METHODS AND ALGORITHMS IN NAVIGATION

# Methods and Algorithms in Navigation

## Marine Navigation and Safety of Sea Transportation

*Editors*

Adam Weintrit & Tomasz Neumann
*Gdynia Maritime University, Gdynia, Poland*

**CRC Press**
Taylor & Francis Group
Boca Raton   London   New York   Leiden

CRC Press is an imprint of the
Taylor & Francis Group, an **informa** business

A BALKEMA BOOK

First issued in hardback 2017

*CRC Press/Balkema is an imprint of the Taylor & Francis Group, an informa business*

© 2011 Taylor & Francis Group, London, UK

Published by: CRC Press/Balkema
P.O. Box 447, 2300 AK Leiden, The Netherlands
e-mail: Pub.NL@taylorandfrancis.com
www.crcpress.com – www.taylorandfrancis.co.uk – www.balkema.nl

ISBN 13: 978-1-138-40221-8 (hbk)
ISBN 13: 978-0-415-69114-7 (pbk)

# List of reviewers

Prof. Yasuo **Arai**, President of Japan Institute of Navigation, Japan,
Prof. Vidal **Ashkenazi**, FRIN, Nottingham Scientific Ltd, UK,
Prof. Andrzej **Banachowicz**, West Pomeranian University of Technology, Szczecin, Poland,
Prof. Marcin **Barlik**, Warsaw University of Technology, Poland,
Prof. Eugen **Barsan**, Master Mariner, Constanta Maritime University, Romania,
Prof. Tor Einar **Berg**, Norwegian Marine Technology Research Institute, Trondheim, Norway,
Prof. Jarosław **Bosy**, Wroclaw University of Environmental and Life Sciences, Wroclaw, Poland,
Prof. Zbigniew **Burciu**, Master Mariner, Gdynia Maritime University, Poland,
Prof. Eamonn **Doyle**, National Maritime College of Ireland, Cork Institute of Technology, Cork, Ireland,
Prof. Andrzej **Fellner**, Silesian University of Technology, Katowice, Poland,
Prof. Wlodzimierz **Filipowicz**, Master Mariner, Gdynia Maritime University, Poland,
Prof. Wieslaw **Galor**, Maritime University of Szczecin, Poland,
Prof. Witold **Gierusz**, Gdynia Maritime University, Poland,
Prof. Marek **Grzegorzewski**, Polish Air Force Academy, Deblin, Poland,
Prof. Andrzej **Grzelakowski**, Gdynia Maritime University, Poland,
Prof. Stanisław **Gucma**, Master Mariner, President of Maritime University of Szczecin, Poland,
Prof. Lucjan **Gucma**, Maritime University of Szczecin, Poland,
Prof. Michal **Holec**, Gdynia Maritime University, Poland,
Prof. Qinyou **Hu**, Shanghai Maritime University, China,
Prof. Jacek **Januszewski**, Gdynia Maritime University, Poland,
Prof. Tae-Gweon **Jeong**, Master Mariner, Secretary General, Korean Institute of Navigation and Port Research,
Prof. Mirosław **Jurdzinski**, Master Mariner, FNI, Gdynia Maritime University, Poland,
Prof. Lech **Kobylinski**, Polish Academy of Sciences, Gdansk University of Technology, Poland,
Prof. Krzysztof **Kolowrocki**, Gdynia Maritime University, Poland,
Prof. Serdjo **Kos**, FRIN, University of Rijeka, Croatia,
Prof. Kazimierz **Kosmowski**, Gdańsk University of Technology, Poland,
Prof. Eugeniusz **Kozaczka**, Polish Acoustical Society, Gdansk University of Technology, Poland,
Prof. Bogumil **Laczynski**, Master Mariner, Gdynia Maritime University, Poland,
Dr. Dariusz **Lapucha**, Fugro Fugro Chance Inc., Lafayette, Louisiana, United States,
Prof. Joong Woo **Lee**, Korean Institute of Navigation and Port Research, Pusan, Korea,
Prof. Andrzej S. **Lenart**, Gdynia Maritime University, Poland,
Prof. Andrzej **Lewinski**, Radom University of Technology, Poland,
Prof. Józef **Lisowski**, Gdynia Maritime University, Poland,
Prof. Vladimir **Loginovsky**, Admiral Makarov State Maritime Academy, St. Petersburg, Russia,
Prof. Artur **Makar**, Polish Naval Academy, Gdynia, Poland,
Prof. Boleslaw **Mazurkiewicz**, Gdańsk University of Technology, Poland,
Prof. Janusz **Mindykowski**, Gdynia Maritime University, Poland,
Prof. Nikitas **Nikitakos**, University of the Aegean, Greece,
Prof. Wiesław **Ostachowicz**, Gdynia Maritime University, Poland,
Prof. Stanisław **Oszczak**, FRIN, University of Warmia and Mazury in Olsztyn, Poland,
Prof. Zbigniew **Pietrzykowski**, Maritime University of Szczecin, Poland,
Prof. Jerzy B. **Rogowski**, MRIN, Warsaw University of Technology, Poland,
Prof. Wladysław **Rymarz**, Master Mariner, Gdynia Maritime University, Poland,
Prof. Roman **Smierzchalski**, Gdańsk University of Technology, Poland,
Prof. Leszek **Smolarek**, Gdynia Maritime University, Poland,
Prof. Jac **Spaans**, Netherlands Institute of Navigation, The Netherlands,
Prof. Cezary **Specht**, Polish Naval Academy, Gdynia, Poland,
Cmdr. Bengt **Stahl**, Nordic Institute of Navigation, Sweden,
Prof. Janusz **Szpytko**, AGH University of Science and Technology, Kraków, Poland,
Prof. Lysandros **Tsoulos**, National Technical University of Athens, Greece,
Prof. Waldemar **Uchacz**, Maritime University of Szczecin, Poland,
Prof. George Yesu Vedha **Victor**, International Seaport Dredging Limited, Chennai, India,
Prof. Peter **Voersmann**, President of German Institute of Navigation DGON, Deutsche Gesellschaft für Ortung und Navigation, Germany,
Prof. Jin **Wang**, Liverpool John Moores University, UK,
Prof. Ryszard **Wawruch**, Master Mariner, Gdynia Maritime University, Poland,
Prof. Adam **Weintrit**, Master Mariner, FRIN, FNI, Gdynia Maritime University, Poland,

Prof. Bernard **Wisniewski**, Maritime University of Szczecin, Poland,

Prof. Adam **Wolski**, Master Mariner, MNI, Maritime University of Szczecin, Poland,

Prof. Hideo **Yabuki**, Master Mariner, Tokyo University of Marine Science and Technology, Tokyo, Japan,

Prof. Homayoun **Yousefi**, MNI, Chabahar Maritime University, Iran

# Contents

Navigational Systems and Simulators. Introduction ......................................................................... 9
*A. Weintrit & T. Neumann*

*Methods and Algorithms* ...................................................................................................... 11

1. Fuzzy–neuron Model of the Ship Propulsion Risk.............................................................. 13
   *A. Brandowski, A. Mielewczyk, H. Nguyen & W. Frackowiak*

2. Kalman-Bucy Filter Design for Multivariable Ship Motion Control ................................ 21
   *M. Tomera*

3. Method of Evaluation of Insurance Expediency of Stevedoring Company's Responsibility
   for Cargo Safety ....................................................................................................................... 33
   *M. Ya. Postan & O. O. Balobanov*

4. Experimental and Numerical Methods for Hydrodynamic Profiles Calculation ............... 37
   *A. Scupi & D. Dinu*

5. Asymptotic Stability of a Class of Positive Continuous-Discrete 2D Linear Systems ...... 41
   *T. Kaczorek*

6. Application of CFD Methods for the Assessment of Ship Manoeuvrability in Shallow Water ....... 45
   *T. Górnicz & J. Kulczyk*

7. Comprehensive Methods of the Minimum Safe Under Keel Clearance Valuation to
   the Restricted Tidal Waters ..................................................................................................... 51
   *G. Szyca*

*Collision Avoidance* ............................................................................................................ 57

8. Knowledge Base in the Interpretation Process of the Collision Regulations at Sea ......... 59
   *P. Banaś & M. Breitsprecher*

9. A Method for Assessing a Causation Factor for a Geometrical MDTC Model
   for Ship-Ship Collision Probability Estimation .................................................................... 65
   *J. Montewka, F. Goerlandt, H. Lammi & P. Kujala*

10. The Sensitivity of Safe Ship Control in Restricted Visibility at Sea................................. 75
    *J. Lisowski*

11. Experimental Research on Evolutionary Path Planning Algorithm with Fitness Function
    Scaling for Collision Scenarios .............................................................................................. 85
    *P. Kolendo, R. Śmierzchalski & B. Jaworski*

12. A New Definition of a Collision Zone For a Geometrical Model For Ship-Ship
    Collision Probability Estimation ............................................................................................ 93
    *J. Montewka, F. Goerlandt & P. Kujala*

13. Uncertainty in Analytical Collision Dynamics Model Due to Assumptions in Dynamic
    Parameters ............................................................................................................................... 101
    *K. Stahlberg, F. Goerlandt, J. Montewka & P. Kujala*

14. Applied Research of Route Similarity Analysis Based on Association Rules.................. 109
    *Zhe Xiang, Ru-ru Liu, Qin-you Hu & Chao-jian Shi*

15. An Analysis the Accident Between M/V Ocean Asia and M/V SITC Qingdao in Hanam
    Canal (Haiphong Port)............................................................................................................ 115
    *Vinh Nguyen Cong*

*Geodetic Problems in Navigational Applications* ................................................................. 121

16. A Novel Approach to Loxodrome (Rhumb Line), Orthodrome (Great Circle) and
    Geodesic Line in ECDIS and Navigation in General ...................................................... 123
    *A. Weintrit & P. Kopacz*

17. Approximation Models of Orthodromic Navigation ...................................................... 133
    *S. Kos & D. Brčić*

18. Solutions of Direct Geodetic Problem in Navigational Applications ............................ 141
    *A.S. Lenart*

*Route Planning in Marine Navigation* ...................................................................... 147

19. Advanced Navigation Route Optimization for an Oceangoing Vessel ......................... 149
    *E. Kobayashi, T. Asajima & N. Sueyoshi*

20. On the Method of Ship's Transoceanic Route Planning .............................................. 157
    *O. D. Pipchenko*

21. Weather Hazard Avoidance in Modeling Safety of Motor-driven Ship for Multicriteria
    Weather Routing .......................................................................................................... 165
    *P. Krata & J. Szlapczynska*

22. Evolutionary Sets of Safe Ship Trajectories: Evaluation of Individuals ...................... 173
    *R. Szlapczynski & J. Szlapczynska*

23. Development of a 3D Dynamic Programming Method for Weather Routing .............. 181
    *S. Wei & P. Zhou*

*Aviation and Air Navigation* ...................................................................................... 189

24. Position Reference System for Flight Inspection Aircraft ........................................... 191
    *M. Kubiš & A. Novák*

25. RNAV GNSS Essential Step for the LUN Implementation and the Chance for the Polish
    General Aviation ......................................................................................................... 199
    *K. Banaszek, A. Fellner, P. Trómiński & P. Zadrąg*

26. Aircraft Landing System Utilizing a GPS Receiver with Position Prediction Functionality .......... 207
    *J. Biały, J. Ćwiklak, M. Grzegorzewski, S. Oszczak, A. Ciećko & P. Kościelniak*

# Navigational Systems and Simulators. Introduction

A. Weintrit & T. Neumann
*Gdynia Maritime University, Gdynia, Poland*

## PREFACE

The contents of the book are partitioned into five parts: methods and algorithms (covering the chapters 1 through 7), collision avoidance (covering the chapters 8 through 15), geodetic problems in navigational applications (covering the chapters 16 through 18), route planning in marine navigation (covering the chapters 19 through 23), aviation and air navigation (covering the chapters 24 through 26). It shows the origins of navigation and how techniques, methods, and systems have been developed. Certainly, this subject may be seen from different perspectives.

The first part deals with methods and algorithms. The contents of the first part are partitioned into seven chapters: Fuzzy-neuron model of the ship propulsion risk, Kalman-Bucy observer design for multivariable ship motion control, Method of evaluation of insurance expediency of stevedoring company's responsibility for cargo safety, Experimental and numerical methods for hydro-dynamic profiles calculation, Asymptotic stability of a class of positive continuous-discrete 2D linear systems, Application of CFD methods for the assessment of ship manoeuvrability in shallow water, and Comprehensive methods of the minimum safe under keel claearance valuation to the tidal restricted waters.

The second part deals with collision avoidance. The principles of anti-collision, and corresponding methods, models and scenarios are described in general and with some particular samples of applications. The contents of the second part are partitioned into eight chapters: Knowledge base in the interpretation process of the collision regulations at sea, A method for assessing a causation factor for a geometrical MDTC model for ship-ship collision probability estimation, The sensitivity of safe ship control in restricted visibility at sea, Experimental research on evolutionary path planning algorithm with fitness function scaling for collision scenarios, A new definition of a collision zone for a geometrical model for ship-ship collision probability estimation, Uncertain-

ty in analytical collision dynamics model due to assumptions in dynamic parameters, Applied research of route similarity analysis based on association rules, and Analysis the accident between m/v Ocean Asia and m/v SITC Qingdao in Hanam Canal (Haiphong Port).

The third part deals with geodetic problems in navigational applications. Last reports and research results in the field of navigational calculations' methods applied in marine navigation are surveyed. The contents of the third part are partitioned into three chapters: A novel approach to loxodrome (rhumb line), orthodrome (great circle) and geodesic line in ECDIS and navigation in general. Approximation models of orthodromic navigation, and Solutions of direct geodetic problem in navigational applications.

The fourth part deals with route planning in marine navigation. Different kinds of ocean route optimization, weather routing and route planning methods are presented. The contents of the forth part are partitioned into fifth chapters: Advanced navigation route optimization for an oceangoing vessel, On the method of ship's transoceanic route planning, Weather hazard avoidance in modeling safety of motor-driven ship for muticriteria weather routing, Evolutionary sets of safe ship trajectories: evaluation of individuals, and Development of a 3D dynamic programming method for weather routing.

The fifth part deals with aviation and air navigation. The contents of this part are partitioned into two chapters: Position reference system for flight inspection aircraft, RNAV GNSS essential step for the LUN implementation and the chance for the Polish general aviation, and Aircraft landing system utilizing a GPS receiver with position prediction functionality.

Methods and Algorithms

# 1. Fuzzy–neuron Model of the Ship Propulsion Risk

A. Brandowski, A. Mielewczyk, H. Nguyen & W. Frackowiak
*Gdynia Maritime University, Poland*

ABSTRACT: A prediction model is presented of the ship propulsion risk, i.e. a risk of the consequences of loss of the ship propulsion capability. This is an expert model based on opinions elicited by the ship power plant operators. The risk level depends, among other things, on the reliability state of the ship propulsion system components. This state is defined by operators in a linguistic form. The formal risk model parameters are determined by means of a neural network. The model may be useful in the ship operation decision processes.

## 1 INTRODUCTION

The risk prediction model consists of a dangerous event (DE) module and the event consequence module. The DE connects the two modules - it initiates consequences of particular causes. In the case of propulsion risk (PR), the event DE is immediate loss of the propulsion capability by the ship, i.e. an immediate catastrophic failure (ICF) of its propulsion system (PS) (Brandowski 2005, Brandowski et al. 2007, 2008, 2009a). The event may be caused by the PS element failures or operator errors.

It is assumed that the model parameter identification will be based on opinions of the ship power plant operators, hereinafter referred to as experts. The opinions will be formulated mainly in a linguistic form, supported to a minimum extent by numerical data.

The ship PS is well developed. In the example of a simple PS presented below, it consists of 11 subsystems (SS) and these of 92 sets of devices (SD) including several hundred devices (D) altogether. The PS sizes, the expert ability to express the opinions necessary to construct a propulsion risk model and the limited number of experts that the authors managed to involve in the study influenced the model form.

The expert investigation methods used in the PR modelling were presented in publications (Brandowski 2005; Brandowski et al. 2007, 2008, 2009a; Nguyen 2009)

## 2 THE PROPULSION RISK PREDICTION MODEL

The PR model form is determined by data that can be obtained from experts. It is assumed that they elicit:

- Annual numbers $N$ of the system ICF type failures;
- System operating time share in the calendar time of the system observation by the expert $t^{(a)}\%$.
- Linguistic estimation of the share of number of PS fault tree (FT) cuts in the failure number $N$ during a year.
- Linguistic estimation of chances or chance preferences of the occurrence of system ICF specific consequences, on the condition that the event itself occurs.

Those opinions are a basis for the construction of a system risk prediction model.

The following assumptions are made as regards the system risk model:

- The system may be only in the active use or stand-by use state. The system ICF type events may occur only in the active use state.
- The formal model of a PS ICF event stream is the Homogeneous Poisson Process (HPP). It is a renewal process model with negligible renewal duration time. This assumption is justified by the expert opinions, which indicate that catastrophic failures (CF) of some systems may occur quite frequently, even several times a year, but in general they cause only a relatively short break in normal system operation. Serious consequences with longer breaks in the system operation are less frequent. Also the exponential time between failures distribution, as in the case of HPP, is

characteristic of the operation of many system classes, including the ship devices (Modarres et al. 1999, Podsiadlo 2008). It is appropriate when defects of the modeled object and the operator errors are fully random, abrupt and no gradual, without wear and/or ageing-type defects. This corresponds with the situation where inspection and renewals are regularly carried out and prevent that type of defects.

The following assumptions were made with reference to the model:
- The HPP parameter is determined in a neural network from data elicited by experts. The network can be calibrated with real data obtained from the system (or a similar systems) operation.
- The failure consequences are determined from data on the chances of occurrence elicited in the expert opinions.
- The operators perform predictions of the system reliability condition and PR, i.e. of the system ICF specific consequences, based on subjective estimations of the analysed system component condition.

For given ICF event a fault tree (FT) is constructed, where the top event is an ICF type PS failure and the basic events are the system minimum cut or cut failures. The notion of minimum cut is generally known. Cut is defined as a set of elements (devices) fulfilling a specific function which loss of that function results in a system ICF. In the case of minimum cut, failures of the same system elements may appear in more than one minimum cuts. Therefore, they are not disjoint events in the probabilistic sense. Besides, obtaining reliable expert opinions on the minimum cut failures is almost unrealistic. Also in the case of a PS ICF event cause decomposition to the minimum cut level the number of basic events in the FT increases considerably - the top event decomposition is deeper. The more basic events it contains, the more data are needed to tune the neural network in a situation when the number of competent experts available is generally very limited. In the case of cuts (not minimum cuts), they can be arranged to form a complete set of events. The failure numbers are then easier to estimate by experts as the cuts include more devices. Such failures are serious events in the ship operation process, very well remembered by the experts. Besides, there are generally fewer cuts than minimum cuts in the FTs.

Cuts have defined reliability structures (RS). If those structures and the number of cut failures within a given time interval are known, then the number of failures of particular devices in the cuts can be determined.

The diagram of a model in Figure 1 illustrates the PR prediction within a period of time $t^{(p)}$. The system operator inputs estimated reliability states of the cut elements (block (1) of the model). The elements are devices (D) of the all system cuts. The estimates are made by choosing the value of the linguistic variable $LV$ = *average annual number of ICF events from the set {minimum, very small, small, medium, large, very large, critical}* for the individual Ds. The operator may be supported in that process by a database.

Having the reliability states of the FT cuts and their RS structures, average numbers $N_{ik}$ of these cut ICF failures are determined by "operator algorithm" (block (2)). The appropriate methods are presented in section 3 of this paper. They are input data to the neural network.

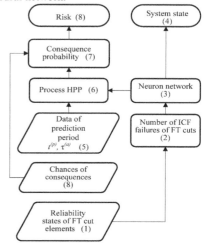

Figure 1. Diagram of the fuzzy-neuron model of risk prediction

The neural network, performing generalized regression, determines the system ICF type failure annual number $N$ in the numerical and linguistic values (block (3)). In the first case, the network determines the respective value of an $LV$ variable singleton membership function, and in the second case - a corresponding linguistic value of that function. In both cases 7 values of the $LV$ were adopted. The network may be more or less complex depending on the number of cuts and the FT structure.

The neural network is built for a specific PS, according to its properties and size. Each cut at the FT lowest level implies an entry to the network. The network error decreases with the increasing amount of data. We are interested in teaching data with errors fulfilling some statistical standards and that depends on the number and appropriate choice of experts.

If there is disproportion between the number of entries and the teaching data lot size, then the system FT may be divided at the lower composition levels and then the component networks "assembled" again. In the ship PR risk prediction example here below, the ship PS was decomposed into subsystems (SS) and those into sets of devices (SD).

The system reliability condition, according to its operator, i.e. annual number $N$ of its ICFs, is presented in a linguistic form by giving the $LV$ value determined in block (3) (block (4)).

Input to the model is risk prediction calendar time $t^{(p)}$ [year] and the modeled PS active use time coefficient $\tau^{(a)}$. The prediction time is chosen as needed, in connection with the planned sea voyages.

The PS active use time coefficient:

$$\tau^{(a)} = \left(t^{(a)} / 100\right)t^{(p)} \tag{1}$$

where $t^{(a)}$ % = propulsion system active use time as a share of prediction calendar time $t^{(p)}$ (approximately equal to the share of ship at sea time).

The value of $\tau^{(a)}$ coefficient is determined by operator from the earlier or own estimates.

The probability of the system ICF event occurrence within the prediction time $t^{(p)}$ is determined by a size $K$ vector (block (6)):

$$P\{ICF_k, t^{(p)}\} = \left[\frac{\left(\lambda^{(a)}\tau^{(a)}t^{(a)}\right)^k}{k!}e^{-\lambda(a)\tau(a)t(p)} : k = 1,2,...,K\right] \tag{2}$$

where $\lambda^{(a)} = N / \tau \, t$ [1/year] = intensity function (rate of occurrence of failures, ROCOF) related to the active use time, where $N$ = number of the system ICFs within $t = 1$ year of observation, with the active use time coefficient $\tau$ determined by neural network; $k$ = number of ICFs.

Vector (2) expresses the probability of occurrence of $k = 1,2,...,K$ system ICFs within the prediction time $t^{(p)}$ interval.

Probability of occurrence of specific consequences on the condition of the analysed system ICF occurrence:

$$P\{C / ICF\} \tag{3}$$

where C = C1 $\cap$ C2 = very serious casualty C1 or serious casualty C2 (IMO 2005).

This probability value is input by the operator from earlier data obtained from expert investigations for a specific ship type, shipping line, ICF type and ship sailing region. The values may be introduced to the prediction program database.

The consequences C are so serious, that they may occur only once within the prediction time $t^{(p)}$, after any of the $K$ analysed system ICFs. The risk of consequence occurrence after each ICF event is determined by vector whose elements for successive $k$-th ICFs are sums of probabilities of the products of preceding ICF events, non-occurrence of consequences C of those events and occurrence of the consequences of $k$-th failure (block (7)):

$$\Re\{C, t^{(p)}\} = \left[P\{C / ICF\}\sum_{k=1}^{x}P\{ICF_x\}\left(1 - P\{C / ICF\}\right)^{x-1} : x = 1,2,..,K\right] \tag{4}$$

Risk (4) is presented in block (8).

# 3 OPERATOR'S ALGORITHM

## 3.1 *Cut models*

The algorithm allows processing of the subjective estimates of numbers of device D failures, creating FT cuts, into numerical values of the numbers of failures of those cuts. They are the neural network input data. The algorithm is located in block (2) of the prediction model. The data are input to the model during the system operation, when devices change their reliability state. Additionally, the algorithm is meant to aid the operator in estimating the system condition.

The numerical values of the numbers of failures in cuts are determined by computer program from the subjective linguistic estimates of the numbers of failures of component devices D. The estimates are made by the system operators and based on their current knowledge of the device conditions. This is simple when cut is a single-element system, but may be difficult with complex RS cuts. The algorithm aids the operator in the estimates. Specifically, it allows converting the linguistic values of D device ICF events into corresponding numerical values of the cuts. The data that may be used in this case are connected with cuts - the universe of discourse ($UD$) of linguistic variables $LV$ of the cut numbers of failures for defined RSs. These numbers are determined from the expert investigations.

Cuts are sets of devices with specific RS - systems in the reliability sense. They may be single- or multi-element systems. They are distinguished in the model because they can cause subsystem ICFs and in consequence a PS failure. Annual numbers of the cut element (device) ICFs change during the operation process due to time, external factors and the operational use.

The conversion problem is presented for the case when in the system FT cuts of subsystems (CSS) are distinguished and in them cuts of sets of devices (CSD). The following CSD notation is adopted:

$$CSD_{ik} = \{e_{ikl}, l = 1,2,...,L_{ikL}\} \tag{5}$$

where $CSD_{ik}$ = $k$-th cut of i-th subsystem, $k = 1.2,...,K$, $i = 1,2,...,I$; $e_{ikl}$ = $l$-th element of $k$-th CSD, $l = 1,2,..., L$.

The CSD cut renewal process parameters, i.e. intensity functions $\lambda$ (ROCOF), are determined from the expert investigations of the system PS. In this case, they are applied only to the ICFs causing the loss of CSD function. Annual numbers of failures $N$, whose functions are intensity functions $\lambda$, are determined. It may be assumed that the numbers elicited by experts are average values in their space of professional experience gained during multi-year seamanship. Then the asymptotic intensity function takes the form (Misra 1992):

$$\lambda^{(a)} \cong \frac{N}{\tau t} \tag{6}$$

where $N$ = average number of the analysed system failures during the observation time $t$; $\tau$ = active use time coefficient; $t = 1$ year = calendar time that the estimate of the number of failures is related to.

We are interested in the dependence on the number of CSD cut ICFs to the number of such failures of the cut elements. It is determined from the formulas of the relation of systems, of specific reliability structures, failure rate to the failure rates of their components. It should be remembered that in the case of a HPP the times between failures have exponential distributions, whose parameter is the modeled object failure rate, in the analysed case equals to the process renewal intensity function $\lambda$. The formulas for the ship system CSD cut reliability structures are given below.

In the case of a single-element structure, the annual numbers of the cut failures and device failures are identical.

$$N_{ik} = N_{ikl}, i \in \{1,2,...,I\}, k \in \{1,2,...,K\}, l = 1 \tag{7}$$

where $N_{ik}$ = annual number of failures of $k$-th cut in $i$-th subsystem; $N_{ikl}$ = annual number of failures of $l$-th device.

In a series RS, the number of system failures is a sum of the numbers of failures of its components.

$$N_{ik} = N_{ik_1} + N_{ik_2} + ... + N_{ik_l} + ... + N_{ik_L} \tag{8}$$

A decisive role in that structure plays a "weak link", i.e. the device with the greatest annual number of failures. The CSD cut number of failures must then be greater than the weak link number of failures.

In a two-element parallel RS, we obtain from the average time between failures formula (Misra, 1992):

$$N_{lk} = \frac{N_{ik_1}^2 N_{ik_2} + N_{ik_1} N_{ik_2}^2}{N_{ik_1} N_{ik_2} + N_{ik_1}^2 N_{ik_2}^2} \tag{9}$$

If one element in that structure fails then it becomes a single element structure. Similar expressions can be easily derived for a three-element parallel structure.

In the structures with stand-by reserve, only part of the system elements are actively used, the other part is a reserve used when needed. The reserve is switched on by trigger or by the operator action. The trigger and the system functional part create the series reliability structure. When the trigger failure rate is treated as constant and only one of the two elements is actively used ($L = 2$), then:

$$N_{ik} = N_{ik}^p + \frac{N_{ik_1} N_{ik_2}}{N_{ik_1} + N_{ik_2}} \tag{10}$$

where $N_{ik}^p$ = annual number of trigger failures.

In the case of a three-element structure ($L = 3$) with two stand-by elements, we obtain:

$$N_{ik} = N_{ik}^p + \frac{N_{ik_1} N_{ik_2} N_{ik_3}}{N_{ik_2} N_{ik_3} + N_{ik_1} N_{ik_3} + N_{ik_1} N_{ik_2}} \tag{11}$$

In the load-sharing structures, as the expert data on the number of failures in the case when entire cut load is taken over by one device are not available, a parallel RS (equation (9)) is adopted.

In operation, the CSD cut elements may become failure and cannot be operated. If in a two-element RS with stand-by reserve one element is non-operational then it becomes a single element structure. If in a three-element RS with stand-by reserve one element is non-operational then it becomes a two-element structure with one element in reserve. If in that structure two elements are non-operational then it becomes a single-element structure. Identical situation occurs in the case of element failures in the parallel RS systems.

## 3.2 Fuzzy approach to the cut failure number estimate problem

Our variables $LV$ are estimates of the average linguistic annual numbers of ICFs failures $N_{ik}$ of cuts $CSD_{ik}$ and $N_{ikl}$ devices $D_{ikl}$, $i = 1,2, ,I, k = 1,2,...,K, l = 1,2, ...,L$. We define those variables and their linguistic term-sets $LT$-$S$. We assume seven-element sets of those values: *minimum, very small, small, medium, high, very high, critical*. We assume that these values represent the *reliability state* of appropriate objects.

From the expert investigations we obtain the universe of discourse values $UD_{ik}$ of individual cuts. Each of those universes is divided into six equal intervals. We assume that the boundary values

$$N_{ik}^1, N_{ik}^2,..., N_{ik}^7$$

of those intervals are singleton member functions of the corresponding linguistic variable values $LV_{ik}$.

The universe of discourse values $UD_{ik}$ are the variability intervals of the numbers of failures of cuts $CSD_{ik}$ appearing on the left hand sides of equations (7) – (11). In the case of a single element RS, parallel RS and with stand-by reserve composed of identical elements in terms of reliability, we can easily determine the minimum and maximum numbers of element failures.

$$N_{ikl}^1, N_{ikl}^7$$

and their universes of discourse $UD_{ikl}$ and then the singleton seven-element member functions:

$$N_{ikl}^1, N_{ikl}^2,..., N_{ikl}^7$$

If all the cut elements remain in the *minimum* state then the cut is also in the *minimum* state. If all the cut elements remain in the *critical* state then the cut is also in the *critical* state. The situation is more difficult when the cut elements are not identical in terms of reliability. Then expert opinion-based heuristic solutions must be applied.

## 4 CASE STUDY

The example pertains to the prediction of a seagoing ship propulsion risk. Determination of the probability of loss of propulsion capability is difficult because of the lack of data on the reliability of PS elements and of operators. This applies in particular to the risk estimates connected with decisions made in the ship operation phase.

The object of investigation was a PS consisting of a low-speed piston combustion engine and a constant pitch propeller, installed in a container carrier operating on the Europe - North America line.

The FT of analysed PS is shown in the Figure 2. For reasons of huge number of SDs the structure of fuel oil subsystem is only described within the lowest FT level. The object was decomposed into subsystems (SS) (propulsion assembly and auxiliary installations necessary for the PS functioning - 11 SSs altogether) and the subsystems into sets of devices ((SD) - 92 sets altogether). Each SS makes the CSS cut and each SD – the SDC cut. In considered case the system FT consists of alternatives of those cuts. In general such FT structure doesn't have to appear in the case of PS.

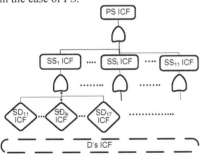

Figure 2. Fault tree of a ship propulsion system ICF
Legend: PS – propulsion system; ICF – immediate catastrophic failure;
SS$_i$ – subsystem, i =1 -fuel oil subsystem, 2 -sea water cooling subs.; 3 – low temperature fresh water cooling subs.; 4 – high temperature fresh water cooling subs.; 5 – startig air subs.; 6 – lubrication oil subs.; 7 – cylider lubrication oil subs.; 8 - electrical subs.; 9 – main engine subs.; 10 – remote control subs.; 11 – propeller + shaft line subs.
SD$_{1k}$ – set of devices; ik = 11 - fuel oil service tanks; 12 – f. o. supply pumps; 13 – f. o. circulating pumps; 14 – f. o. heaters; 15 –filters; 16 – viscosity control arrangement; 17 - piping's heating up steam arrangement.

The FT allowed the building the neural network. The sets of input signals for the network were assigned.

Using the code (IMO, 2005), five categories of ICF consequences were distinguished, including *very serious casualty C1, serious casualty C2 and three incident categories*. Consequences of the alternative of first two events were investigated ($C = C1 \cap C2$).

The consequences are connected with losses. They may involve people, artifacts and natural environment. They are expressed in physical and/or financial values. Detailed data on losses are difficult to obtain, particularly as regards rare events like the C1 and C2 type consequences. They cannot be obtained from experts either, as most of them have never experienced that type of events. In such situation, the risk was related only to the type C consequences of an ICF event.

### 4.1 *Acquisition and processing of expert opinions*

The experts in the ICF event investigation were ship mechanical engineers with multi-year experience (50 persons). Special questionnaires were prepared for them, containing definition of the investigated object, SS and SD schemes, precisely formulated questions and tables for answers. The questions asked pertained to the number of ICF type events caused by equipment failures or human errors within one year and the share of time at sea in the ship operation time (PS observation time by expert). These were the only questions requiring numerical answers.

Other questions were of a linguistic character and pertained to the share of ICF type failures of individual SSs in the annual number of the PS ICF type events and the share of ICF failures of individual SD sets in the annual numbers of SS failures. In both presented cases the experts chose one of five values of the linguistic variables: *very great, great, medium, small, very small*. The elicited linguistic opinions were compared in pairs and then processed by the AHP method (Saaty 1980; Nguyen 2009). The obtained distribution of subsystem shares complies with the engineering knowledge. The greatest shares are due to the main engine and the electric power and fuel supply systems and the smallest - due to the propeller with shaft line.

The experts in the ICF event consequence field were ship mechanical engineers and navigation officers (37 in number). A similar questionnaire was prepared with questions about preferences of 5 possible consequences (*C1 - very serious casualty, C2 - serious casualty and 3 types of incidents*) of the ICF type event occurrence. The casualty types were defined in accordance with the code (IMO, 2005). The experts could choose from the following preferences: *equivalence, weak preference, significant preference, strong preference, absolute preference, and inverse of these preferences* (Saaty, 2005; Nguyen 2009). After processing of the so obtained data by the AHP method, a normalized vector of shares of the ICF type event consequences was obtained.

### 4.2 Some results

The PR model was subjected to a broad range of tests. Some of the results are presented below. Figures 3 and 4 present the probability of the occurrence of defined numbers ICF type events of PS in dependence on the prediction time, when PS is in excellent and critical reliability states. The number of ICF events from 1 to 5 was adopted for each of those states. The probability was performed for the prediction time $t^{(p)}$ = 1, 3 and 6 months. The diagrams 3 and 4 indicate that the occurrence of ICF events and their numbers are significantly greater when PS is in the critical state than in excellent state.

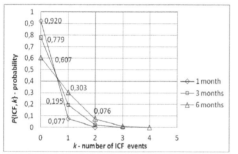

Figure.3. Probability of the ICF type events versus the numbers of those events for the selected times of risk prediction. PS reliability state is excellent.

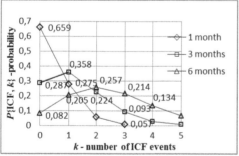

Figure 4. Probability of the ICF type events versus the number of those events for the selected times of risk prediction. PS reliability state is critical.

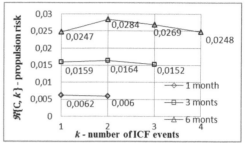

Figure 5. Propulsion risk versus the numbers of ICF events for selected prediction times. PS reliability state excellent.

Figures 5 and 6 presents the PR risk, i.e. the risk of type C consequences after occurrence of an ICF event, for the prediction times $t^{(p)}$ = 1, 3 and 6 months, when PS is in the excellent and critical for states. The diagrams show increased risk with deteriorating PS reliability.

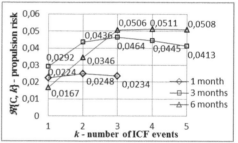

Figure 6. Propulsion risk versus the numbers of the ICF events for selected times of prediction. PS reliability state is critical

## 5 SUMMARY

A fuzzy-neural model of risk prediction has been developed, based on the knowledge acquired from experts. It is a model of homogeneous Poisson renewal process, where parameters are determined by means of a neural network. The model parameter estimation data were acquired from experts - the modeled system operators. Their opinions were elicited in a numerical form as regards the events observed by them many times and in a linguistic form in the cases where their knowledge might be less precise. The neural network was tuned with the elicited opinions. The network may be calibrated with data collected in the system operation process. In this way the Homogeneous Poisson Process can be adapted to real operating conditions - it becomes non-homogeneous in steps. The model allows prediction of the risk of dangerous events consequences, which may occur due to different systems.

In the expert investigations we have to rely on data obtained from experts and models are constructed from that data. The adequacy and type of obtained information depends on the form and adequacy of the data. The expert competence level must not be exceeded. In the case reported here, it might have happened in the estimates of occurrence of the ICF event consequences. In the authors' opinion, the competence level was not exceeded as the remaining data are concerned, as the choice of experts was careful.

The expert-elicited data have an impact on the level of adequacy of models used in the investigations - like data like model. A number of simplifying assumptions had to be made. Some of them are the following: two states of the use of modeled objects, failures possible only in the active use state, homogeneity of the Poisson renewal process, the cut notion, definition of the ICF event consequences etc.

Results of the propulsion risk estimates quoted in section 4 are not questionable as regards the order of magnitude of the numbers. Events from the subset of C consequences occur at present in about 2% of the ship population (20 ships out of 1000 in a year). This applies to ships above 500 GT. There are at present about 50 thousand such ships (Graham, 2009; Podsiadlo 2008). The results are also adequate in terms of trends of changes in the investigated values, which are in compliance with the character of the respective processes.

It has to be taken into account that results of a subjective character may be (but not necessarily) subject to greater errors than those obtained in a real operating process. The adequacy of such investigations depends on the method applied, and particularly on the proper choice of experts, their motivation, as well as the type of questions asked. In the expert investigations the fuzzy methods are especially useful, as they allow the experts to express their opinions in a broader perspective.

In the authors' opinion, the main difficulty in the neural network application for modeling is the necessity of having a considerable amount of input and output data for tuning the models. In the prospective investigations the data are generally in short supply. They may be gathered after some time in the operating process of the respective objects, but that may appear to be too late.

There is a chance of further developing and using the risk prediction program, developed under the project, aboard ships and not only for the propulsion systems. It could be coupled with the existing equipment renewal management or operating management programs.

The investigations presented in the paper were supported by Ministry of Science and Higher Education in the frame of a study project.

REFERENCES

IMO, MSC-MEPC.3/Circ.1. 2005. Casualty-related matters. Reports on marine casualties and incidents. Revised harmonized reporting procedures – Reports required under SOLAS regulation I/21 and MARPOL 73/78, articles 8 and 12.

Brandowski A. 2005. *Subjective probability estimation in risk modeling* (in Polish). Problemy Eksploatacji 3/2005 (58). Zeszyty Naukowe Instytutu Technologii Eksploatacji Radom.

Brandowski A., Frackowiak W., Mielewczyk A. 2007. Subjective reliability estimation of a seagoing ships. Proceedings of ESREL2007 Conferece. Stravanger.

Brandowski A., Frackowiak W., Nguyen H., Podsiadlo A. 2008. *Subjective propulsion risk of a seagoing ship*. Proceedings of ESREL2008 Conference. Valencia.

Brandowski A., Frackowiak W., Nguyen H., Podsiadlo A. 2009. *Risk estimation of a sea-going ship casualty as the consequence of propulsion loss*. Proceedings of ESREL2009 Conference. Prague.

Brandowski A. 2009. *Estimation of the probability of propulsion loss by a seagoing ship based on expert opinions*. Polish Maritime Research 1/2009. Gdańsk University of Technology. Gdańsk.

Saaty T.L. et al.1980. *The Analytic Hierarchy Process*. New York.McGraw-Hill.

Nguyen H. 2009. *Application of AHP method in the risk estimation of ship systems* (in Polish). Polish Maritime Research 1/2009. Gdańsk 2009.

Modarres M., Kaminskiy M., Krivtsov. 1999. *Reliability Engineering and Risk Analysis*. New York, Basel: Marcel Dekker, Inc.

Misra K. B. 1992. *Reliability Analysis and Prediction*. A Methodology Oriented Treatment. ELSEVIER. Amsterdam, Oxford, New York, Tokyo.

Piegat A. 1999. *Fuzzy modeling and contol* (in Polish). Akademicka Oficyna Wydawnicza EXIT. Warszawa.

Graham P. 2009. *Casualty and World Fleet Statistics as at 31.12.2008*. IUMI Facts & Figures Committee..

Podsiadło A. 2008. *Analysis of failures in the engines of main ship propulsion* (in Polish). Internal study of Gdynia Maritime University. Gdynia.

# 2. Kalman-Bucy Filter Design for Multivariable Ship Motion Control

M. Tomera

*Gdynia Maritime University, Faculty of Marine Electrical Engineering, Department of Ship Automation, Poland*

ABSTRACT: The paper presents a concept of Kalman-Bucy filter which can be used in the multivariable ship motion control system. The navigational system usually measures ship position coordinates and the ship heading, while the velocities are to be estimated using an available mathematical model of the ship. The designed Kalman-Bucy filter has been simulated on a computer model and implemented on the training ship to demonstrate the filtering properties.

## 1 INTRODUCTION

Modern ships are equipped with complicated ship motion control systems, the goals of which depend on tasks realised by an individual ship. The tasks executed by the control system include, among other actions, controlling the ship motion along the course or a given trajectory (path following and trajectory tracking), dynamical positioning and reduction of ship rolls caused by waves. Figure 1 presents basic components of the ship motion control system.

The guidance system generates a required smooth reference trajectory, described using given positions, velocities, and accelerations. The trajectory is generated by algorithms which make use of the required and current ship positions, and the mathematical model with complementary information on the executed task and, possibly, the weather.

The control system processes the motion related signals and generates the set values for actuators to reduce the difference between the desired trajectory and the current trajectory. The controller can have a number of operating modes depending on the executed tasks. On some ships and in some operations the required control action can be executed in several ways due to the presence of a number of propellers. Different combinations of actuators can generate the same control action. In those cases the control system has also to solve the control allocation problem, based on the optimisation criteria (Fossen, 2002).

The navigation system measures the ship position and the heading angle, collects data from various sensors, such as GPS, log, compass, gyro-compass, radar. The navigation system also checks the quality of the signal, passes it to the observer system in which the disturbances are filtered out and the ship state variables are calculated. Stochastic nature of the forces generated by the environment requires the use of observers for estimating variables related with the moving ship and for filtering the disturbances in order to use the signals in the ship motion control systems.

Filtering and estimating are extremely important properties in the multivariable control systems. In many cases the ship velocity measurements are not directly available, and the velocity estimates are to be calculated from the position and heading values measured by the observer. Unfortunately, these measurements are burdened with errors generated by environmental disturbances like wind, sea currents and waves, as well as by sensor noise.

One year after publishing his work on a discrete filter (Kalman, 1960), Rudolph Kalman, this time together with Richard Bucy, published the second work in which they discussed the problem of continuous filtering (Kalman & Bucy, 1961). This work has also become the milestone in the field of optimal filtering. In the present article the continuous Kalman filter is derived based on the discrete Kalman filter, assuming that the sampling time tends to zero. A usual tendency in numerical calculations is rather reverse: starting from continuous dynamic equations, which are digitised to arrive at the discrete difference equations being the approximates of the initial continuous dynamics. In the Kalman filter idea the discrete equations are accurate as they base on accurate difference equations of the model of the process.

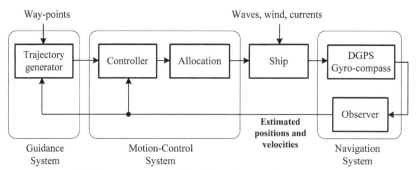

Way-points                       Waves, wind, currents

Figure 1. Basic components of modern ship motion control system (Fossen, 2002).

The dynamic positioning systems have been developed since the early sixties of the last century. The first dynamic control systems were designed using conventional PID controllers working in cascade with low-pass filters or cut-off filters to separate the motion components connected with the sea waves. However, those systems introduce phase delays which worsen the quality of the control (Fossen, 2002).

From the middle of 1970s more advanced control techniques started to be used, which were based on optimal control and the Kalman filter theory. The first solution of this type was presented by (Balchen et al., 1976). It was then modified and extended by Balchen himself and other researchers: (Balchen et al., 1980a; Balchen et al., 1980b; Fung and Grimble, 1983; Saelid et al. 1983; Sorensen et al., 1996; Strand et al. 1997). The new solutions made use of the linear theory, according to which the kinematic characteristics of the ship were to be linearized in the form of sets of predefined ship heading angles, with an usual resolution of 10 degrees. After the linearization of the nonlinear model, the observer based on such a model is only locally correct. This is the disadvantage of the Kalman filter. The Kalman filter can make use of measurements done by different sensors at different accuracy levels, and calculate ship velocity estimates which are not measured in the majority of ship positioning applications.

The main goal of the article is designing and testing the observer for the ship motion velocity estimation.

## 2  DISCRETE MODEL OF THE PROCESS

Discussed are time-dependent discrete processes, which are recorded by sampling continuous processes at discrete times. Let us assume that the continuous process is described by the following equation:

$$\dot{\mathbf{x}}(t) = \mathbf{A}(t)\mathbf{x}(t) + \mathbf{G}(t)\mathbf{u}(t) \qquad (1)$$

where $\mathbf{u}$ is the input vector having the form of white noise. The state transition matrix for equation (1) takes the form:

$$\mathbf{x}(t) = e^{\mathbf{A}(t_0)(t-t_0)}\mathbf{x}(t_0) + \int_{t_0}^{t} e^{\mathbf{A}(\tau)(t-\tau)}\mathbf{G}(\tau)u(\tau)d\tau \qquad (2)$$

For the discrete model, the objects of analysis are process samples recorded at times $t_0$, $t_1$, ..., $t_k$, .... Equation (2) written for a single sampling interval can be presented as

$$\mathbf{x}(t_{k+1}) = \mathbf{F}(t_{k+1},t_k)\mathbf{x}(t_k) + \int_{t_k}^{t_{k+1}} \mathbf{F}(t_{k+1},\tau)\mathbf{G}(\tau)u(\tau)d\tau \qquad (3)$$

which can be briefly written as

$$\mathbf{x}_{k+1} = \mathbf{F}_k\mathbf{x}_k + \mathbf{w}_k \qquad (4)$$

where $\mathbf{F}_k$ is the state transition matrix for the step between times $t_k$ and $t_{k+1}$ at the absence of the excitation function

$$\mathbf{F}_k = \mathbf{F}(t_{k+1},t_k) = e^{\mathbf{A}(t_k)T} = \mathbf{I} + \mathbf{A}(t_k)\cdot T \qquad (5)$$

and $\mathbf{w}_k$ is the excited response at time $t_{k+1}$ due to the presence of the white noise at the input in the time interval $(t_k, t_{k+1})$, i.e. accidental disturbances affecting the process

$$\mathbf{w}_k = \int_{t_k}^{t_{k+1}} \mathbf{F}(t_{k+1},\tau)\mathbf{G}(\tau)u(\tau)d\tau \qquad (6)$$

The white noise is a stochastic signal having the mean value equal to zero and finite variance. The matrix elements $\mathbf{w}_k$ can reveal non-zero cross correlation at some times $t_k$. The covariance matrix connected with $\mathbf{w}_k$ is denoted as

$$E\left\{\mathbf{w}_k\mathbf{w}_i^T\right\} = \mathbf{Q}_k \qquad (7)$$

The covariance matrix $\mathbf{Q}_k$ can be determined using the formula written in the following integral form

$$\mathbf{Q}_k = E\left\{\mathbf{w}_k\mathbf{w}_i^T\right\}$$

$$= E\left\{\left[\int_{t_k}^{t_{k+1}} \mathbf{F}(t_{k+1},\xi)\mathbf{G}(\xi)u(\xi)d\xi\right]\left[\int_{t_k}^{t_{k+1}} \mathbf{F}(t_{k+1},\eta)\mathbf{G}(\eta)u(\eta)d\eta\right]^T\right\}$$

$$= \int_{t_k}^{t_{k+1}}\int_{t_k}^{t_{k+1}} \mathbf{F}(t_{k+1},\xi)\mathbf{G}(\xi)E\left[\mathbf{u}(\xi)\mathbf{u}^T(\eta)\right]\mathbf{G}^T(\eta)\mathbf{F}^T(t_{k+1},\eta)d\xi d\eta \tag{8}$$

The matrix $E[\mathbf{u}(\xi)\mathbf{u}^T(\eta)]$ is the matrix of the Dirac delta function, well known from continuous models.

## 3 DISCRETE KALMAN FILTER

Briefly, the Kalman filter tries to estimate, in an optimal way, the state vector of the controlled process modelled by the linear and stochastic difference equation having the form given by formula (4). Observations (measurements) of the process are done at discrete times and meet the following linear relation

$$\mathbf{y}_k = \mathbf{H}_k\mathbf{x}_k + \mathbf{v}_k \tag{9}$$

where $\mathbf{x}_k$ is the state vector of the process at time $t_k$, $\mathbf{y}_k$ is the vector of the values measured at time $t_k$, $\mathbf{H}_k$ is the matrix representing the relation between the measurements and the state vector at time $t_k$, and $\mathbf{v}_k$ represents the measurement errors. It is assumed that the signals $\mathbf{v}_k$ and $\mathbf{w}_k$ have the mean value equal to zero and there is no correlation between them.

The covariance matrix for the vector $\mathbf{w}_k$ is given by formula (7), while that for $\mathbf{v}_k$ is defined in the following way

$$E\left\{\mathbf{v}_k\mathbf{v}_k^T\right\} = \mathbf{R}_k \tag{10}$$

$$E\left\{\mathbf{w}_k\mathbf{v}_k^T\right\} = 0, \quad \text{for all } k \text{ and } i \tag{11}$$

It is assumed that the initial values of the process estimates are known at the beginning time $t_k$ and that these estimates until time $t_k$ base on the knowledge about the process. Such an estimate is denoted as $\overline{\mathbf{x}}_k$ where the bar means that this is the best estimate at time $t_k$ before the measurement. The estimation error is defined as:

$$\overline{\mathbf{e}}_k = \mathbf{x}_k - \overline{\mathbf{x}}_k \tag{12}$$

and the related error covariance matrix is

$$\overline{\mathbf{P}}_k = E\left\{\overline{\mathbf{e}}_k\overline{\mathbf{e}}_k^T\right\} = E\left\{(\mathbf{x}_k - \overline{\mathbf{x}}_k)(\mathbf{x}_k - \overline{\mathbf{x}}_k)^T\right\} \tag{13}$$

At sampling times $t_k$ at which the measurement $\mathbf{y}_k$ is done, the possessed estimate $\overline{\mathbf{x}}_k$ is corrected using the following relation (Brown & Hwang, 1997; Franklin et al. 1998)

$$\hat{\mathbf{x}}_k = \overline{\mathbf{x}}_k + \mathbf{L}_k\left(\mathbf{y}_k - \mathbf{H}_k\overline{\mathbf{x}}_k\right) \tag{14}$$

where $\hat{\mathbf{x}}_k$ is the estimate updated by the performed measurement, and $\mathbf{L}_k$ is the scaling amplification. The task is to find the vector amplifications $\mathbf{L}_k$

which update the estimate in the optimal way. For this purpose the minimisation of the mean square error is done. Then, the covariance matrix is determined for the error relating to the estimate updated by the performed measurement.

$$\mathbf{P}_k = E\left\{\mathbf{e}_k\mathbf{e}_k^T\right\} = E\left\{(\mathbf{x}_k - \hat{\mathbf{x}}_k)(\mathbf{x}_k - \hat{\mathbf{x}}_k)^T\right\} \tag{15}$$

In time intervals between the sampling times, the estimates are calculated using the following formula

$$\overline{\mathbf{x}}_{k+1} = \mathbf{F}_k\hat{\mathbf{x}}_k \tag{16}$$

Firstly, the covariance matrix $\overline{\mathbf{P}}_{k+1}$ is calculated using formula (13) after correcting it by one sample ahead

$$\overline{\mathbf{P}}_{k+1} = E\left\{\left(\mathbf{x}_{k+1} - \overline{\mathbf{x}}_{k+1}\right)\left(\mathbf{x}_{k+1} - \overline{\mathbf{x}}_{k+1}\right)^T\right\} \tag{17}$$

After placing relations (4) and (16) into formula (17) we get (Brown & Hwang, 1997)

$$\overline{\mathbf{P}}_{k+1} = E\left\{\left[\mathbf{F}_k(\mathbf{x}_k - \hat{\mathbf{x}}_k) + \mathbf{w}_k\right]\left[\mathbf{F}_k(\mathbf{x}_k - \hat{\mathbf{x}}_k) + \mathbf{w}_k\right]^T\right\}$$

$$= \mathbf{F}_k E\left\{(\mathbf{x}_k - \hat{\mathbf{x}}_k)(\mathbf{x}_k - \hat{\mathbf{x}}_k)^T\right\}\mathbf{F}_k^T + \mathbf{F}_k E\left\{(\mathbf{x}_k - \hat{\mathbf{x}}_k)\mathbf{w}_k^T\right\}$$

$$+ E\left\{\mathbf{w}_k(\mathbf{x}_k - \hat{\mathbf{x}}_k)^T\right\}\mathbf{F}_k^T + E\left\{\mathbf{w}_k\mathbf{w}_k^T\right\} \tag{18}$$

No correlation is assumed between the estimation error signals $\mathbf{e}_k$ and the disturbances $\mathbf{w}_k$. After placing the covariances defined by formulas (7) and (15) into relation (18) we get the required error covariance matrix between the sampling times

$$\overline{\mathbf{P}}_{k+1} = \mathbf{F}_k\mathbf{P}_k\mathbf{F}_k^T + \mathbf{Q}_k \tag{19}$$

The estimation error covariance matrix $\mathbf{P}_k$ is calculated for sampling times in the similar way. After placing relations (9) and (14) into formula (15) we get

$$\mathbf{P}_k = E\left\{\left[(\mathbf{x}_k - \overline{\mathbf{x}}_k) - \mathbf{L}_k\mathbf{H}_k(\mathbf{x}_k - \overline{\mathbf{x}}_k) - \mathbf{L}_k\mathbf{v}_k\right]\right.$$

$$\left. \cdot \left[(\mathbf{x}_k - \overline{\mathbf{x}}_k) - \mathbf{L}_k\mathbf{H}_k(\mathbf{x}_k - \overline{\mathbf{x}}_k) - \mathbf{L}_k\mathbf{v}_k\right]^T\right\} \tag{20}$$

And after similar algebra operations as in formula (21) we get the following solution

$$\mathbf{P}_k = \overline{\mathbf{P}}_k - \mathbf{L}_k\mathbf{H}_k\overline{\mathbf{P}}_k - \overline{\mathbf{P}}_k\mathbf{H}_k^T\mathbf{L}_k^T$$

$$+ \mathbf{L}_k\left(\mathbf{H}_k\overline{\mathbf{P}}_k\mathbf{H}_k^T + \mathbf{R}_k\right)\mathbf{L}_k^T \tag{21}$$

The final task is to calculate the optimal values of the amplification matrix $\mathbf{L}_k$. This task is realised by finding such values of the vector $\mathbf{L}_k$. which minimise the trace of the matrix $\mathbf{P}_k$ being the sum of the mean square errors of the estimates of all state vector elements. The trace of the matrix $\mathbf{P}_k$ is differentiated with respect to $\mathbf{L}_k$ and made equal to zero. What can be easily noticed, the second and third term of equation (21) are linear with respect to $\mathbf{L}_k$, while the fourth term is quadratic and the trace of $\mathbf{L}_k\mathbf{H}_k\overline{\mathbf{P}}_{k}$ is equal to the trace of its transposition $\mathbf{L}_k\mathbf{H}_k\overline{\mathbf{P}}_k$ $\overline{\mathbf{P}}_k\mathbf{H}_k^T\mathbf{L}_k^T$. We get (Brown & Hwang, 1997)

$$\frac{d(\text{trace } \mathbf{P}_k)}{d\mathbf{L}_k} =$$
$$-2\left(\mathbf{H}_k\overline{\mathbf{P}}_k^T\right)^T + 2\mathbf{L}_k\left(\mathbf{H}_k\overline{\mathbf{P}}_k\mathbf{H}_k^T + \mathbf{R}_k\right) = 0 \quad (22)$$

and after some transformations we arrive at the required form of the matrix $\mathbf{L}_k$, bearing the name of Kalman amplification:

$$\mathbf{L}_k = \overline{\mathbf{P}}_k\mathbf{H}_k^T\left(\mathbf{H}_k\overline{\mathbf{P}}_k\mathbf{H}_k^T + \mathbf{R}_k\right)^{-1} \quad (23)$$

Now we remove the matrix $\mathbf{L}_k$ from equation (21) by placing the relation (23) to get

$$\mathbf{P}_k = \overline{\mathbf{P}}_k - \overline{\mathbf{P}}_k\mathbf{H}_k^T\left(\mathbf{H}_k\overline{\mathbf{P}}_k\mathbf{H}_k^T + \mathbf{R}_k\right)^{-1}\mathbf{H}_k\overline{\mathbf{P}}_k \quad (24)$$

The obtained recurrent calculation algorithm, based on equations (14), (16), (19) (23) and (24), is widely known as the Kalman filter. Equation (24) can be presented in another form taking into account the $\mathbf{L}_k$ relation given by the formula (23) which gives

$$\mathbf{P}_k = \overline{\mathbf{P}}_k - \mathbf{L}_k\mathbf{H}_k\overline{\mathbf{P}}_k = (\mathbf{I} - \mathbf{L}_k\mathbf{H}_k)\overline{\mathbf{P}}_k \quad (25)$$

## 4 CONVERSION FROM DISCRETE TO CONTINUOUS EQUATIONS DESCRIBING THE KALMAN FILTER ALGORITHM

The general form of a continuous process is already given by equation (1), while the equation describing the measuring model for a continuous system is

$$\mathbf{y} = \mathbf{C}\mathbf{x} + \mathbf{v} \quad (26)$$

Consequently, by analogy with the discrete model it is assumed that the vectors $\mathbf{u}(t)$ and $\mathbf{v}(t)$ are the vectors of the stochastic process bearing the name of the white noise with the zero cross correlation value.

$$E\left\{\mathbf{u}(t)\mathbf{u}^T(\tau)\right\} = \mathbf{Q}\delta(t - \tau) \quad (27)$$

$$E\left\{\mathbf{v}(t)\mathbf{v}^T(\tau)\right\} = \mathbf{R}\delta(t - \tau) \quad (28)$$

$$E\left\{\mathbf{u}(t)\mathbf{v}^T(\tau)\right\} = 0 \quad (29)$$

The covariance parameters $\mathbf{Q}$ and $\mathbf{R}$ play a similar role to that played by the parameters $\mathbf{Q}_k$ (7) and

$\mathbf{R}_k$ (10) in the discrete filter, but have different numerical values.

To do the conversion from discrete to continuous form, let us first define the relations between $\mathbf{Q}_k$ and $\mathbf{R}_k$, on the one hand, and the corresponding values of $\mathbf{Q}$ and $\mathbf{R}$ for small sampling intervals $\Delta t$. Based on formula (5) we can notice that for small $\Delta t$ values the discrete transition matrix $\mathbf{F}_k = \mathbf{I}$. After applying this conclusion to the matrix $\mathbf{Q}_k$ given by formula (8) we get (Brown & Hwang, 1997).

$$\mathbf{Q}_k = \iint_{\Delta t \to 0} \mathbf{G}(\xi)E\left[\mathbf{u}(\xi)\mathbf{u}^T(\eta)\right]\mathbf{G}^T(\eta)d\xi d\eta \quad (30)$$

Then placing the equation (27) into the equation (30) and integrating for a small time interval $\Delta t$ we get

$$\mathbf{Q}_k = \mathbf{G}\mathbf{Q}\mathbf{G}^T\Delta t \quad (31)$$

Deriving the equation relating $\mathbf{R}_k$ to $\mathbf{R}$ is not straightforward. In the continuous model the signal $\mathbf{v}(t)$ is the white noise and direct sampling returns the measuring noise with infinite variance. In order to get equivalent values at the discrete and continuous times it is assumed that the continuous measurements are averaged over the time interval $\Delta t$ in the sampling process. The state variables $\mathbf{x}$ are not "white" and can be approximated as constant along this interval. Hence

$$\mathbf{y}_k = \frac{1}{\Delta t}\int_{t_{k-1}}^{t_k} y(t)dt = \frac{1}{\Delta t}\int_{t_{k-1}}^{t_k} \left[\mathbf{C}\mathbf{x}(t) + \mathbf{v}(t)\right]dt$$

$$\approx \mathbf{H}_k\mathbf{x}_k + \frac{1}{\Delta t}\int_{t_{k-1}}^{t_k} \mathbf{v}(t)dt \quad (32)$$

where $\mathbf{H}_k = \mathbf{C}(t_k)$. This way a new equivalent is obtained which defines the relation between the continuous time and discrete time domains

$$\mathbf{v}_k = \frac{1}{\Delta t}\int_{\Delta t \to 0} \mathbf{v}(t)dt \quad (33)$$

From formula (10) we get

$$E\left[\mathbf{v}_k\mathbf{v}_k^T\right] = \mathbf{R}_k = \frac{1}{\Delta t^2}\iint_{\Delta t \to 0} E\left[\mathbf{v}(u)\mathbf{v}^T(v)\right]dudv \quad (34)$$

Placing equation (28) into equation (34) and integrating we get the required relation

$$\mathbf{R}_k = \frac{\mathbf{R}}{\Delta t} \quad (35)$$

The amplification of the discrete Kalman filter is given by formula (23). After placing relation (35) into this formula we get

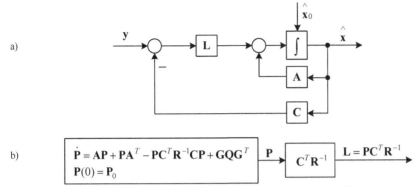

a)

b)

$$\dot{\mathbf{P}} = \mathbf{AP} + \mathbf{PA}^T - \mathbf{PC}^T\mathbf{R}^{-1}\mathbf{CP} + \mathbf{GQG}^T$$
$$\mathbf{P}(0) = \mathbf{P}_0$$

$\mathbf{P}$ → $\boxed{\mathbf{C}^T\mathbf{R}^{-1}}$ → $\mathbf{L} = \mathbf{PC}^T\mathbf{R}^{-1}$

Figure 2. Block diagram of the continuous Kalman filter.

$$\mathbf{L}_k = \overline{\mathbf{P}}_k\mathbf{H}_k^T\left(\mathbf{H}_k\overline{\mathbf{P}}_k\mathbf{H}_k^T + \mathbf{R}/\Delta t\right)^{-1}$$

$$\approx \overline{\mathbf{P}}_k\mathbf{H}_k^T\mathbf{R}^{-1}\Delta t \qquad (36)$$

as the second term inside the parentheses is dominating in formula (36)

$$\mathbf{R}/\Delta t \gg \mathbf{H}_k\overline{\mathbf{P}}_k\mathbf{H}_k^T \qquad (37)$$

When passing from the discrete form to the continuous form, the error covariance matrix between sampling times, given by formula (19), nears to $\overline{\mathbf{P}}_{k+1} \approx \overline{\mathbf{P}}_k$ when $\Delta t \to 0$. Therefore only one error matrix $\mathbf{P}$ is in force in the continuous filter. Equation (36) takes the form

$$\mathbf{L}_k = \mathbf{PC}^T\mathbf{R}^{-1}\Delta t \qquad (38)$$

And, finally, when $\Delta t \to 0$, we arrive at the formula for determining the amplification $\mathbf{L}$ in the continuous Kalman filter (Brown & Hwang, 1997)

$$\mathbf{L} = \frac{\mathbf{L}_k}{\Delta t} = \mathbf{PC}^T\mathbf{R}^{-1} \qquad (39)$$

As the next step, the equation describing the estimation error covariance matrix is to be derived. Placing relation (25) into equation (19) we get

$$\overline{\mathbf{P}}_{k+1} = \mathbf{F}_k\left(\mathbf{I} - \mathbf{L}_k\mathbf{H}_k\right)\overline{\mathbf{P}}_k\mathbf{F}_k^T + \mathbf{Q}_k$$

$$= \mathbf{F}_k\overline{\mathbf{P}}_k\mathbf{F}_k^T - \mathbf{F}_k\mathbf{L}_k\mathbf{H}_k\overline{\mathbf{P}}_k\mathbf{F}_k^T + \mathbf{Q}_k \qquad (40)$$

Then the discrete transition matrix $\mathbf{F}_k$ in equation (40) is substituted by its approximate given by formula (5) (Brown & Hwang, 1997)

$$\overline{\mathbf{P}}_{k+1} = \left(\mathbf{I} + \mathbf{A}\Delta t\right)\overline{\mathbf{P}}_k\left(\mathbf{I} + \mathbf{A}\Delta t\right)^T$$

$$- \left(\mathbf{I} + \mathbf{A}\Delta t\right)\mathbf{L}_k\mathbf{H}_k\overline{\mathbf{P}}_k\left(\mathbf{I} + \mathbf{A}\Delta t\right)^T + \mathbf{Q}_k$$

$$= \overline{\mathbf{P}}_k + \mathbf{A}\overline{\mathbf{P}}_k\Delta t + \overline{\mathbf{P}}_k\mathbf{A}^T\Delta t + \mathbf{A}\overline{\mathbf{P}}_k\mathbf{A}^T\Delta t^2$$

$$- \mathbf{L}_k\mathbf{H}_k\overline{\mathbf{P}}_k - \mathbf{A}\mathbf{L}_k\mathbf{H}_k\overline{\mathbf{P}}_k\Delta t - \mathbf{L}_k\mathbf{H}_k\overline{\mathbf{P}}_k\mathbf{A}^T\Delta t$$

$$- \mathbf{A}\mathbf{L}_k\mathbf{H}_k\overline{\mathbf{P}}_k\mathbf{A}^T\Delta t^2 + \mathbf{Q}_k \qquad (41)$$

It can be seen from equation (38), than in equation (41) the matrix $\mathbf{L}_k$ is of an order of $\Delta t$. And, after removing all terms containing $\Delta t$ of and order higher than one from equation (41) we get the simplified form (Brown & Hwang, 1997):

$$\overline{\mathbf{P}}_{k+1} = \overline{\mathbf{P}}_k + \mathbf{A}\overline{\mathbf{P}}_k\Delta t + \overline{\mathbf{P}}_k\mathbf{A}^T\Delta t - \mathbf{L}_k\mathbf{H}_k\overline{\mathbf{P}}_k + \mathbf{Q}_k \,(42)$$

After substituting formula (38) for $\mathbf{L}_k$ and formula (31) for $\mathbf{Q}_k$ in equation (42) we arrive at

$$\overline{\mathbf{P}}_{k+1} = \overline{\mathbf{P}}_k + \mathbf{A}\overline{\mathbf{P}}_k\Delta t + \overline{\mathbf{P}}_k\mathbf{A}^T\Delta t$$

$$- \overline{\mathbf{P}}_k\mathbf{H}_k^T\mathbf{R}^{-1}\mathbf{H}_k\overline{\mathbf{P}}_k\Delta t + \mathbf{GQG}^T\Delta t \qquad (43)$$

Based on equation (43) we can get the following difference

$$\frac{\overline{\mathbf{P}}_{k+1} - \overline{\mathbf{P}}_k}{\Delta t} = \mathbf{A}\overline{\mathbf{P}}_k + \overline{\mathbf{P}}_k\mathbf{A}^T$$

$$- \overline{\mathbf{P}}_k\mathbf{H}_k^T\mathbf{R}^{-1}\mathbf{H}_k\overline{\mathbf{P}}_k + \mathbf{GQG}^T \qquad (44)$$

Then, after limiting $\Delta t \to 0$ and removing all subscripts and bars over matrices we get the matrix differential equation

$$\dot{\mathbf{P}} = \mathbf{AP} + \mathbf{PA}^T - \mathbf{PC}^T\mathbf{R}^{-1}\mathbf{CP} + \mathbf{GQG}^T \qquad (45)$$

with the initial condition

$$\mathbf{P}(0) = \mathbf{P}_0 \qquad (46)$$

The last remaining step is to derive the state estimation equation given by formula (14). Placing the below given relation (47),

$$\overline{\mathbf{x}}_k = \mathbf{F}_{k-1}\hat{\mathbf{x}}_{k-1} \qquad (47)$$

25

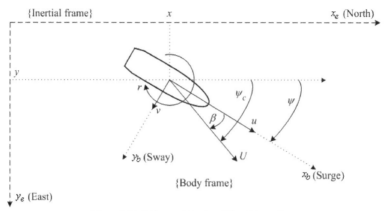

Figure 3. Variables used for describing ship motion.

derived from equation (16), into formula (14) makes it possible to arrive at the following form of equation (17)

$$\hat{\mathbf{x}}_k = \mathbf{F}_{k-1}\hat{\mathbf{x}}_{k-1} + \mathbf{L}_k\left(\mathbf{y}_k - \mathbf{H}_k\mathbf{F}_{k-1}\hat{\mathbf{x}}_{k-1}\right) \qquad (48)$$

Here again, the matrix $\mathbf{F}_{k-1}$ is approximated by formula (4)

$$\hat{\mathbf{x}}_k = \hat{\mathbf{x}}_{k-1} + \mathbf{A}\hat{\mathbf{x}}_{k-1}\Delta t + \mathbf{L}_k\left(\mathbf{y}_k - \mathbf{H}_k\hat{\mathbf{x}}_{k-1} - \mathbf{H}_k\mathbf{A}\hat{\mathbf{x}}_{k-1}\Delta t\right) (49)$$

After removing all terms containing $\Delta t$ of an order higher than one from equation (50) and observing that $\mathbf{L}_k = \mathbf{L}\Delta t$, the equation takes the form

$$\hat{\mathbf{x}}_k - \hat{\mathbf{x}}_{k-1} = \mathbf{A}\hat{\mathbf{x}}_{k-1}\Delta t + \mathbf{L}\Delta t\left(\mathbf{y}_k - \mathbf{H}_k\hat{\mathbf{x}}_{k-1}\right) \qquad (50)$$

Finally, after dividing by $\Delta t$,

$$\frac{\hat{\mathbf{x}}_k - \hat{\mathbf{x}}_{k-1}}{\Delta t} = \mathbf{A}\hat{\mathbf{x}}_{k-1} + \mathbf{L}\left(\mathbf{y}_k - \mathbf{H}_k\hat{\mathbf{x}}_{k-1}\right) \qquad (51)$$

reducing by $\Delta t \rightarrow 0$, and removing all subscripts and bars over variable matrices we get the matrix differential equation of continuous state estimation

$$\dot{\hat{\mathbf{x}}} = \mathbf{A}\hat{\mathbf{x}} + \mathbf{L}\left(\mathbf{y} - \mathbf{C}\hat{\mathbf{x}}\right) \qquad (52)$$

Equations (39), (45), (46) and (52) compose the continuous Kalman filter. They are shown in Fig. 1. Figure 1(a) shows a block diagram illustrating the principle of operation of the continuous Kalman filter. The input signal for the filter is the measured noised output signal of the object. Figure 1(b) presents the method of determining the optimal amplification $\mathbf{L}$.

## 5 APPLYING THE KALMAN-BUCY FILTER FOR THE ESTIMATION OF SHIP MOTION VELOCITIES

The algorithm of the continuous Kalman-Bucy filter described in the previous section was applied for the estimation of ship motion parameters. The tests were performed on the training ship *Blue Lady* owned by the Foundation of Sailing Safety and Environment Protection in Ilawa. *Blue Lady* is the physical model, in scale 1:24, of a tanker designed for transporting crude oil. The overall length of *Blue Lady* is $L = 13.75$ m, the width is $B = 2.38$ m, and the mass is $m = 22.934 \times 10^3$ [kg].

The ship sailing on the surface of the water region is considered a rigid body moving in three degrees of freedom. The ship position $(x, y)$ and the ship course $\psi$ in the horizontal plane with respect to the stationary, inertial coordinate system $\{x_e, y_e\}$ are represented by the vector $\boldsymbol{\eta} = [x, y, \psi]^T$. The second coordinate system $\{x_b, y_b\}$ is connected with the moving ship and fixed to its centre of gravity. Velocities of the moving ship are represented by the vector $\mathbf{v} = [u, v, r]^T$, where $u$ is the longitudinal ship velocity (surge), $v$ is the lateral velocity (sway), and $r$ is the angular speed (yaw). These variables are shown in Fig. 3.

The position coordinates $(x, y)$ are measured by DGPS (Differential Global Positioning System), while the ship course $\psi$ is measured by the gyrocompass. These three measured state variables are collected in the vector $\boldsymbol{\eta} = [x, y, \psi]^T$. The three remaining state variables, composing the vector $\mathbf{v} = [u, v, r]^T$, are to be estimated.

The ship motion equations simply express the Newton's second law of motion in three degrees of freedom. These equations, formulated in the stationary coordinate system connected with the map of the water region, have the following form (Clarke, 2003).

$$m\ddot{x} = X \qquad (53)$$

$$m\ddot{y} = Y \qquad (54)$$

$$I_z\ddot{\psi} = N \qquad (55)$$

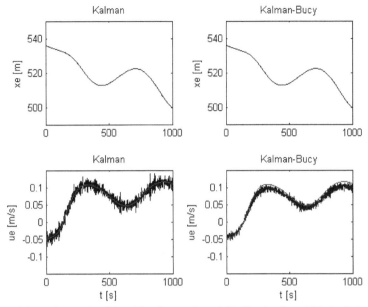

Figure 4. Simulation study: actual position with estimate and actual (black) and estimated (blue) velocity $u$ in surge.
Left-hand column – discrete Kalman filter, right-hand column – continuous Kalman-Bucy filter.

where $X$ and $Y$ are forces acting along the $x_b$ and $y_b$ axes, respectively, $N$ is the torque, $m$ is the mass of the ship, and $I_z$ is the moment of inertia along the lateral axis directed downwards.

The above differential equations can be presented as three sets of dynamic equations having the following general form

$$\dot{\mathbf{x}} = \mathbf{A}\mathbf{x} + \mathbf{B}\mathbf{u} \qquad (56)$$

$$y = \mathbf{C}\mathbf{x} \qquad (57)$$

For each degree of freedom the matrices $\mathbf{A}$, $\mathbf{B}$, $\mathbf{C}$ are identical and take the form

$$\mathbf{A} = \begin{bmatrix} 0 & 0 \\ 1 & 0 \end{bmatrix}, \qquad \mathbf{B} = \begin{bmatrix} 1 \\ 0 \end{bmatrix}, \qquad \mathbf{C} = \begin{bmatrix} 0 & 1 \end{bmatrix} \qquad (58)$$

while the state vectors for consecutive states of freedom are the following

$$\mathbf{x}_1 = \begin{bmatrix} u_x \\ x \end{bmatrix}, \qquad \mathbf{x}_2 = \begin{bmatrix} u_y \\ y \end{bmatrix}, \qquad \mathbf{x}_3 = \begin{bmatrix} r \\ \psi \end{bmatrix} \qquad (59)$$

where $u_x = dx/dt$, $v_y = dy/dt$, $r = d\psi/dt$ are velocities in the stationary coordinate system. The velocity vector $\mathbf{v} = [u,v,r]^T$ expressed in the moving coordinate system $\{x_b, y_b\}$, can be calculated based on the velocities determined in the stationary coordinate system $\{x_e, y_e\}$ and making use of the relation

$$\begin{bmatrix} u \\ v \\ r \end{bmatrix} = \begin{bmatrix} \cos\psi & \sin\psi & 0 \\ -\sin\psi & \cos\psi & 0 \\ 0 & 0 & 1 \end{bmatrix} \cdot \begin{bmatrix} u_x \\ v_y \\ r \end{bmatrix} \qquad (60)$$

The continuous Kalman-Bucy filter implemented for estimating the motion parameters on the training ship Blue Lady worked based on the system and equations shown in Fig. 2. Moreover, for the purposes of the algorithm of the Kalman-Bucy filter, relevant values of the coefficients in the matrices $\mathbf{G}$, $\mathbf{Q}$ and $\mathbf{R}$ were selected. For the first and second degree of freedom the following values were adopted:

$$\mathbf{G}_x = \mathbf{G}_y = \begin{bmatrix} 1 & 0 \\ 0 & 0.01 \end{bmatrix}, \quad \mathbf{Q}_x = \mathbf{Q}_y = \begin{bmatrix} 0.1 & 0 \\ 0 & 0.2 \end{bmatrix},$$

$$R_x = R_y = 0.01 \qquad (61)$$

while for the third degree of freedom:

$$\mathbf{G}_\psi = \begin{bmatrix} 0.2 & 0 \\ 0 & 0.01 \end{bmatrix}, \quad \mathbf{Q}_\psi = \begin{bmatrix} 1 & 0 \\ 0 & 1 \end{bmatrix}, \quad R_\psi = 0.1 \qquad (62)$$

The covariances of the position coordinates measured by GPS were equal to $R_x = R_y = 0.01$ while the covariances of the ship course measurement were equal to $R_\psi = 0.1$ and were determined based on the experimental tests done on the training ship *Blue Lady*.

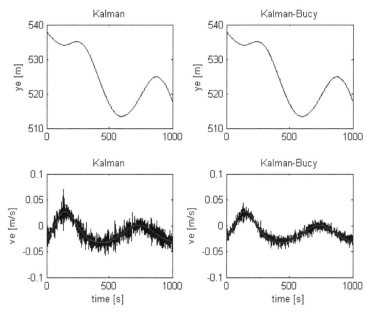

Figure 5. Simulation study: actual position with estimate and actual (black) and estimated (blue) velocity $v$ in sway. Left-hand column – discrete Kalman filter, right-hand column – continuous Kalman-Bucy filter.

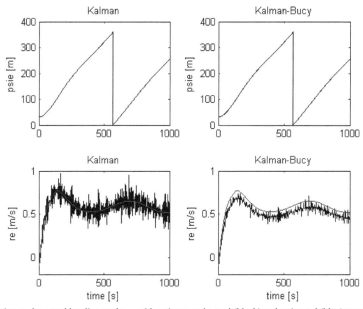

Figure 6. Simulation study: actual heading angle $\psi$ with estimate and actual (black) and estimated (blue) angular rate $r$ in yaw. Left-hand column – discrete Kalman filter, right-hand column – continuous Kalman-Bucy filter.

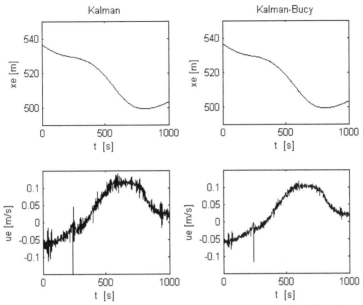

Figure 7. Experimental data: measured position with estimate and estimated velocity $u$ in surge.
Left-hand column – discrete Kalman filter, right-hand column – continuous Kalman-Bucy filter.

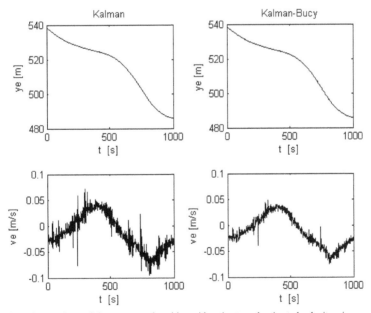

Figure 8. Experimental data: measured position with estimate and estimated velocity $v$ in sway.
Left-hand column – discrete Kalman filter, right-hand column – continuous Kalman-Bucy filter.

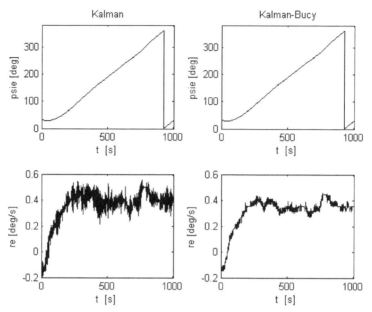

Figure 9. Experimental data: measured heading angle $\psi$ with estimate and estimated angular rate $r$ in yaw. Left-hand column – discrete Kalman filter, right-hand column – continuous Kalman-Bucy filter.

Initially, the simulation investigations were performed in the calculating environment Matlab/Simulink based on the mathematical model of the training ship *Blue Lady*, described in detail by (Gierusz, 2001; Gierusz. 2005). These investigations were performed at the presence of measurement noises, which were added to the positions and heading measurements in simulations and at the presence of the external disturbances. In simulation study assumed that on ship was acting wind with average speed equal 2 m/s and in direction 0 degrees. The simulation results are shown in Figs. 4-6. The actual and estimated velocities in surge, sway and yaw are shown in the bottom of plots.

After simulation tests, the algorithm of the discrete Kalman-Bucy filter was implemented on the training ship *Blue Lady* sailing on the lake Silm near Ilawa. The performed experimental tests aimed at testing the quality of filter operation and its resistance to disturbances. In order to provide good opportunities for comparison, the tests of the training ship Blue Lady were also performed using the discrete Kalman filter described by Tomera (Tomera, 2010).

The results recorded in the experimental tests are shown in Figs. 7 – 9. The diagrams in the left-hand column present the results obtained for the discrete Kalman filter while the time histories in the right-hand column refer to the investigations performed using the continuous Kalman-Bucy filter.

The presented diagrams reveal that the estimates of the position coordinates and the course are identi-cal as the measured values. On the other hand, the correspondence between the time-histories of the estimated velocities is much worse, as their curves are not smooth and are burdened with relatively large errors. All inaccuracies in the measured values are reflected in the estimated velocity values.

## 6  REMARKS AND CONCLUSIONS

The article describes the method of deriving the algorithm of the continuous Kalman-Bucy filter based on the discrete Kalman filter. This algorithm does not take into account the model of ship dynamics, but only bases on the position coordinates $x$, $y$ measured by GPS and the ship course angle $\psi$ measured by the gyro-compass. The estimates of ship position coordinates $x$, $y$ and ship course angle $\psi$ shown in Figs. 7, 8 and 9 well correspond to the measured values. The correspondence is worse for velocity estimates determined for the moving coordinate system fixed to the ship and collected in the vector $\mathbf{v}=[u,v,r]^T$. The determined time-histories of these velocities are burdened with a noise of relatively high level, lower for the Kalman-Bucy controller than for the discrete Kalman filter. The shapes of the velocity estimate curves indicate that they need additional smoothing.

Additional difficulties in velocity estimation may appear when the instantaneous ship position coordinates measured by GPS reveal rapid changes. In this situation additional oscillations can be observed in the estimated velocities. Fortunately, in the sample results of investigations shown in Figs. 7 through 9 these difficulties were not recorded.

## REFERENCES

Balchen J.G., Jenssen N.A. & Saelid S. 1976. Dynamic positioning using Kalman filtering and optimal control theory. *In IFAC/IFIP Symposium on Automation in Offshore Oil Field Operation*:183-186. Bergen, Norway.

Balchen, J. G., Jenssen, N. A. & Saelid, S. 1980a. Dynamic positioning of floating vessels based on Kalman filtering and optimal control, *Proceedings of the 19th IEEE Conference on Decision and Control*:852-864. New York.

Balchen, J. G., Jenssen, N. A., Mathiasen, E. & Saelid, S. 1980b. A dynamic positioning system based on Kalman filtering and optimal control, *Modeling, Identification and Control*, 1(3):135-163.

Brown, R.G. & Hwang, P.Y.C. 1997. *Introduction to Random Signals and Applied Kalman Filtering with Matlab Exercises and Solutions*. Third Edition. John Wiley & Sons, Inc.

Clarke, D. 2003. The foundations of steering and maneouvering. *Proceedings of the IFAC Conference Manoeuvering and Control Marine Crafts*:10-25. Girona. Spain.

Franklin, G.F., Powell, J.D. & Workman, M. 1998. *Digital Control of Dynamic Systems*. Third Edition. Addison Wesley Longman.

Fung, P. & Grimble, M. 1983. Dynamic Ship Positioning Using Self-Tuning Kalman Filter. *IEEE Transactions on Automatic Control*, 28(3):339-349.

Gierusz, W. 2001. Simulation model of the shiphandling training boat Blue Lady. *Proceedings of Control Applications in Marine Systems*. Glasgow. Scotland. UK.

Gierusz, W. 2005. *Synthesis of multivariable systems of precise ship motion control with the aid of selected resistant system design methods*. Publishing House of Gdynia Maritime University. (in Polish).

Kalman, R.E. 1960. A New Approach to Linear Filtering and Prediction Problems. *Transactions of the ASME – Journal of Basic Engeenering*. 82(D):35-45.

Kalman, R.E. & Bucy R.S. 1961. New Results in Linear Filtering and Prediction Theory. *Transactions of the ASME – Journal of Basic Engeenering*. 83(D):95-108.

Saelid, S., Jensen, N. A. & Balchen, J. G. 1983. Design and analysis of a dynamic positioning system based on Kalman filtering and optimal control, *IEEE Transactions on Automatic Control*, 28(3):331-339.

Sorensen, A. J., Sagatun, S. I. & Fossen, T. I. 1996. Design of a Dynamic Position System Using Model-based Control., *Control Engeenering Practice*, 4(3):359-368.

Strand, J. P., Sorensen, A. J. & Fossen, T. I. 1997. Modelling and control of thruster assisted position mooring systems for ships, *Procedings of IFAC Manoeuvering and Control of Marine Craft (MCMC'97)*:160-165. Brijuni, Croatia.

Tomera, M. 2010. Discrete Kalman filter design for multivariable ship motion control: experimental results with training ship. *Joint Proceedings of Gdynia Maritime Academy & Hochschule Bremerhaven*:26-34. Bremerhaven.

# 3. Method of Evaluation of Insurance Expediency of Stevedoring Company's Responsibility for Cargo Safety

M. Ya. Postan & O. O. Balobanov
*Odessa National Maritime University, Ukraine*

ABSTRACT: The method of insurance expediency of stevedoring company's responsibility for safety of containers under their transshipment at port's terminal is proposed. This method is based on representation of terminal as a queueing system of GI/G/m type and on comparison of the stevedoring company's insurance expenditures and random value of transshipped containers' total damage (sum insured) for a given period of time.

## 1 INTRODUCTION

In operational activity of stevedoring companies, the many cases may occur related to the situations of risk. The main of them are listed below:
– damage of a ship's hull or equipment during the loading/unloading;
– cargo package's damage in result of violation of loading/unloading rules or rules of port's mechanisms exploitation;
– damage of cargo in result of violations of rules of its storage at warehouse;
– failures of port's equipment;
– exceeding of a ship's laytime.

Appearance of above events leads to some additional expenditure for the stevedoring company, charterer or cargo owner. As shows the international commercial practice, many stevedoring companies which operate in big seaports insure their responsibility for safe and qualitative transshipment of cargo (within the framework of contract responsibility) [1].

When the managers of stevedoring company make decision concerning an insurance of its responsibility for safe transshipment of cargo it is useful and even necessary to apply the methods of probability theory and actuarial mathematics [2, 3]. At the same time the standard methods of quantitative evaluation of risk proposed by mathematical risk theory are mainly aimed at insurance companies' profit but not at protection of commercial interests of insurants. Therefore, specifics of seaports operational activity and interrelations between stevedoring company's managers and its clients demand the special methods for actuarial calculations.

The purpose of our paper is working out a method of a risk evaluation of containers damage under their transshipment at a seaport terminal and substantiation of insurance expediency of this risk by a stevedoring company.

## 2 MAIN RESULTS

Our approach is based on representation of port's container terminal as a queueing system of GI/G/m type ($m$ identical servers in parallel, infinite waiting room, service discipline is FIFO).

We denote:

$\omega(t)$ be a random number of served ships in time interval $(0, t)$;

$\gamma_k$ be a random number of containers transshipped on/from the $k$th ship served in time interval $(0, t)$;

$v_k$ be a random number of damaged containers during loading/unloading of the $k$th ship;

$\xi_{ki}$ be a random value of damage caused to the $i$th container loaded on or unloaded from the $k$th ship (estimated in money).

It is assumed that:

1 the random variables $\gamma_1, \gamma_2, \ldots$ are independent and identically distributed (i.i.d.) with the discrete distribution

$$\pi_M = \Pr\{\gamma_1 = M\}, M = 1, 2, \ldots, \sum_{M \geq 1} \pi_M = 1; \quad (1)$$

2 $v_1, v_2, \ldots$ are the i.i.d. random variables with the conditional binomial distribution

$$\Pr\{\underset{k}{\chi} = n \mid \gamma_k = M\} = C_M^n p^n q^{M-n} \quad (q = 1-p), \tag{2}$$
$$n = 0,1,\dots,M,$$

where $p$ is the probability that a damage is caused to arbitrary container through a stevedoring company's fault;

$\xi_{11}, \xi_{12},\dots, \xi_{21},\dots$ are the i.i.d. random variables with the distribution function (d.f.)

$$D(x) = \Pr\{\xi_{11} \le x\}; \tag{3}$$

the sequences of random variables $\nu_1, \nu_2,\dots$ and $\xi_{11}, \xi_{12},\dots$ are mutually independent.

If $\tau$ denotes the constant loading/unloading time of one container, than service time of the $k$th served ship is the random variable $\gamma_k \tau$. We shall consider the steady-state regime of our queueing system functioning and assume that the following stability condition holds true

$$\lambda < m /(\tau\, E\gamma_1), \tag{4}$$

where $\lambda^{-1}$ is the mean interarrival time of the ships.

Let us evaluate the total damage in time interval $(0, t)$ caused to containers by stevedoring company $(\Delta(t))$. Using the above designations we can write

$$\Delta(t) = \sum_{k=1}^{\omega(t)} \sum_{i=1}^{\nu_k} \xi_{ki}. \tag{5}$$

The financial managers of a stevedoring company face the dilemma: to insure or not to insure the possible total damage (4) with the gross risk premium rate $c$ (we assume that the sum insured is $\Delta(t)$). Note that $t$ we consider as the period of insurance policy action.

The simplest criterion of insurance expediency is: the average profit of a stevedoring company in result of the total damage insurance must be positive, i.e.

$$E(\Delta(t) - ct) > 0. \tag{6}$$

Taking into account relations (1)-(5) and applying to right-hand side of (5) theorem of total mathematical expectation, from (6) we have

$$E\omega(t) E\nu_1 E\xi_{11} > ct, \tag{7}$$

where

$$E\xi_{11} = \int_0^\infty x\, dD(x) < \infty,\quad E\nu_1 = p\sum_{n\ge 1} n\pi_n < \infty. \tag{8}$$

For ergodic queue (see (4)) $E\omega(t) = \lambda t$ [4]. Therefore, from (7) we obtain

$$\lambda\, E\nu_1 E\xi_{11} > c. \tag{8}$$

More precise criterion than (8) is

$$\Pr\{\Delta(t) > ct\} \ge 1 - \varepsilon, \tag{9}$$

where $\varepsilon$ is a given small probability. For application of criterion (9) we need to determine the d.f. of stochastic process $\Delta(t)$.

For the sake of simplicity, we suppose that $\gamma_k = N, k = 1,2,\dots,$ where $N$ may be interpreted as hold capacity of a ship (in TEU). In other words, we assume that each ship arrives for loading/unloading of exactly $N$ containers. Then by theorem of total probability, taking into account (5), mutual independence of $\omega(t)$ and $\nu_1, \nu_2,\dots$, we can write

$$F(x,t) = \Pr\{\Delta(t) \le x\} = \Pr\{\omega(t) = 0\} + \sum_{k=1}^\infty \Pr\{\omega(t) = k\} \times$$
$$\times \sum_{n_1=0}^N \cdots \sum_{n_k=0}^N \prod_{i=1}^k C_N^{n_i} p^{n_i} q^{N-n_i} D^{(n_1+\dots+n_k)}(x), \tag{10}$$

where $D^{(n)}(x)$ is $n$-multiple convolution of d.f. $D(x)$ with itself, $D^{(0)}(x) \equiv 1$.

Due to the formula (10) the criterion (9) takes the form

$$F(ct,t) = \Pr\{\omega(t) = 0\} + \sum_{k=1}^\infty \Pr\{\omega(t) = k\} \times \tag{11}$$
$$\times \sum_{n_1=0}^N \cdots \sum_{n_k=0}^N \prod_{i=1}^k C_N^{n_i} p^{n_i} q^{N-n_i} D^{(n_1+\dots+n_k)}(ct) \le \varepsilon.$$

In practice, $N$ may be considered as large and $p$ as small quantities. Therefore, the binomial terms in (11) may approximately be substituted for the Poisson distribution. From (11), it follows

$$\Pr\{\omega(t) = 0\} + \sum_{k=1}^\infty e^{-ak}\Pr\{\omega(t) = k\} \times$$
$$\times \sum_{n_1=0}^\infty \cdots \sum_{n_k=0}^\infty \prod_{i=1}^k \frac{a^{n_i}}{n_i!} D^{(n_1+\dots+n_k)}(ct) \le \varepsilon, \tag{12}$$

where $a = Np$.

The Laplace-Stieltjes transform of d.f. (10) on variable $x$ is given by

$$\int_0^\infty e^{-sx} d_x F(x,t) = \Pr\{\omega(t)=0\} +$$

$$+ \sum_{k=1}^\infty \Pr\{\omega(t)=k\} \sum_{n_1=0}^N \cdots \sum_{n_k=0}^N \prod_{i=1}^k C_N^{n_i} [p\delta(s)]^{n_i} q^{N-n_i} =$$

$$= \Pr\{\omega(t)=0\} + \sum_{k=1}^\infty \Pr\{\omega(t)=k\}[p\delta(s)+q]^{kN} =$$

$$= \Phi([p\delta(s)+q]^N, t), \ \text{Re}\, s \geq 0, \qquad (13)$$

where

$$\delta(s)=\int_0^\infty e^{-sx} dD(x) \text{ and } \Phi(y,t)=\sum_{k=1}^\infty y^k \Pr\{\omega(t)=k\} \text{ is}$$

the generating function of stochastic process' $\omega(t)$ distribution, $|y| \leq 1$.

In particular, from (13) we find

$$E\Delta(t) = -\frac{\partial}{\partial s}\, \Phi([p\delta(s)+q]^N, t)\Big|_{s=0} = Np\mathrm{E}\,\xi_1 \mathrm{E}\omega(t),$$

$$\mathrm{Var}\Delta(t) = \frac{\partial^2}{\partial s^2}\, \Phi([p\delta(s)+q]^N, t)\Big|_{s=0} - (\mathrm{E}\Delta(t))^2 =$$

$$= Np[\mathrm{E}\,\xi_1^2 - p(\mathrm{E}\,\xi_1)^2]\mathrm{E}\omega(t) + (Np\mathrm{E}\,\xi_1)^2 \mathrm{Var}\omega(t).$$

One more simplification of criterion (9) may be done by application of the Chebyshev's inequality. Applying this inequality, taken in modified form [5], we obtain (under condition (6))

$$\Pr\{\Delta(t) > ct\} \geq \frac{(\mathrm{E}\Delta(t) - ct)^2}{\mathrm{E}\Delta^2(t)}. \qquad (14)$$

Hence, the criterion (9) may be reduced to the simple inequality

$$\frac{(\mathrm{E}\Delta(t) - ct)^2}{\mathrm{E}\Delta^2(t)} \geq 1 - \varepsilon.$$

For application of the criteria (11),(12),(14) it is necessary to find the probabilistic distribution of process $\omega(t)$. It may be found by the methods of queueing theory [6]. Below, will be considered two particular cases of queue GI/G/m for which this distribution is known.

1 Queue of M/D/∞ type, i.e. with infinite number of servers, the Poisson input with the rate $\lambda$, and constant service time. Such queueing system is good approximation to multi-server queue if $\lambda\tau N \ll m$. As it was shown in [7], for such system (in equilibrium)

$$\Pr\{\omega(t) = k\} = \frac{(\lambda t)^k}{k!} e^{-\lambda t}, \ k = 0,1,2,\ldots \qquad (15)$$

and, consequently, $\mathrm{Var}\omega(t) = \mathrm{E}\omega(t) = \lambda t$. In this case the condition (11) takes the following form

$$e^{-\lambda t} \sum_{k=0}^\infty \frac{(\lambda t)^k}{k!} \sum_{n_1=0}^N \cdots \sum_{n_k=0}^N \prod_{i=1}^k C_N^{n_i} p^{n_i} q^{N-n_i} D^{(n_1+\ldots+n_k)}(ct)] \leq \varepsilon. \qquad (16)$$

From (15), it follows also that

$$\Phi([p\delta(s)+q]^N, t) = \exp\{-\lambda t[1-(p\delta(s)+q)^N]\}.$$

For inversion of this expression the known numerical methods of the Laplace transform inversion may be used [8].

The criterion (16) is too complex for calculations. Note that in this case $\Delta(t) - ct$ is the compound Poisson process with the drift $c$ [9]. Therefore if $t \to \infty$, we can apply the central limit theorem for such kind of stochastic processes [9]. Hence, instead of (9), we have as $t \to \infty$

$$\Pr\{\Delta(t) - ct \leq 0\} \approx \mathrm{N}(R\sqrt{t}) \leq \varepsilon, \qquad (17)$$

where

$$R = (c/\mathrm{E}\,\xi_1 - \lambda Np) \times$$

$$\times \sqrt{\lambda Np[\mathrm{E}\,\xi_1^2 + (N-1)p(\mathrm{E}\,\xi_1)^2]};$$

$N(x)$ is the standard normal distribution with zero mean and variance equals to unity.

2 One-server queue of M/D/1 type, i.e. with the Poisson input and constant service time. For such system the following result is valid [6]:

$$\Phi^*(y,\theta) \equiv \int_0^\infty e^{-\theta t} \Phi(y,t)dt =$$

$$= \Phi_0^*(y,\theta)\left[1 + \frac{\lambda(1-z_0)(1-e^{-\theta N\tau})}{\theta(1-ye^{-\theta N\tau})}\right] - \frac{\lambda(1-e^{-\theta N\tau})}{\theta^2(1-ye^{-\theta N\tau})} \times$$

$$\times\left[(1-y)e^{-\theta N\tau} + \frac{1}{y}(1-\rho)(1-z_0)\frac{z_0 - ye^{-\lambda(1-z_0)N\tau}}{z_0 - e^{-\lambda(1-z_0)N\tau}}\right] +$$

$$+ \frac{\rho}{\theta}, \qquad (18)$$

where $\rho = \lambda N\tau < 1$;

$$\Phi_0^*(y,\theta) = \frac{1-\rho}{\theta(\theta+\lambda-\lambda z_0)} \times$$

$$\times\left[\theta + \frac{\lambda(1-z_0)ye^{-\lambda(1-z_0)N\tau}(1-e^{-\theta N\tau})}{e^{-\lambda(1-z_0)N\tau} - z_0}\right];$$

$z_0$ is the unique root of the equation

$$z_0 = y \exp[-(\theta + \lambda - \lambda z_0)N\tau]$$

in the domain $|y| \leq 1$, $\operatorname{Re} \theta > 0$.

With the help of relation (18) we can determine $\operatorname{Var}\omega(t)$ and then use the criterion (14).

## 3 NUMERICAL RESULTS

Let us demonstrate the application of criterion (17) for real initial data. Put $\lambda = 5$ ships per month, $t = 25$ months, $p = 10^{-3}$, and assume that $E\Delta_{11}^2 = 2(E\Delta_{11})^2$. The results of calculations of probability in the formula (17) for different values of $Np$ and ratio $c/E\Delta_{11}$ are given in the Table.

Table

| $c/E\Delta_{11}$ / $Np$ | 0,15 | 0,20 | 0,25 | 0,30 | 0,35 |
|---|---|---|---|---|---|
| 0,10 | 0,0436 | 0,0721 | 0,1112 | 0,1635 | 0,2327 |
| 0,15 | 0,0092 | 0,0150 | 0,0244 | 0,0384 | 0,0582 |
| 0,20 | 0,0021 | 0,0035 | 0,0058 | 0,0092 | 0,0140 |
| 0,25 | 0,0006 | 0,0009 | 0,0015 | 0,0024 | 0,0037 |

From these results, it follows the expedience of insurance, for example, if $Np = 0,1$, $c/E\Delta_{11} \leq 0,2$ or $Np > 0,1$, $c/E\Delta_{11} \leq 0,3$ because probability in (17) is sufficiently small in these cases.

## 4 CONCLUSIONS

The real problems of risk-management concerning the port operator's (or stevedoring company's) activity may be formulated and solved with application of mathematical risk theory. The main feature of above problems is: first of all they must be aimed at the protection of financial state of stevedoring company but not an insurance firm. In most cases these problems may not be solved by standard theoretical methods and require the use of combination of different fields of applied probability, for example, ruin theory, queueing and reliability theories, theory of storage processes, etc. This is necessary for modeling the port's operational activity side by side with the corresponding financial processes [10].

For practical applications of results obtained it is necessary to use the corresponding statistical data concerning the cases of containers damage and values of damage, moments of ships' arrival, etc. for a previous period. Such information must be accumulated in the data base of a stevedoring company.

## REFERENCES

1. Brown RH (1985-1993) Marine Insurance: Vol. I. Principles and Basic Practice; Vol.II.Cargo Practice; Vol. III. Hull Practice, London: Witherby Publishers
2. Grandell J (1992) Aspects of Risk Theory, Springer, Berlin Heidelberg New York
3. Asmussen S (2001) Ruin Probabilities, World Scientific, Singapore New Jersey London Hong Kong
4. Harris R (1974) The expected number of idle servers in a queueing system. Operations Research 22, 6: 1258-1259
5. Feller W (1971) An Introduction to Probability Theory and Its Applications. Vol.II. 2nd Ed. Jhon Wiley & Sons, Inc., New York London Sydney Toronto
6. Jaiswall NK (1968) Priority Queues, Academic Press, New York London
7. Mirasol N (1963) The output of an M/G/∞ queueing system is Poisson. Operations Research 11, 2: 282-284
8. Krylov VI, Skoblya NS (1968) Handbook on Numerical Inversion of Laplace Transform, Nauka i Tehnika, Minsk (in Russian)
9. Prabhu NU (1997) Stochastic Storage Processes: Queues, Insurance Risk, Dams, and Data Communications. 2nd Ed. Springer, Berlin Heidelberg New York
10. Postan MYa (2006) Economic-Mathematical Models of Multimodal Transport, Astroprint, Odessa (in Russian)

# 4. Experimental and Numerical Methods for Hydrodynamic Profiles Calculation

A. Scupi & D. Dinu
*Constanta Maritime University*

ABSTRACT: The calculation of a hydrodynamic profile for a fluid that flows around mainly consists in determining the variation of drag force and lift force. Thus, for NACA 6412 profile, we will calculate and compare the changes of values of the coefficient forces mentioned above. The calculation will be done both experimentally in a naval wind tunnel and with a computational fluid dynamics - CFD (ANSYS 13). These experimental and numerical approaches can be used to study finite scale naval profiles such as the rudder.

## 1 INTRODUCTION

The main objectives of the paper are: to see how the experimental and numerical calculations of a hydrodynamic profile match, to identify the reasons for which data are not concurring and to also see whether we can use numerical methods for designing or pre-designing purposes.

Calculation of hydrodynamic profile belongs to the engineering field, where we can use three main directions of investigation: an experimental method, a numerical method and an analytical method. In our case, it is very difficult to use the analytical method because the fluid flow as described by Navier-Stokes equations has not been yet solved analytically.

Therefore calculating the forces acting on a hydrodynamic profile can be solved using one of the two methods mentioned above: the experimental and the numerical method. The results in this case are not very precise because in problem statement some simplifying assumptions, specific to our domain, have been considered by default. (OANȚĂ, 2009)

Numerical methods most generally used by computational software are: the finite element method, the finite difference method, the boundary element method and the finite volume method. ANSYS 13 uses finite element and finite volume method.

Since software using numerical methods for solving engineering problems of varying difficulty and providing satisfactory results, have emerged in the past 15 years, most problems have been solved by the experimental method. Therefore the approach proposed in this paper by comparing the two methods try to present more clearly the physical phenomenon investigated and the differences between the two methods.

To study the coefficients $C_x$ (drag coefficient) and $C_y$ (lift coefficient), we must remark at first that in the phenomenon of fluid flow around a wing, one of physical quantities, i.e. the force (lift force or drag force), is a variable size depending on the incidence angle $\alpha$. Therefore, it can be said that the process under study is a nonlinear one. The $\Pi$ theorem applies both to linear phenomena and nonlinear phenomena.

Let's analyze the similarity of the simple nonlinear process (one size variable), described by the implicit function (DINU, 1994):

$$f(\rho, v, c, l, R, \Gamma, \tau, p, \alpha) = 0 \qquad (1)$$

where the force R, is a function of $\alpha$ :

$$R = f_1(\alpha) \qquad (2)$$

In relation (1):
$\rho$ - fluid density;
$v$ - fluid velocity;
$\Gamma$ - velocity circulation,
$p$ - fluid pressure;
$\tau$ - period of swirl separation;
$c$ - chord length;
$l$ - wing span.

Nonlinear dependence expressed by equation (1) is a curve obtained experimentally. Its equation is obtained by putting the condition that the power polynomial has the form:

$$R = k_0 + k_1\alpha + k_2\alpha^2 + ... + k_n\alpha^n \qquad (3)$$

and the polynomial should be verified by some experimental points. The experimental points can be determined both experimentally and by using a fluid flow modeling program.

## 2 WORKING PARAMETERS

We considered a NACA 6412 profile with a relative elongation 6 with the following characteristics:
– the length of the chord equal to 0.080 [m];
– the wing span equal to 0.480[m].

The profile (Fig. 1) is located in an air stream with a velocity of 15 m/s. The Reynolds number calculated with formula (4) has the value of 85.000.

$$Re = \frac{v \cdot c}{\upsilon} \qquad (4)$$

where:
$v$ = fluid velocity m/s;
$c$ = chord length m;
$\upsilon$ = cinematic viscosity m²/s.

The angles of incidence are (Fig. 1):
- $-10^0 \div 0^0$ step $2^0$;
- $+0^0 \div 15^0$ step $3^0$;

The forces that are acting upon a hydrodynamic and aerodynamic profile are: the lift force and the friction force or force due to boundary layer detachment. These forces give a resultant force $R$ which decomposes by the direction of velocity at infinity and by a direction perpendicular to it. $R_x$ component is called drag force and $R_y$ component is called lift force.

$R$ force can also be decomposed by the direction of chord (component $R_t$- called tangential force) and by the direction perpendicular to the chord ($R_n$ component – called normal force). (DINU, 2010)

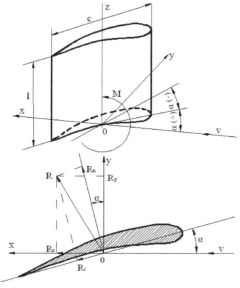

Figure 1. General representation of the profile

## 3 EXPERIMENTAL DETERMINATION OF THE AERODYNAMIC FORCES

Experiments were made in a naval aerodynamic tunnel. Airflow was uniform on a section of $510 \times 580$ mm.

A tensometric balance was used to determine the forces acting upon the wing. Results of tests are given in Table 1.

Table 1. Results of experiments

| Results | $R_x$ | $C_x$ | $R_y$ | $C_y$ |
|---|---|---|---|---|
| Incidence angles | N | | N | |
| $\alpha = -10^0$ | 1.2134 | 0.2293 | -1.80246 | -0.3406 |
| $\alpha = -8^0$ | 0.8567 | 0.1619 | -1.69714 | -0.3207 |
| $\alpha = -6^0$ | 0.6620 | 0.1251 | -1.31877 | -0.2492 |
| $\alpha = -4^0$ | 0.5069 | 0.0958 | -0.95679 | -0.1808 |
| $\alpha = -2^0$ | 0.4212 | 0.0796 | 0.16511 | 0.0312 |
| $\alpha = 0^0$ | 0.4503 | 0.0851 | 2.16284 | 0.4087 |
| $\alpha = +3^0$ | 0.6884 | 0.1301 | 5.432238 | 1.0265 |
| $\alpha = +6^0$ | 0.9679 | 0.1829 | 8.198366 | 1.5492 |
| $\alpha = +9^0$ | 1.4558 | 0.2751 | 10.10931 | 1.9103 |
| $\alpha = +12^0$ | 2.0125 | 0.3803 | 11.37568 | 2.1496 |
| $\alpha = +15^0$ | 2.7528 | 0.5202 | 12.2255 | 2.3102 |

In Fig. 2 and Fig. 3 we have represented the graphics of the function $C_y(\alpha)$ and $C_x(\alpha)$, respectively.

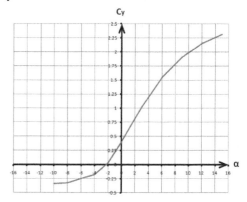

Figure 2. Graphic of $C_y$ experimentally obtained

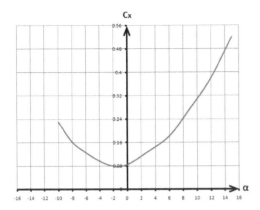

Figure 3. Graphic of $C_x$ experimentally obtained

## 4 DETERMINATION OF THE AERODYNAMIC FORCES USING CFD

### 4.1 *NACA 6412 profile*

Using Design Modeler v. 13.0, we were able to accurately reproduce the NACA 6412 profile, as represented in figure 4. Airflow was uniform on a bigger section 980 × 511 mm.

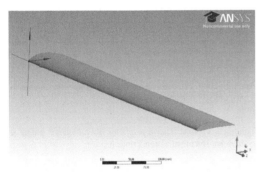

Figure. 4 Geometric representation of the NACA 6412 profile

### 4.2 *Profile discretization*

After the geometric representation of NACA profile, we went to its discretization, as shown in Fig. 5.

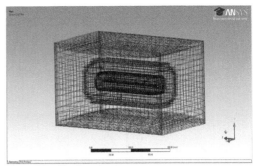

Figure 5. Profile discretization

We discretized the NACA profile in more than 10 million cells, of which 9 million are hexahedrons, 55.000 are wedges, 35.000 are polyhedral, 1500 are pyramids and only 400 are tetrahedrons. The mesh has also over 30 million faces and 11 million knots.

### 4.3 *Calculation of the aerodynamic forces*

Using Fluent program version 13.0, we set the boundary conditions as follows:
- The profile is attacked with a velocity of 10 m/s, under different angles, namely -10°, -8°, -6°, -4°, -2°, 0°, +3°, +6°, +9°, +12°, +15°;
- Behind the profile, atmospheric pressure is equal to 101325 Pa.
- The fluid motion is turbulent with a Prandtl number equal to 0.667.
- The air density is considered constant and it is equal to 1.225 kg/m³;
- The air dynamic and cinematic viscosity are also considered constant and are equal to $1.7894 \times 10^{-5}$ kg/ms, 0.0001460735 m²/s, respectively;
- The turbulence viscosity ratio is set to 10.

Process has stabilized after 208 iterations allowing us to visualize the values of drag and lift forces and their coefficients, presented in table 2.

Table 2. Results using CFD

| Results | $R_x$ | $C_x$ | $R_y$ | $C_y$ |
|---|---|---|---|---|
| Incidence angles | N | | N | |
| $\alpha = -10^0$ | 0.4536 | 0.0857 | -1.5145 | -0.2862 |
| $\alpha = -8^0$ | 0.3606 | 0.0681 | -0.8234 | -0.1556 |
| $\alpha = -6^0$ | 0.2819 | 0.0532 | -0.4900 | -0.0926 |
| $\alpha = -4^0$ | 0.2616 | 0.0494 | -0.1756 | -0.0332 |
| $\alpha = -2^0$ | 0.2529 | 0.0477 | 0.1582 | 0.0299 |
| $\alpha = 0^0$ | 0.2683 | 0.0506 | 2.1501 | 0.4063 |
| $\alpha = +3^0$ | 0.3402 | 0.0642 | 5.3861 | 1.0178 |
| $\alpha = +6^0$ | 0.4604 | 0.0869 | 8.0507 | 1.5213 |
| $\alpha = +9^0$ | 0.6160 | 0.1164 | 9.4658 | 1.7887 |
| $\alpha = +12^0$ | 0.8447 | 0.1596 | 10.4375 | 1.9723 |
| $\alpha = +15^0$ | 1.1043 | 0.2086 | 11.3418 | 2.1432 |

In the Fig. 6 and Fig. 7 we have represented the graphics of the function $C_y(\alpha)$ and $C_x(\alpha)$, respectively.

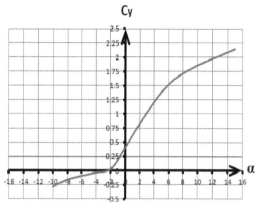

Figure 6. Graphic of $C_y$ obtained using CFD

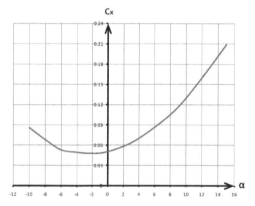

Figure 7. Graphic of $C_x$ obtained using CFD

## 5 CONCLUSIONS

Comparing the $C_y$ coefficients values obtained by experiment and using CFD, we can make the observation that they are very similar in a the field of the incidence angles $[-2^0, 6^0]$. Also, comparing the $C_x$ graphic, we remark that the graphics are very similar, but between the values there are some differences.

These differences are due to experimental errors (errors of measurement devices), numerical errors (rounding errors), and also discretization errors.

Also, the CFD programme doesn't take into account the induce resistance in the case of finite span wings. As a consequence an induce angle $\alpha_i$ will appear which thus decreases the incidence angle $\alpha$. The alteration of direction and value of velocity bring about the corresponding alteration of lift force, which is perpendicular on the direction of stream velocity. (DINU, 1999)

In order to reduce these differences, it is recommended that the object of study be discretized into a larger number of cells. It is also advisable to leave out some simplifying conditions, and to impose various other conditions that simulate reality to a better precision (e.g. energy equations, air compressibility).

CFD can replace the experiment within certain limits, being a good method for pre-designing.

REFERENCES

Dinu, D. 1994. Trecerea coeficienţilor $C_x$ şi $C_y$ de la model la natură în teoria similitudinii la două scări a aripilor hidrodinamice. *Buletin Tehnic RNR, 3-4:10-12.* Bucureşti: RNR

Dinu, D. 1999. *Hydraulics and Hydraulic Machines.* Constanta Sigma Trading Metafora.

Dinu, D. 2010. *Mecanica fluidelor pentru navalişti.* Constanta: Nautica

Oanţă, E. 2009. *Proiectarea aplicaţiilor software în ingineria mecanică.* Constanţa : Nautica

Iorga, V., Jora, B., Nicolescu, C., Lopătan, I., Fătu, I. 1996. *Programare numerică.* Bucureşti: Teora

Rădulescu, V. 2004. *Teoria profilelor hidrodinamice izolate şi în reţea.* Bucureşti : Bren-Printech.

# 5. Asymptotic Stability of a Class of Positive Continuous-Discrete 2D Linear Systems

T. Kaczorek

*Faculty of Electrical Engineering, Bialystok University of Technology, Poland*

ABSTRACT: New necessary and sufficient conditions for the asymptotic stability of a class of positive continuous-discrete 2D linear systems are presented. Effectiveness of the stability tests are demonstrated on numerical examples.

## 1 INTRODUCTION

In positive systems inputs, state variables and outputs take only nonnegative values. A variety of models having positive systems behavior can be found in engineering, management science, economics, social sciences, biology and medicine etc.. An overview of state of the art in positive systems is given in the monographs [6, 9]. The positive continuous-discrete 2D linear systems have been introduced in [8], positive hybrid linear systems in [10] and the positive fractional 2D hybrid systems in [11]. Different methods of solvability of 2D hybrid linear systems have been discussed in [14] and the solution to singular 2D hybrids linear systems has been derived in [16]. The realization problem for positive 2D hybrid systems has been addressed in [12]. Some problems of dynamics and control of 2D hybrid systems have been considered in [5, 7]. The problems of stability and robust stability of 2D continuous-discrete linear systems have been investigated in [1-4, 17-19]. The stability of positive continuous-time linear systems with delays has been addressed in [13]. Recently the stability and robust stability of Fornasini-Marchesini type model and of Roesser type model of scalar continuous-discrete linear systems have been analyzed by Buslowicz in [2-4].

In this note new necessary and sufficient conditions for asymptotic stability of positive continuous-discrete 2D linear systems will be presented.

The following notation will be used: $\Re$ - the set of real numbers, $Z_+$ - the set of nonnegative integers, $\Re^{n\times m}$ - the set of $n \times m$ real matrices, $\Re_+^{n\times m}$ - the set of $n \times m$ matrices with nonnegative entries and $\Re_+^n = \Re_+^{n\times 1}$, $I_n$ - the $n \times n$ identity matrix.

## 2 PRELIMINARIES

Consider the linear continuous-discrete 2D system [8, 9]

$$\dot{x}(t, i+1) = A_1\dot{x}(t, i) + A_2 x(t, i+1) + Bu(t, i),$$
$$t \in \Re_+, \ i \in Z_+ \quad (1)$$

where

$$\dot{x}(t, i) = \frac{\partial x(t, i)}{\partial t}, \ x(t, i) \in \Re^n, \ u(t, i) \in \Re^m$$

$$A_1, A_2 \in \Re^{n\times n}, \ B \in \Re^{n\times m}.$$

**Definition 1.** The linear continuous-discrete 2D system (1) is called (internally) positive if $x(t, i) \in \Re_+^n$, $t \in \Re_+$, $i \in Z_+$ for any input $u(t, i) \in \Re_+^m$ and all initial conditions

$$x(0, i) \in \Re_+^n, \ i \in Z_+, \ x(t, 0) \in \Re_+^n, \ \dot{x}(t, 0) \in \Re_+^n,$$
$$t \in \Re_+ \quad (2)$$

**Theorem 1.** [8, 9] The linear continuous-discrete 2D system (1) is positive if and only if

$$A_2 \in M_n, \ A_1 \in \Re_+^{n\times n} \text{ and } B \in \Re_+^{n\times m} \quad (3)$$

where $M_n$ is the set of $n \times n$ Metzler matrices (with nonnegative off-diagonal entries).

**Theorem 2.** The positive system (1) is unstable if at least one diagonal entry of the matrix $A_2$ is nonnegative.

**Proof.** This follows immediately from (1) for $B = 0$ since for positive system (1) $A_1 \in \Re_+^{n\times n}$ [9]. $\square$

**Definition 2.** The point $x_e$ is called equilibrium point of the asymptotically stable system (1) if

$$0 = A_1 x_e + A_2 x_e + 1_n, \ 1_n = [1 \ \ldots \ 1]^T \in \Re_+^n \qquad (4)$$

From (4) we have

$$x_e = -[A_1 + A_2]^{-1} 1_n = -A^{-1} 1_n, \ A = A_1 + A_2 \qquad (5)$$

**Remark 1.** The positive system (1) is asymptotically stable only if diagonal entries of the matrix $A = A_1 + A_2$ are negative. Therefore, in (5) the matrix $A$ is a Metzler matrix with negative diagonal entries and $-A^{-1} \in \Re_+^{n \times n}$ [9].

## 3 MAIN RESULT

**Theorem 3.** The positive continuous-discrete 2D system (1) is asymptotically stable if and only if there exists a strictly positive vector $\lambda \in \Re_+^n$ such that

$$A\lambda < 0, \ A = A_1 + A_2 \qquad (6)$$

In what follows it is assumed that the matrix $A_2$ is asymptotically stable Metzler matrix (all its diagonal entries are negative).

**Proof.** Integrating the equation (1) with $B = 0$ in the interval $(0, +\infty)$ for $i \to +\infty$ we obtain

$$x(+\infty, +\infty) - x(0, +\infty) = (A_1 + A_2) \int_0^{+\infty} x(\tau, +\infty) d\tau \qquad (7)$$

If the system is asymptotically stable then $x(+\infty, +\infty) = 0$ and from (7) we obtain (6) for $\lambda = \int_0^{+\infty} x(\tau, +\infty) d\tau$. Note that $\lambda > 0$ for every $x(0, +\infty) > 0$. By Remark 1 the positive system (1) is asymptotically stable only if diagonal entries of the Metzler matrix $A$ are negative. In this case there exists a strictly positive vector $\lambda > 0$ satisfying (6). $\square$

**Remark 2.** As the strictly positive vector $\lambda$ we may choose the equilibrium point $x_e$, i.e. $\lambda = x_e$.

Substitution of (5) into (6) yields

$$A\lambda = (A_1 + A_2) x_e = -[A_1 + A_2][A_1 + A_2]^{-1} 1_n = -1_n \ (8)$$

**Example 1.** Consider the positive system (1) with the matrices

$$A_1 = \begin{bmatrix} 0.2 & 0.1 \\ 0.1 & 0.3 \end{bmatrix}, \quad A_2 = \begin{bmatrix} -0.5 & 0.1 \\ 0.2 & -0.6 \end{bmatrix} \qquad (9)$$

In this case the matrix

$$A = A_1 + A_2 = \begin{bmatrix} -0.3 & 0.2 \\ 0.3 & -0.3 \end{bmatrix} \qquad (10)$$

is asymptotically stable Metzler matrix and from (5) we obtain the equilibrium point

$$x_e = -A^{-1} 1_n = \begin{bmatrix} -0.3 & 0.2 \\ 0.3 & -0.3 \end{bmatrix}^{-1} \begin{bmatrix} 1 \\ 1 \end{bmatrix} = \frac{1}{0.03} \begin{bmatrix} 0.5 \\ 0.6 \end{bmatrix} \qquad (11)$$

For $\lambda = x_e$ we have

$$(A_1 + A_2)\lambda = \begin{bmatrix} -0.3 & 0.2 \\ 0.3 & -0.3 \end{bmatrix} \frac{1}{0.03} \begin{bmatrix} 0.5 \\ 0.6 \end{bmatrix} = -\begin{bmatrix} 1 \\ 1 \end{bmatrix} \qquad (12)$$

Therefore, by Theorem 3 the positive system (1) with (9) is asymptotically stable.
From Theorem 2 and 3 the following theorem follows.

**Theorem 4.** The positive continuous-discrete 2D linear system (1) is asymptotically stable if and only if the matrix $A = A_1 + A_2$ is asymptotically stable (Hurwitz Metzler matrix).

To test the asymptotic stability of the positive system (1) the following theorem is recommended.

**Theorem 5.** The positive continuous-discrete 2D linear system (1) is asymptotically stable if and only if one of the following equivalent conditions is satisfied:
1 all coefficients $a_0, \ldots, a_{n-1}$ of the polynomial

$$\det[I_n s - A] = s^n + a_{n-1} s^{n-1} + \ldots + a_1 s + a_0 \qquad (13)$$

are positive,

2 the diagonal entries of the matrices

$$A_{n-k}^{(k)} \text{ for } k = 1, \ldots, n-1 \qquad (14)$$

are negative, where $A = A_1 + A_2$ and

$$A_n^{(0)} = A = \begin{bmatrix} a_{11}^{(0)} & \cdots & a_{1,n}^{(0)} \\ \vdots & \cdots & \vdots \\ a_{n,1}^{(0)} & \cdots & a_{n,n}^{(0)} \end{bmatrix} = \begin{bmatrix} A_{n-1}^{(0)} & b_{n-1}^{(0)} \\ c_{n-1}^{(0)} & a_{n,n}^{(0)} \end{bmatrix},$$

$$A_{n-1}^{(0)} = \begin{bmatrix} a_{11}^{(0)} & \cdots & a_{1,n-1}^{(0)} \\ \vdots & \cdots & \vdots \\ a_{n-1,1}^{(0)} & \cdots & a_{n-1,n-1}^{(0)} \end{bmatrix}$$

$$b_{n-1}^{(0)} = \begin{bmatrix} a_{1,n}^{(0)} \\ \vdots \\ a_{n-1,n}^{(0)} \end{bmatrix}, \quad c_{n-1}^{(0)} = [a_{n,1}^{(0)} \quad \cdots \quad a_{n,n-1}^{(0)}]$$

$$A_{n-k}^{(k)} = A_{n-k}^{(n-1)} - \frac{b_{n-k}^{(k-1)} c_{n-k}^{(k-1)}}{a_{n-k+1,n-k+1}^{(k-1)}} =$$

$$= \begin{bmatrix} a_{11}^{(k)} & \cdots & a_{1,n-k}^{(k)} \\ \vdots & \cdots & \vdots \\ a_{n-k,1}^{(k)} & \cdots & a_{n-k,n-k}^{(k)} \end{bmatrix} = \begin{bmatrix} A_{n-k-1}^{(k)} & b_{n-k-1}^{(k)} \\ c_{n-k-1}^{(k)} & a_{n-k,n-k}^{(k)} \end{bmatrix},$$

$$b_{n-k-1}^{(k)} = \begin{bmatrix} a_{1,n-k}^{(k)} \\ \vdots \\ a_{n-k-1,n-k}^{(k)} \end{bmatrix},$$

$$c_{n-k-1}^{(k)} = [a_{n-k,1}^{(k)} \quad \cdots \quad a_{n-k,n-k-1}^{(k)}]$$

for $k = 0,1,\ldots,n-1$.

**Proof.** By Theorem 4 the positive system (1) is asymptotically stable if and only if the Metzler Matrix $A$ is asymptotically stable and this holds if and only if all coefficients of the polynomial (13) are positive [9]. In [15] was shown that the Metzler matrix $A$ is asymptotically stable if and only if the condition ii) is satisfied. □

**Example 2.** Using Theorem 5 check the stability of the positive system (1) with (9). The polynomial (13) for the matrix (10) has the form

$$\det[I_n s - A] = \begin{vmatrix} s+0.3 & -0.2 \\ -0.3 & s+0.3 \end{vmatrix} = s^2 + 0.6s + 0.03 \ (16)$$

All coefficients of the polynomial (16) are positive and by Theorem 5 the positive system is asymptotically stable.
Using (14) for $n = 2$ we obtain

$$A_1^{(1)} = a_{11}^{(0)} - \frac{a_{12}^{(0)} a_{21}^{(0)}}{a_{22}^{(0)}} = -0.3 + \frac{0.2*0.3}{0.3} = -0.1 < 0$$

The condition ii) of Theorem 5 is satisfied and the positive system (1) with (9) is asymptotically stable.

# 4 CONCLUDING REMARKS

New necessary and sufficient conditions for asymptotic stability of a class of positive continuous-discrete 2D linear systems have been established (Theorems 3-5). The effectiveness of the new stability tests have been demonstrated on numerical examples. The considerations can be also extended for fractional positive 2D continuous-discrete linear systems.

## ACKNOWLEDGMENT

This work was supported by Ministry of Science and Higher Education in Poland under work S/WE/1/11.

## REFERENCES

[1] Bistritz Y., "A stability test for continuous-discrete bivariate polynomials", *Proc. Int. Symp. on Circuits and Systems* 3, 682-685 (2003).
[2] Busłowicz M., "Improved stability and robust stability conditions for a general model of scalar continuous-discrete linear systems", *Measurement Automation and Monitoring*, (submitted for publication).
[3] Busłowicz M., "Stability and robust stability conditions for a general model of scalar continuous-discrete linear systems", *Measurement Automation and Monitoring*, no. 2, 2010, 133-135.
[4] Busłowicz M., "Robust stability of the new general 2D model of a class of continuous-discrete linear systems", *Bull. Pol. Acad. Sci. Techn.* vol. 57, no. 4, 2010.
[5] Dymkov M., I. Gaishun, E. Rogers, K. Gałkowski and D. H. Owens, "Control theory for a class of 2D continuous-discrete linear systems", *Int. J. Control* 77 (9), 847-860 (2004).
[6] Farina L. and Rinaldi S., *Positive Linear Systems; Theory and Applications*, J. Wiley, New York 2000.
[7] Gałkowski K, Rogers E., Paszke W. and Owens D. H., "Linear repetitive process control theory applied to a physical example", *Int. J. Appl. Math. Comput. Sci.* 13 (1), 87-99 (2003).
[8] Kaczorek T., "Reachability and Minimum energy control of positive 2D continuous-discrete systems", *Bull. Pol. Acad. Sci. Techn.* vol. 46, no. 1, 1909, 85-93.
[9] Kaczorek T., *Positive 1D and 2D Systems*, Springer-Verlag, London, 2002.
[10] Kaczorek,T., "Positive 2D hybrid linear systems", *Bull. Pol. Acad. Sci. Tech.* vol. 55, no. 4, 2007, 351-358.
[11] Kaczorek T., "Positive fractional 2D hybrid linear systems", *Bull. Pol. Acad. Tech.* 56 (3), 273-277 (2008).
[12] Kaczorek T., "Realization problem for positive 2D hybrid systems", *COMPEL* 27 (3), 613-623 (2008).
[13] Kaczorek T., "Stability of positive continuous-time linear systems with delays", *Bul. Pol. Acad. Sci. Techn.* vol. 57. no. 4, 2009, 395-398.
[14] Kaczorek T., Marchenko V. and Sajewski Ł., "Solvability of 2D hybrid linear systems - comparison of the different methods", *Acta Mechanica et Automatica* 2 (2), 59-66 (2008).
[15] Narendra K.S. and Shorten R., "Hurwitz stability of Metzler matrices", *IEEE Trans. Autom. Contr.* vol. 55, no. 6 June 2010, 1484-1487.

[16] Sajewski Ł., "Solution of 2D singular hybrid linear systems", *Kybernetes* 38 (7/8), 1079-1092 (2009).

[17] Xiao Y., "Stability test for 2-D continuous-discrete systems", *Proc. 40th IEEE Conf. on Decision and Control* 4, 3649-3654 (2001).

[18] Xiao Y., "Stability, controllability and observability of 2-D continuous-discrete systems:, *Proc. Int. Symp. on Circuits and Systems* 4, 468-471 (2003).

[19] Xiao Y., "Robust Hurwitz-Schur stability conditions of polytopes of 2-D polynomials", *Proc. 40th IEEE Conf. on Decision and Control* 4, 3643-3648 (2001).

# 6. Application of CFD Methods for the Assessment of Ship Manoeuvrability in Shallow Water

T. Górnicz & J. Kulczyk
*Wrocław University of Technology, Poland*

ABSTRACT: Safety in water transport plays a significant role. One way to increase safety in the waterways is to ensure that ships have proper manoeuvrability. Evaluation of manoeuvring properties performed at an early stage of design can detect problems that later would be difficult to solve. To make such an analysis the numerical methods can be used. In the paper the numerical method to evaluate the ship manoeuvrability on the shallow water is presented. Additionally author shows the procedure of determining hydrodynamic coefficients on a basis of CFD calculation. The simulation results of inland ship was compared with experimental data.

## 1 INTRODUCTION

For a long time ship manoeuvring characteristics were considered to be of marginal importance. Nevertheless, from the operational point of view of inland vessels, manoeuvring characteristics play a crucial role. Operating in restricted waterways substantially increases the risk of accidents and disasters as a result of manoeuvring errors. To increase level of safety on the waterways, it is important to ensure that vessels have appropriate manoeuvring characteristics. For this purpose it is necessary to develop research methods for determining the manoeuvring properties of inland ships.

Currently, one of the more commonly used research methods in ship hydrodynamics are numerical methods. With improvements in computing power and increasingly more accurate CFD methods it is possible to simulate more complicated cases. This paper presents a numerical method for prediction of ship manoeuvrability. It is based on a mathematical model of the ship equations of motion, described in section 3. This model requires knowledge of the coefficients of the hydrodynamic forces acting on the ship. In section 4 there is presented a method for determining the hydrodynamic coefficients using CFD.

On inland waters the most restrictive rules are described in Rhine Regulations. The severity of the Rhine Regulations is the result of difficult conditions on the Rhine: a strong current and high traffic intensity. Vessels that fulfil these requirements are allowed to operate on other waterways in Europe. These regulations specify requirements for: a minimum speed of the vessel, the ability to stop, the abil-

ity to move backwards and turning manoeuvrability. In addition, there is specified an additional manoeuvre - "evasive action" , which is analogous to zigzag test and is a specific requirement for inland waters. The main characteristic of this evasive action [1] is the moment of switching the rudder to the opposite side, which occurs when the ship reaches the desired angular velocity.

## 2 MATHEMATICAL MODEL OF SHIP MOTION

Motion simulation is based on two coordinate systems. The first one is the global, stationary coordinate system. The ship position and orientation is determined in the global coordinate system. The second one is the local coordinate system. The centre of the local coordinate system is located at the centre of gravity of the ship. Equations of motion of the ship are written in the local coordinate system. Figure 1 shows the two coordinate systems used in the paper. It also shows the velocity components of the vessel.

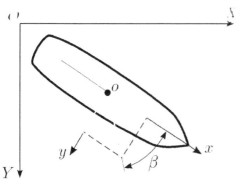

Figure 1. Coordinate systems and velocity components of the vessel.

The movement of the ship is considered to be planar motion (limited to three degrees of freedom).

This study only analysed small velocities at which sinkage and trimming of the ship is minimal. Therefore, these phenomena are not considered in the research..

Interaction between the hull, rudder and propellers are only simulated by appropriate coefficients.

In inland vessels the centre of gravity is far lower than in marine ships. In addition, the centre of gravity is located near the centre of buoyancy. As a result of this , the roll of the vessel is relatively small and could also be neglected.

Wavelength on inland waterways is relatively small in proportion to the length of the vessel. In the study the influence of waves on the trajectory of the ship was not taken into consideration.

The equations of motion (1) were written for the centre of gravity of the ship. The left side of equations describes the ship as a rigid body. On the right side of equations there are hydrodynamic external forces (X, Y) and the hydrodynamic moment (N) acting on the vessel.

$$m\dot{u} - mvr = X$$
$$m\dot{v} + mur = Y \qquad (1)$$
$$I_{zz}\dot{r} = N$$

where:

$m$ – vessel mass; $u$ –longitudinal velocity; $v$ – transversal velocity; $r$ – yaw rate; ˙- dot, the derivative of a variable over time; $I_{zz}$ – moment of inertia of the vessel; $X,Y,N$ – hydrodynamic force and moment acting on the vessel according to the axes of local coordinate system. Detailed information about model can be found in [2].

External hydrodynamic forces and the hydrodynamic moment acting on the ship were written like in the MMG model [3], as a sum of the components.

$$X = X_H + X_P + X_R$$
$$Y = Y_H + Y_R \qquad (2)$$
$$N = N_H + N_P + N_R$$

where:

$X_H$, $Y_H$, $N_H$ – hydrodynamic forces and moment acting on the bare hull; $X_P$, $N_P$- force and moment induced by the operation of propellers, $X_R$, $Y_R$, $N_R$ – forces and moment induced by flow around the rudder.

### 2.1 Hull

To determine the hydrodynamic forces induced by flow around the hull the following mathematical model was used [3]:

$$X_H = -m_x\dot{u} - R_T(u) + X_{vv}v^2 + (X_{vr} + m_y)vr +$$
$$\qquad + X_{rr}r^2 + X_{vvvv}v^4$$
$$Y_H = -m_y\dot{v} + Y_vv + (Y_r - m_xu)r + Y_{vvv}v^3 + Y_{vvr}v^2r +$$
$$\qquad + Y_{vrr}vr^2 + Y_{rrr}r^3 \qquad (3)$$
$$N_H = -j_{zz}\dot{r} + N_vv + N_rr + N_{vvv}v^3 + N_{vvr}v^2r +$$
$$\qquad + N_{vrr}vr^2 + N_{rrr}r^3$$

where:

$m_x$, $m_y$ – added mass coefficients, in $x$ and $y$ direction; $j_{zz}$ – added inertia coefficient; $R_T(u)$ – hull resistance; $X_{vv}$, $X_{vr}$,..., $Y_v$, $Y_r$,... - coefficient of hydrodynamic forces acting on the hull; $N_v$, $N_r$, ... – coefficients of hydrodynamic moment acting on the hull.

### 2.2 Rudders

Hydrodynamic forces induced by rudder laying can be calculated on the basis of equations (4). The model was taken from[4].

$$X_R \quad = -(1-t_R)F_N\sin\delta$$
$$Y_R \quad = -(1+a_H)F_N\cos\delta \qquad (4)$$
$$N_R \quad = -(x_R + a_Hx_H)F_N\cos\delta$$

where:

$t_R$ - coefficient of additional drag; $a_H$ - ratio of additional lateral force; $x_R$ – x-coordinate of application point of $F_N$; $x_H$ - x-coordinate of application point of additional lateral force; $\delta$ rudder angle; $F_N$ – normal hydrodynamic force acting on the rudder.

The value of normal force $F_N$ was determined on the basis of model (5).

$$F_N = 0{,}5{\cdot}\rho A_R C_N U_R^2$$

$$U_R^2 = \left(1-w_R\right)^2 (1+C{\cdot}g(s))$$

$$g(s) = \eta K \frac{\left(2-(2-K){\cdot}s\right)s}{\left(1-s\right)^2}$$

$$\eta = \frac{D}{h_R}$$

$$K = 0{.}6\frac{1-w_P}{1-w_R} \tag{5}$$

$$s = 1-(1-w_P)U\frac{\cos\beta}{n{\cdot}P}$$

$$w_R = w_{R0}\frac{w_P}{w_{P0}}$$

$$\alpha_R = \delta-\gamma{\cdot}\beta_R'$$

$$\beta_R' = \beta-2x_R'{\cdot}r'$$

where:

$\rho$ – water density; $A_R$ – rudder area; $C_N$ – normal force coefficient; $U_R$ - effective rudder inflow velocity; $C$ – coefficient, dependent on rudder angle sense ($C{\approx}1.0$); $D$ – propeller diameter; $h_R$ – height of rudder; $w_P$ - wake fraction at propeller location; $w_R$ - wake fraction at rudder location; $w_{R0}$ - effective wake fraction at rudder location, in straight ahead ship motion; $U$ – total velocity of vessel; $\beta$ – drift angle; $n$ – rotational speed of propeller; $P$ – propeller pitch; $x'_R$ – non-dimensional x-coordinate of application point of $F_N$; $r'$ – non-dimensional yaw rate.

## 2.3 Propellers

This paper uses a mathematical model of hydrodynamic forces generated by two propellers.

$$X_P = (1-t)(T_1+T_2)$$

$$N_P = (1-t)(T_1-T_2)d \tag{6}$$

where:

$t$ - thrust deduction factor; $T_1, T_2$ – thrust generated by propellers; $d$- distance from the axis of propeller to symmetry plane of the vessel, in $y$ direction.

Propeller thrust was determined on the basis of the relation (6).

$$T = n^2 D^4 K_T(J)$$

$$K_T(J) = a_0 + a_1 J + a_2 J_2$$

$$J = U\cos\beta_P\frac{1-w_P}{nD} \tag{7}$$

$$w_P = w_{P0}{\cdot}\exp(-4.0\beta_P^2)$$

$$\beta_P = \beta - x_P'{\cdot}r'$$

where:

$T$ – propeller thrust; $J$ - advance coefficient; $a_0$, $a_1$, $a_2$ – $K_T$ polynomial coefficients; $w_{P0}$ - effective

wake fraction in straight ahead ship motion; $x'_P$ – non-dimensional x-coordinate of propeller..

Values of thrust coefficients used in the calculations were derived from experimental research.

## 3 DETERMINING OF HYDRODYNAMIC COEFFICIENTS

### 3.1 The coefficients of the hydrodynamic forces acting on the hull

The model of the hydrodynamic forces described in section 2.1 requires knowledge of the values of hydrodynamic coefficients: $X_{vv}$, $X_{vr}$,..., $Y_v$, $Y_r$,..., $N_v$, $N_r$, .... The literature, including [5], describes the empirical formulas derived for marine ships. Due to the complicated nature of these forces, the results can be insufficiently accurate. For this reason, preference is given to other, more accurate method to determine the coefficients of hydrodynamic forces. One of these methods is the CFD calculation. In addition to the providing accurate results, numerical calculations enable the model to take into account specific operational conditions of the inland ship, for example, that the impact of shallow water on the hydrodynamic forces acting on the hull.

o determine the coefficients of hydrodynamic interactions it is necessary to have a database containing the values of the hydrodynamic forces and the corresponding to them values of velocity ($u,v,r$) and the acceleration ($\dot{u},\dot{v},\dot{r}$) of the ship. One way to obtain such a database is to simulate with help of CFD software the series of tests (manoeuvres): yaw , yaw with drift, sway test. Figure (2) shows the trajectory of a ship during these manoeuvres.

The values of the added mass ($m_x$ ,$m_y$) coefficients , added inertia ($j_{zz}$) coefficient and the ship resistance ($R_T(u)$) can be obtained on the basis of empirical methods or CFD calculations.

During the yaw manoeuvre the ship transverse velocity $v$ and acceleration $\dot{v}$ are zero. The equations of the hydrodynamic forces acting on the hull can be simplified to the following form:

$$X_H = -m_x\dot{u} - R_T(u) + X_{rr}r^2$$

$$Y_H = -m_x ur + Y_r r + Y_{rrr}r^3 \tag{7}$$

$$N_H = -j_{zz}\dot{r} + N_r r + N_{rrr}r^3$$

In the results of CFD simulation of the yaw manoeuvre the relation between hydrodynamic forces and moment ($X_H$, $Y_H$ , $N_H$) and velocity $u$, $r$, and acceleration $\dot{u}$ is obtained. The least squares method can be used on the results from yaw simulation to approximate the following coefficients: $X_{rr}$, $Y_r$, $Y_{rrr}$, $N_r$, $N_{rrr}$ .

During the sway manoeuvre the speed $r$ and the acceleration $\dot{r},\dot{u}$ are zero. The equations of hydrodynamic forces acting on the hull can be simplified to the following form:

$$X_H = -R_T(u) + X_{vv}v^2 + X_{vvvv}v^4$$
$$Y_H = -m_y\dot{v} + Y_vv + Y_{vvv}v^3 \quad (8)$$
$$N_H = N_vv + N_{vvv}v^3$$

When simulating the sway manoeuvre using CFD the data showing the relation between hydrodynamic forces acting on the hull ($X_H$, $Y_H$, $N_H$) and velocities $u$ and $v$ is obtained. The least squares method is used on the results from sway simulation to approximate the following coefficients: $X_{vv}$, $X_{vvvv}$, $Y_v$, $Y_{vvv}$, $N_v$, $N_{vvv}$.

The CFD simulation of the yaw with drift manoeuvre provides the calculations for the rest of hydrodynamic coefficients: $X_{vr}$, $Y_{vvr}$, $Y_{vrr}$, $N_{vvr}$, $N_{vrr}$.

More information about determining hydrodynamic coefficients can be found[6]

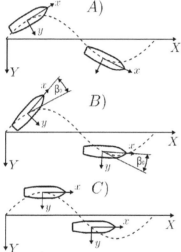

Figure 2. The manoeuvres for determining the hydrodynamic coefficients: A) yaw manoeuvre B) yaw with drift, C) sway manoeuvre.

## 3.2 *Rudder characteristic.*

A mathematical model of hydrodynamic forces acting on the rudders, described in section 2.2, requires knowledge of the normal force coefficient ($C_N$). The characteristics of isolated rudders have been described in many sources, for example [7]. Working conditions of rudders installed under the hull of the inland vessels may be significantly different than these from a single rudder. This is due to the presence of wake and propeller streams as well as the impact of the limited depth of the waterway. In order to determine the correct characteristics of the rudder, more accurate methods should be used.

In studies to determine the characteristics of the rudders CFD methods were used. The calculations were carried out in two ways. Firstly, the approach shown in Figure 3, was based on calculations of a rudder located in the propeller stream. The geometry of propeller was replaced by the disk with a pressure jump. The value of the pressure jump was equivalent to propeller thrust. Restriction of flow around a rudder at the top and bottom edge was simulated by two flat plates. The symbols $c$ and $d$ on the scheme denotes distance between rudder and plates. The main advantage of this approach is the low complexity of the discrete model, which significantly accelerates the calculations.

The second method used in the study was to build a full geometric model of the entire hull with propellers and rudders. This solution required the usage of discrete models with a much larger number of elements. CFD calculations of the entire hull give more accurate results but require large computing power. Figure 4 shows an example of a discrete model of the stern of the ship with the rudders and simplified models of propellers.

Additional information on the rudder force numerical calculations can be found in [8].

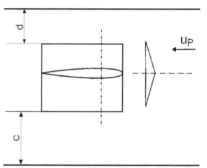

Figure 3. Scheme of model for rudder force calculations.

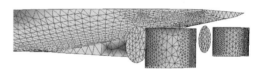

Figure 4. Example of discrete model of the stern of the ship with propellers and rudders.

## 4 CFD METHODS

In the studies the Ansys Fluent commercial CFD software was used. It is based on a finite volume method. To solve the three-dimensional turbulent flow the RANS method was used. The turbulence model *k-ε-Realizable* was used. The boundary layer was calculated using the *Enhanced Wall Treatment* model. The result of research presented in [9] shows that this model works best for flows with large pressure gradients and a separation phenomenon.

The calculations mainly used a structured mesh with hexagonal elements. Unstructured (tetragonal) grids were only used in the calculation of the hull with rudders and propellers, due to the complicated geometries involved.

To simulate the yaw, yaw with drift, sway manoeuvre the moving mesh technology was technique was utilised. Parameters of mesh motion were defined in an additional batch program (UDF) to the system Ansys Fluent (UDF). The program is written in C language.

## 5 RESULTS

Numerical calculations were performed for a model of inland transport vessel (ship A). The scale model was $\lambda = 21.81$.

For the same scale a physical model was created and tests in the towing tank were performed. All tests were performed in the shallow water. In this paper the results for h/T=1.89 are presented. Table 1 contains the main parameters of the ship A.

Table 1. The main parameters of the ship A

| Parameter | unit | ship | value |
| --- | --- | --- | --- |
| LPP | m | 85.50 | 3.920 |
| B | m | 11.45 | 0.525 |
| T | m | 2.65 | 0.122 |
| CB | - | | 0.0853 |
| LCB | m | 43.71 | 2.004 |
| Number of propellers | - | | 2 |
| Number of rudders | - | | 2 |

| Parameters of the propeller (model) | | |
| --- | --- | --- |
| Parameter | unit | value |
| D | m | 0.08 |
| P/D | - | 1.1102 |
| $A_E/A_0$ | - | 0.7474 |
| Z | - | 4 |

| Parameters of the rudder (model) | | |
| --- | --- | --- |
| Parameter | unit | value |
| $h_R$ | m | 0.092 |
| $A_R$ | m² | 0.0103 |
| Profile | - | NACA0012 |

### 5.1 Direct Comparison

CFD calculations are characterized by a number of restrictions and simplifying assumptions. In the numerical calculations of yaw, yaw with drift and sway test, which are necessary for determining the value of hydrodynamic coefficients, a problem with calculation stability occurred . After disabling two-phase flow (air-water) the problem disappeared. Therefore, the calculation of coefficients of hydrodynamic forces acting on the hull $X_H$, $Y_H$, and $N_H$ does not include the impact of the free surface. This problem didn't appear in the calculation of the resistance of the ship. In the subsequent research only situations with low ship speed (Fr = 0.12) were analysed. When the ship speed is relatively low the influence of free surface on the hydrodynamic force is negligible.

The figures 5,6,7 shows a comparison of results obtained from CFD calculations and the experimental test performed in the towing tank. The charts show the course of the hydrodynamic forces $X_H$, $Y_H$ and hydrodynamic moment $N_H$ during the yaw and sway test. During the tests in towing tank the PMM mechanism was used. The worst comparison between CFD and experiment is for $X_H$. This force is relatively small and it is very susceptible for different factors and disturbances. The comparison for hydrodynamic moment $N_H$ is much better. The hydrodynamic moment depends mainly on huge difference of pressure on both sides of a hull. And the pressure field was accurately predicted by CFD calculations.

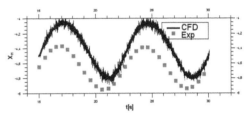

Figure 5. Comparison of the results of the experiment and CFD calculations for the characteristic of $X_H$ forces obtained during yaw test.

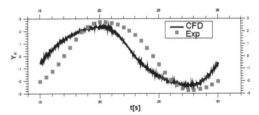

Figure 6. Comparison of the results of the experiment and CFD calculations for calculations for $Y_H$ force obtained during yaw test.

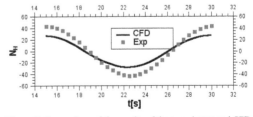

Figure 7. Comparison of the results of the experiment and CFD calculations for the $N_H$ moment obtained during the sway test.

## 5.2 Indirect Comparison

In the towing tank the model tests of ship A were performed. The following tests were performed: yaw, yaw with drift and sway. On the basis of the experimental results the coefficients of hydrodynamic forces were calculated. Similarly, on the basis of the CFD calculations the results of a second set of hydrodynamic forces coefficients was determined. For both sets of coefficients the simulation of standard manoeuvres was performed.

The manoeuvre results were obtained from the author's program to simulate the motion of the ship.

The following figures show the comparison of the results obtained from experimental tests and computer simulation. The figures 8,9,10 illustrate the characteristic charts for the turning circle manoeuvre, evasive action and spiral test.

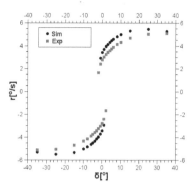

Figure 10. The comparison of characteristic chart of the spiral test.

## 6 SUMMARY

The paper presents a numerical method for evaluating manoeuvring characteristics. Research on a model of the inland vessel showed good agreement between numerical calculations and experimental results.

Due to problems with the stability of the CFD code only the mono-fluid calculations were performed. Further studies will be carried out to calculate the free surface effects, and taking into account phenomena such as sinkage and trim.

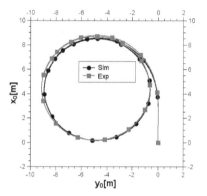

Figure 8. Comparison of trajectory for turning circle manoeuvre, rudder $\delta = 35\,°$.

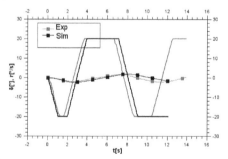

Figure 9. The comparison of characteristic chart of the evasive action.

## BIBLIOGRAPHY

[1] COMMISSION DIRECTIVE 2008/126/EC European Parliament and of the Council laying down technical requirements for inland waterway vessels, Official Journal of the European Union L 32/1.
[2] Właściwości manewrowe statku śródlądowego na wodzie ograniczonej, Tabaczek T., Górnicz T., Kulczyk J., Zawiślak M, Raporty Instytutu Konstrukcji i Eksploatacji Maszyn, SPR, Politechnika Wrocławska, 2010.
[3] The Specialist Committee on Esso Osaka Final Report and Recommendations, Conference Proceedings of the 23rd ITTC, Volume II, Venice, 2002.
[4] Kijima, K., Tanaka, S., Furukawa, Y., Hori, T.On prediction Method of Ship Maneuvering Characteristics, MARSIM'93, St. John's, Newfoundland, Canada, 1993.
[5] The Manoeuvring Committee Final Report and Recommendations, Conference Proceedings of the 23rd ITTC, Volume II, Venice, 2002
[6] Marine Rudders and Control Surfaces, Anthony F. Molland and Stephen R. Turnock, Elsevier, 2007.
[7] Metoda oceny właściwości manewrowych statku śródlądowego. Górnicz T, Raporty Instytutu Konstrukcji i Eksploatacji Maszyn, PRE, Politechnika Wrocławska, 2010.
[8] Badanie wpływu głębokości wody na siły hydrodynamiczne działające na ster statku śródlądowego. Górnicz Jan, Kulczyk Jan, Tabaczek Tomasz, XXV. Plavebné dni, 2008,
[9] Computional Fluid Dynamics A practical Approach, Jiyuan Tu, Guan Heng Yeoch and Chaoqun Liu, Elsevier, 2008.

# 7. Comprehensive Methods of the Minimum Safe Under Keel Clearance Valuation to the Restricted Tidal Waters

G. Szyca

*Associated British Ports, Humber Estuary Service, Pilotage Department, England*

ABSTRACT: The main purpose of the paper is familiarization with the matter of most crucial aspects of the minimum under keel reserve for the sea-going ships navigating on the restricted tidal waters. For the purpose of this paper, river Humber was used as good example of high tidal range in conjunction with variety weather conditions encountering in this area affecting the tides. The Author made use of his research and job experience as the pilot, closely co-operating with other authorities, e.g. Vessel Traffic Service ( VTS ) Humber, local Hydrographic Department as well as Maritime Coastguard Agency ( MCA ) and Maritime Accident Investigation Bureau ( MAIB ). The extensive statistics of groundings and near miss situations, in connection with in-depth analysis will be due to presented with conclusions and proposals of sorting out some problems. The new concept of Dynamic Under Keel Clearance DUKC® software with the trials assessments will be widely put forward in compare with currently utilizing tools and software by VTS and Ships' Traffic Centre.

## 1 INTRODUCTION

### 1.1 *Criteria of the subject choice*

It is a common knowledge that statistical, groundings are on the top of the list of all reported ships' accidents, incidents and near miss situations. Upon UK Maritime Accident Investigation Bureau (MAIB) reports, groundings represented 73.2% of all noted ships accidents last year and over 64% for the latest decade. Additionally, groundings are in strict conjunction with secondary hazards and accidents, appearing shortly after that, e.g. oil pollution, lost of the stability, damage to the hull and finally collision with other ships in extreme cases. The above mentioned hazards classified in conjunction with the groundings represent 7.5% of the total number listed accidents. Significant numbers ( over 55% roughly – precise estimations are still under the count) of all groundings had happened on the restricted waters, such as rivers, estuaries, bays and creeks. The average tidal range for these cases was above 5 meters.

Due to the careful study and analysis, it was concluded that the main reasons of the majority of the total groundings were insufficient under keel clearance, errors in squat assumption and wrongly calculated tidal figures in connection to the passage plans.

The groundings are relatively easy to classify, so any precautionary measures should be selected very carefully using all available traditional methods with modern software and sophisticated tools.

### 1.2 *River Humber as an example of restricted tidal area*

The Humber Estuary is one of the busiest trade routes in Britain and represents about 11% of overall seaborne trade.

Numerous sandbanks, swift tidal currents, dense traffic, dynamic changing of bathymetric data and unprecedented elsewhere silltation, do the River Humber very difficult for navigation and conservancy. Since 1972 the VTS Humber likely others in Great Britain continuously developing and self evaluating, under supervising Maritime Coastguard Agency (MCA) and Maritime Accident Investigation Bureau (MAIB), fulfilling SOLAS Convention, Chapter V Regulation 12, Maritime Safety Committee (MSC) Circular 1065, International Maritime Organization (IMO) Resolution A.857(20) and corresponding with International Association of Marine Aids to Navigation and Lighthouse ( IALA ) VTS Manual 2008 placing emphasis on IALA Model courses and Training Guidelines.

Despite of every effort was made by the local Hydrographic Department Hull in the scope of survey and dredging, the dynamic and unpredicted silltation of the river due to daily migration of the huge masses of the mud and the other riverbed materials ( mainly on the upper part of the river ) is considered as a main reason of the massive ships' groundings, posing highest percentage of overall marine casualties on the River Humber. Grounding incidents divided onto to two groups: first when ships were re-floated on the same tide and the second, when ships not re-floated on the same tide ( means: re-floated on the next tide or under external assistance). Generally speaking, it is quite obvious that the measured two kinds of groundings showing very similar figures. In presented hereunder results, about 50% of them were caused by wrongly calculated tide values ( time and high ), or inadequate prediction (tide fluctuation). In 40% of overall cases, the measured accidents were brought out by pilots with 2 years practice or less. Over 25% of total groundings concerned sudden "tide cut" either height or time, in the areas called "no point to return", where was not enough space to turn around the ship and abort the passage[1].

Below diagram illustrates the figures of the total groundings of sea-going ships on the river Humber within the space of five years:

Diagram 1.1 The statistics of the vessels groundings with re-

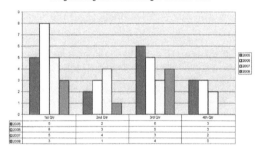

V/ls grounding but not refloating in the same Tide

| | 1st Qtr | 2nd Qtr | 3rd Qtr | 4th Qtr |
|---|---|---|---|---|
| 2005 | 5 | 2 | 6 | 3 |
| 2006 | 8 | 3 | 5 | 3 |
| 2007 | 5 | 4 | 3 | 2 |
| 2008 | 3 | 1 | 4 | 0 |

Although every single case of the grounding has been detailed  evaluated and analyzed both by MAIB and Humber Estuary Service, at the present stage, it is very difficult to find out about any links between the casualty, the professionalism of the VTS staff, ship's crew negligence and  pilot's lack of knowledge. MAIB's grounding investigations revealed that over 50% of all accidents were caused by insufficient under keel clearance and in over 30% cases the official soundings were not covered with actual depths found while grounding or shortly after that.[2] MAIB strictly recommends to apply by VTS sophisticated and efficient software which allows to assess a require under keel clearance and increase ruling Under Keel Clearance ( UKC ) for restricted waters, meeting assumed goals.

Execution of  increasing the minimum under keel clearance, in practice is very difficult, because the

---

[1] Annual Maritime and Coastguard Agency Report, London 2009
[2] Harbour Master Humber Annual Report, Hull 2009

52

period of flood tide in the rivers Trent and Ouse is 3 hours only, so high demands would lead to too frequent cancellations of the  ships and discontinuation upper river traffic. The most reasonable solutions is applying highly, advanced software, particularly designed  for specific waters, conformed with existing systems and flexible collaborated with tide gauges in many locations. Practically such a tool should consist of two independent modules: one for the lower river, second for the upper, therefore UKC requirements are quite different for those two parts of the river. One additional module should be designated for very large ships  (VLS), which are subject to completely different UKC and passage criteria. Another additional obstacle making difficult execution for the required UKC, is the fact, that upper river is not covered by radar stations and all monitoring of the traffic is via Automatic Identification Systems (AIS), Radiotelephones (VHF) and Closed-Circuit Television (CCTV). One alternative option could be using the Portable Operation Approaching and Docking Support System (POADSS), by all transiting ships. That conception will be widely presented in the next chapters.

## 2  ACTUALLY APPLY UNDER KEEL CLEARANCE STANDARDS TO THE RIVER HUMBER

The precise determination of the clear-cut under keel clearance figures still remains as a open issue and  the subject to the dispute between safety measures and commercial interests. The relatively huge differences of the surveys in the short intervals is still the main obstacle to work out uniformed UKC figures.

Table 2.1  Illustration of the dynamic changes to the depths of the River Ouse (upper Humber)  for the period less than one month

| ABP-PORT OF GOOLE, | REACHES DATE: 18/05/09 | ABP-PORT OF GOOLE, | REACHES DATE: 10/06/09 |
|---|---|---|---|
| CHANNELS | DEPTH (METERS ) | CHANNELS | DEPTH (METERS ) |
| Goole Reach | | Goole Reach | |
| EAST SIDE TOP | 1.70 | EAST SIDE TOP | 1.70 |
| EAST SIDE LOWER END | 1.70 | EAST SIDE LOWER END | 1.60 |
| WEST SIDE | 1.40 | WEST SIDE | 1.30 |
| Swinefleet Reach | 1.60 | Swinefleet Reach | 1.80 |
| 72a.33 | | 21b.33 | |
| 21 BANK HOUSE | 1.80 | 21 BANK HOUSE | 2.20 |
| LIGHT - LIGHT | 1.40 | LIGHT - LIGHT | 1.10 |
| JETTY - ORCHARD | 1.90 | JETTY - ORCHARD | 1.70 |
| BANK HOUSE HALE STAITH / SALTMARSHE | 2.70 | BANK HOUSE HALE STAITH / SALTMARSHE | 2.70 |
| Yokefleet Reaches | 1.00 | Yokefleet Reaches | 1.10 |
| M6 70 PILE TO PILE | | M5/8 PILE TO PILE | |
| LIGHT - LIGHT | 1.20 | LIGHT - LIGHT | 1.80 |

Depth of water below the Zero mark at the Tide Gauge Board. Whilst every care is taken compiling this data from actual soundings. Associated British Ports does not accept responsibility for its accuracy due to the unpredictable fluidity of the silt on the river bed.

Hydrographic surveys presented above, for the specific reaches of the river Ouse, in duration of  the 23 days show average difference about 0.20 m and in extreme cases **even 0.40 and 0.50 m.** The commonly established under keel clearance standards for the rivers Ouse, Trent and upper Humber are 0.20m during the day and 0.30m at night (not excluding ships' Companies higher standards), it means that the difference between fault given depth is almost twice more than actual under keel clearance and directly leads towards potential marine hazard. Another idea to establish higher UKC standard may cause, that required minimum would be (e.g. for value 0.50m )equivalent for a ship with the draught of 5.0m, for lower Humber ( UKC is 10% of max. draught ). In the cases of less drafted ships, the UKC for upper river would be greater than for lower. Where max. draught for Ouse and Trent is 5.5m in highest spring tide and average draught is 3.80m, such a solution would be absolutely pointless and leads to nowhere. At the present stage it is seeking for the compromise between demanded safe level of navigation and keeping the waterways fully navigable working on the higher standards but that major, essential problem is not still sorted out and it is the subject to further consultations and advanced trials both by Humber Estuary Service and MCA.

## 3 STATIC METHODS OF THE DETERIMNATION UNDER KEEL RESERVE

Generally speaking, the main strategic assumption in the calculating process of estimation a required under keel clearance (UKC) is available water level at the destination referred to the actual ship's draught. Applying this method, consists of the several variations and derivatives but mainly basis on the tide tables for the specific location, date and time upon drew up harmonic curves and math algorithm. Unfortunately mentioned method does not take into consideration changing hydro-met condition affecting desired tide level and leading straight away to apply additional corrections or decisive modifications current passage plan. All factors must be take into the consideration while unexpected conditions are being encountered to complete safe passage of the ship, including sufficient water level when piloting act is aborted ( return passage ).

### 3.1 Analytic estimation of the demand height of the tide.

The analytic calculation of the predicted tidal level at the port of destination generally basis on the tide tables worked out for the specific ports and it is the part of preliminary process preparing ship' passage plan. Mentioned method may be recognized as a estimated only, because the all tides given in the tide tables are referred to the High Water for the specific location, not providing the heights for intermediate periods. The manual height interpolation of the tide gives the errors about 7 to 12%, there is 0.35 and 0.5m respectively for the height of the tide 5m,

which is unacceptable for 0.3m of the UKC. Besides, relying on the recalculating figures only, given in the Tide Tables without taking into consideration seasonal changes and specifications of the river bed increases the error to the additional 10%, in extremes. Only right, correct action should be applied additional other support or/and alternative reliable methods for double check.

### 3.2 Using remote gauges for the current tidal valuation

The river Humber is fitted with several tide gauges throughout the navigation traffic routes. The average distance between the gauges is 5-7 Nm, which gives to the navigator current information about tidal condition for the specific location via VTS or internet connection.

The tide prediction is not made of each gauge location ( current tide height remotely reading only). Presuming the ship's average speed of 10 knots, bearing mind changing of the datum and assessing variation of the reading for the respective tide gauges the navigator is able to extrapolate the demanded UKC for the specific location, time and height of the tide with necessary margin of the error using following empirical obtained formulas:

$$UKC = [ Dr - ( ( R1 - R2 )/2 + dD ) + Dth ] + 10\%$$
while sailing upriver on rising tide $\qquad$ (1)

$$UKC = [ Dr - (( R1 - R2 )/2 - dD) + Dth ] \text{ while}$$
sailing downriver on rising tide $\qquad$ (2)

$$UKC = [ Dr - (( R1 - R2 )/2 - dD ) + Dth ] + 15\%$$
while sailing down river on falling tide $\qquad$ (3)

where: UKC – under keel clearance [m]
$\qquad$ Dr – ship's draught [m]
$\qquad$ R1 – reading from passing tide gauge [m]
$\qquad$ R2 – reading from the next nearest tide gauge [m]
$\qquad$ dD – difference in the datum between the gauges [m]
$\qquad$ Dth – actual depth for specified point below chart datum [m]

For the double check purpose of the UKC may be used the below table, drew up in the over 15 years period basis on statistical observation, referred to the HW Albert Dock and recalculated for the significant location. This table includes all observed seasonal changes of the tide as well as other fluctuations and there is verified and updated annually.

Table 3.1 Height of the tides referred to Albert Dock HW 8.0 metres

| 8.0 metres (HW Albert-Book) | | | | | | | | | | | |
|---|---|---|---|---|---|---|---|---|---|---|---|
| | 5hrs | 4hrs | 3hrs | 2hrs | 1hr | HW | 1hr | 2hrs | 3hrs | 4hrs | 5hrs |
| Spurn Point | 2.4 | 3.6 | 4.9 | 6.0 | 6.4 | 6.3 | 5.6 | 4.4 | 3.1 | 2.2 | 1.7 |
| Immingham | 2.2 | 3.4 | 4.7 | 5.9 | 6.7 | 6.7 | 6.1 | 5.0 | 3.7 | 2.6 | 1.9 |
| Albert Dock | 1.6 | 2.7 | 4.0 | 5.3 | 6.3 | 6.8 | 6.3 | 5.3 | 4.1 | 2.9 | 2.0 |
| Humber Bridge | 1.6 | 2.4 | 3.4 | 4.6 | 5.7 | 6.2 | 6.0 | 5.0 | 4.0 | 3.0 | 2.1 |
| Brough | 1.1 | 1.3 | 2.1 | 3.6 | 4.8 | 5.6 | 5.5 | 4.6 | 3.5 | 2.6 | 1.9 |
| WW Dykes | 1.0 | 1.0 | 1.6 | 3.0 | 4.5 | 5.3 | 5.4 | 4.4 | 3.5 | 2.6 | 2.0 |
| Blacktoft | 0.4 | 0.1 | 0.7 | 2.2 | 3.7 | 4.6 | 4.8 | 3.9 | 2.9 | 2.1 | 1.5 |
| Goole Docks | 1.0 | 0.8 | 0.7 | 1.3 | 2.8 | 4.3 | 4.8 | 4.2 | 3.3 | 2.5 | 2.0 |
| Burton Stather | 0.0 | 0.0 | 0.3 | 1.7 | 3.1 | 4.2 | 4.4 | 3.6 | 2.7 | 1.8 | 1.2 |
| Flixborough | 0.3 | 0.0 | 0.0 | 1.4 | 2.6 | 4.0 | 4.4 | 3.6 | 2.6 | 1.8 | 1.2 |
| Grove | 0.2 | 0.0 | 0.0 | 0.7 | 2.0 | 3.5 | 4.1 | 3.4 | 2.5 | 1.8 | 1.2 |
| Keadby | 0.0 | 0.0 | 0.0 | 0.3 | 2.3 | 3.3 | 3.8 | 3.3 | 2.4 | 1.6 | 1.0 |

### 3.3 *Comparing the tide fluctuation to the secondary location*

Seeking of the secondary ports, with similar characteristic of the sea/river bed, the datum and not far away located from defined area where is more likely any fluctuation tide data may be used for the extrapolating tidal condition on the site in our interests.

There are several basic assumptions should be made:
- the occurrence of the High Water must be not earlier than 3 hours and later than 7,
- no significant seasonal changes should be affecting the observations,
- mutual changes with the datum not exceeding 1 meter,
- the secondary location has to be situated in the north of defined area,

The North Shields was chosen as a secondary location for the River Humber upon the years of careful observations and wide-ranging analysis. So far there is not any math algorithm allowing precisely described mutual correlation between such a huge main and secondary locations. The average accuracy applying above method are vary and oscillating between 65 and 72%.

## 4 DYNAMIC EVALUATION OF THE AVAILABLE UNDER KEEL CLEARANCE

Dynamic changes of the shipping conditions, such as tidal stream sets and rates, weather conditions, available depths, ships' traffic density or common technical difficulties with the service of the locks or berths, leading towards arising the potential threats and near miss situations for shipping safety in relation to the execution of the original passage plans. VTS operators make every efforts to update any essential data, affecting shipping, either currently or at the periodical broadcasts, but it contents only major information and not included any minor and dynamic developing potential endangers for the specific area. All navigators
( pilots, masters, pilot exemption certificate (PEC) holders ) should have the access to any online nautical, hydro-meteorological, bathymetric and traffic information, covering their passage areas. Such possibilities offers newly working out into the practice, Portable Operational and Docking Support System, commonly known as POADSS.

### 4.1 *Usage of the Portable Operational and Docking Support System (POADSS).*

The POADSS project successfully culminated in live demonstration in Lisbon, last October 2008 proving its complete suitability[3].

The POADSS unit consists of three main elements, two onboard units and the shore one. One onboard unit is an Instrument Unit and the other is a laptop displaying all relevant information for receiving and transmitting data to and from the shore based unit by means of mobile broadband. That information exchange ashore by POADSS Ground Server Station, which sources data from VTS, tide gauges and AIS transmitters. Such own stored data gives to the navigator overview ship's static and dynamic information details as well as surrounding traffic image and environmental conditions in comprehensive overview of all necessary parameters of the particular ship on her passage. Distinct from mostly applied pilotage units the POADSS monitors vertical position ( 3D ) and all dynamic motions. There are four main new applications:
- Internal Measurement Unit with Global Navigation Satellite System ( GNSS ) for determination all dynamic ship's movement,
- Wireless broadband to exchange information in real time ( Web or local map service ).
- Dynamic high density bathymetric and survey data displayed on electronic chart including true dynamic safety contour.
- Dynamic Under Keel Clearance (DUKC®) software.

The above mentioned applications efficiently reduce voice radio communications and maximizes the usability of fairway and enhances the efficiency of the traffic flow. Interoperability with VTS centre is a key element and by using Web Map Service the overall VTS traffic image can be overlaid on the POADSS Electronic Navigation Chart. If the broadband connection lost then AIS information remains available by pilot's plug connection.

However benefits of usage the POADSS are obvious some restrictions and inconveniences still exist:
- if specialized docking system is deployed, this might take up to 15 minutes to set up it,
- still some vessels, such barges or yachts are not fitted with AIS causing POADSS not effective as expected,
- not approved operating and training standards,

---

[3] The Pilot No.296, United Kingdom Maritime Pilot's Association, January 2009

- by using the POADSS in conjunction with Dynamic Passage Planning DPP the maximum draft could be considerably increased and the tidal windows widened without compromising the safety or efficiency of other traffic,
- development of Fibre Optic Gyro's and Micro Electronic Motion Sensors MEMS
- presently are not advanced enough and not offer sufficient accuracy and reliability. It is expected to reach those goal in the next five years.
- the coverage of wireless broadband is still unsatisfactory.
- present stage of development of E-Navigation is not fully capable to be integrated with all POADSS applications.
- The usage of the POADSS is pretty limited at its functions in very narrow river channels and restricted fairways ( upper Humber, rivers Ouse and Trent for instance ).

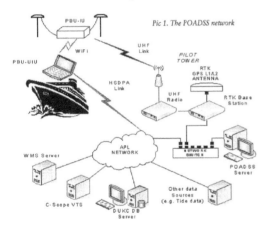

Pic 1. The POADSS network

## 5 METEOROLOGICAL EFFECTS ON TIDES

All meteorological conditions more or less change the tide figures, next affecting available depth at the port of destination closely linked with under keel clearance on the ship's passage. Meteorological condition which differ from the average will cause corresponding differences between predicted and actual tide. Some of the effects are discussed below.

### 5.1 *The effect of the wind*

There was observed that winds blowing longer than 24 hours with the force above 7B from the north directions causing drop of the tide about 30-40cm, paradoxically strong southerly winds don not affecting significantly the height of the tide. After long wind blowing periods ( probability more than 50%) there is more likely that the tide will be above prediction in proportion to the windy period. The new algorithm about wind effect to the tide for river

Humber is under the progress by the Author of hereby Paper .

### 5.2 *Barometric pressure*

The tide tables are computed for average barometric pressure, any significant changes in the atmospheric pressure immediately are reflected in the tidal data. A difference from the average of 34 hPa can cause a difference in height of about 0.3m. That aspect is frequently passed over but is still so essential for navigation on the margin UKC. During predominant of low pressure, for the stationary low, the increase in elevation can be found by the formula[4]:

$$R = [0.01(1010\text{-}P)] \times 0.3; \ [m]$$

R – increase in elevation [m]
P – actual atmospheric pressure [hPa]

For the moving low, the increase in elevation is given by the formula:

$$R1 = \{R / [\ 1\text{-}(\ C / gh\ )]\} \times 0.3 \ [m]$$

C – rate of low motion [ m/s]
g – acceleration to the gravity [ 32.2 m/s]
h – depth of the water [ m]

The number of the British VTS are equipped with modern software applying changes in barometric pressure for tide predictions. So, far VTS Humber is not fitted with such a software.

### 5.3 *Icing*

The rivers Ouse and Trent were frozen since 1962. During 2 months period relatively thick ice on the both rivers made difficulties with the shipping as well as seriously affected the tidal conditions. In the extreme cases the congested ice caused decreasing the height of the tide to figure of 60cm.

### 5.4 *The Aegir*

The "Aegir" is the local word means the head of the significant tidal wave breaking through the river bars and creeks. The aegir occurs while high spring tides around the time of high water ( about half an hour before ). That phenomena seriously interferes predicted height of the tide causing numerous eddies, top runs and significantly gaining tidal streams. Early indication and proper monitoring of the aegir allow to avoid unnecessary risks to navigation. Unfortunately, only visual observations and usage of probability methods can made at the present stage.

---

[4] N.Bowditch, The American Practical Navigator, National Imagery and Mapping Agency L., Publication No.9, 2002

Gathering all information can allow in future to develop more effective statistical methods.

## 6 SUMMARY

The huge numbers of groundings as a result of insufficient under keel clearance are still at unacceptable high level for last several years. Every effort should made to seek out a reasonable compromise between securing minimum under keel clearance requirements for keeping the waterways navigable and rigid standards to the ships' safety. There is no doubt that disturbing statistics should prompt for implementation of the new solutions and different approach the subject of the minimum safe UKC taking into consideration that the human factor still plays major role. The matter of semi liquid sea/river bed being the pattern of any measures and surveys remains still open shall be without any delay sorted out by relevant authorities as well as by the local Safety Navigation River Committees ( SNRC).
So far there is no one defined universal formula for UKC determination for the such extensive area as Humber estuary in the wide space of time. Presently, the main existing "tool" is the probabilistic method, commonly known as dynamic prediction using with co-relation with specialized software. The optimistic prospects after series of advanced trials are placed in the POADSS. At the present stage, the other methods as a deterministic or a stochastic can be used as a support utilities only.

REFERENCES

N.Bowditch, The American Practical Navigator, National Imagery and Mapping Agency L., Publication No.9, 2002
L.Gucma, M.Schoeneich, Probabilistic model of under keel clearance of LNG carriers to given Polish terminals, Proc. of Transport XXI Conference
Dynamic under keel clearance, Information Booklet, OMC International Marine Service Department Australia, 2003
L.Gucma, M.Schoeneich , Probabilistic model of under keel clearance in decision making process of port captain. Proc of Trans – Nav Conference Gdynia 2007
SOLAS 74/78 Convention with further Amendments, London 2007.
IALA VTS Manual 2008, London 2008.
Harbour Master Humber Annual Report, Hull, January 2009.
Hans-Herman Diestel, Compendium on Seamanship and Sea Accidents, Seehafen Verlag, 2005.
Annual Maritime and Coastguard Agency Report, London 2009,
Maritime and Investigations Bureau, Humber Reports, 2005-2009.
Humber Pilot Handbook, Hull 2008..
IALA Recommendation V-103, Standards for the Training and Certification of VTS Personnel.
The Pilot No.296, United Kingdom Maritime Pilot's Association, January 2009.
G.Szyca, Additional Precautionary Measured Upgrading Shipping Safety in Restricted Areas and Adjacent Waters Covered by VTS and Support Decision Centres, Conference materials, Malmo 2009.
I.Jagniszczak, E.M. Łusznikow, Bezpieczeństwo Nawigacji, Gdańsk 2009
Strandings and Their Causes, Cahill, Richard, Third Edition, Nautical Institiute
Role of the Human Element – Assesment of the impact and effectiveness of implementation of the ISM Code, IMO, MSC 81/17

Collision Avoidance

# 8. Knowledge Base in the Interpretation Process of the Collision Regulations at Sea

P. Banaś & M. Breitsprecher
*Institute of Marine Technology, Maritime University of Szczecin, Szczecin, Poland*

ABSTRACT: The article presents the problem of transforming knowledge contained in the provisions of the International Regulations for Preventing Collisions at Sea, and the so called good seamanship in computer applications. Some methods of knowledge representation in decision support in avoidance of collision situations are compared and examined. Acquisition, representation and sharing of knowledge are taken into consideration from the viewpoint of supplementing the knowledge database and computational complexity.

## 1 INTRODUCTION

Widely introduced on ships and land-based facilities, navigational information systems are designed to assist users in the decision-making process. Their main task is collecting and presenting information necessary for safe navigation. Currently available technologies give wider opportunities to assist navigators in interpretation of navigational situations and to generate suggestions of possible solutions. This is connected with the observed development of these information systems in the direction of decision support systems. These systems are based on the use of knowledge, which includes: the existing legal regulations, procedures of conduct, principles of good seamanship, navigational theories. The point is to appropriately acquire that knowledge and create its representation, which enables its efficient and effective use. The problem of decision support also applies to the interpretation of the navigational situation in accordance with the International Regulations for Preventing Collisions at Sea (COLREGS). Due to the fact that in certain areas local law may apply, it is also important to take it into account. Assuming that decision support also includes propositions of solutions (e.g., manoeuvres), a knowledge base should contain the principles of good seamanship.

The knowledge engineering is a branch of information technology which deals with issues of knowledge acquisition, representation and sharing. The methods and tools of this new discipline open way to the construction of systems facilitating the interpretation of the sea route regulations and the application of the principles of good seamanship. The knowledge base built in this way can be used as one of the expert system elements, which may be part of a larger decision support system.

## 2 ASSUMPTION

There are many methods of knowledge representation that can be used in the navigational decision support process. The most important are:
- decision trees, as a representation of possible paths of decision-making depending on the existing and changing conditions;
- logical rules, presented in a simple or complex form, contain the premises and conclusions resulting from them;
- frames that are the base of object-oriented representation of knowledge.

One of the main problems which appears during the design of the knowledge base consists in choosing the proper method of knowledge representation. The decision to choose the proper ways of reasoning is as important as the explanation method and future expansion of knowledge. The selection of these methods is strongly determined by the specificity of the application field, therefore it is important to identify assumptions for a knowledge base and further for an expert system.

The knowledge base should satisfy the following requirements:
- ability to reproduce and verify the rules arising from regulations contained in COLREGS;
- possibility of supplementing the knowledge with special regulations issued by competent authority;
- opportunity to submit informal knowledge contained in the principles of good seamanship;

- open database that allows easy expansion of knowledge;
- simple presentation of the knowledge stored in the database to make it understandable to people who are not systems designers, and easy to verify.

The expert system using the above knowledge base should satisfy the following requirements:
- generation of explicit answers resulting from the knowledge contained in the database;
- interpretation of COLREGS regulations and special rules;
- generation of proposals of actions to be taken arising from the principles of good seamanship;
- explanation of the system response by quoting relevant rules or principles.

An example of the implementation of COLREGS is, designed at the Maritime University of Szczecin, the knowledge base included in the Navigational Decision Support System (NDSS) (Pietrzykowski et al. 2009, Wołejsza 2005). The method of knowledge representation used in this system has a form of decision trees. They give a possibility of checking the subsequent conditions and, on that basis, determining the response of the system. Their main disadvantage is the difficulty of extension, which may cause considerable complexity of the system, and thus the difficulties in the verification of correctness and prolonged time to a response.

Knowledge representation based on logical rules (rule-based knowledge base) is open to change, enables easy verification of accumulated knowledge and makes it possible to offer explanations by simple methods. Due to data security and the systems application field, designers decided that the expert system based on this knowledge base would not be a self-learning system.

## 3 DECISION SUPPORT SYSTEMS

Knowledge bases are an integral part of decision support systems. According to the theory of information systems, it is possible to distinguish several classes:
- transaction systems, whose main task is to collect full information from trustworthy sources and through simple models to allow the basic analysis and compilation;
- management information systems, which aggregate data from other systems, among them transaction systems; they use cross trend analysis using equation models, that operate on deterministic data, but there may occur shortcomings and contradictions;
- decision support systems, which have an extended data analysis mechanism for probabilistic, often contradictory, incomplete and incorrect; when they assist the decision making process, they use

knowledge bases, and optimization and simulation models;
- expert systems, built with the use of knowledge and skills of people who are experts in the appropriate field, which are given as logic and heuristic models;
- artificial intelligence systems, which use a wide range of artificial intelligence methods to perform modelling and analyses.

Because of the wide use of Decision Support Systems, in addition to the foregoing division, they can be characterized by taking into consideration many other factors and criteria. For example, the division can be based on decision-making model or decision-making process modelling (description, prescription). The division by type of controlled system or process comes down to the determination whether the model used is deterministic, statistical or fuzzy. The number of steps in decision-making process (in one step or in n steps), the number decision-makers involved in decision-making process (individual decision or group decision) and the time factor (how the time span is determined to take actions and decisions) are further factors. The next factor characterizing the decision support system is the manner it works – static, when it stops just after giving the answer or dynamic – system that at a given time discretization and at the occurrence of certain events (inputs) operates continuously and appropriately adapts to the existing conditions. Despite the divisions and classifications, which in fact may reflect many aspects, decision support systems, in general, deal with data acquisition, convert information into knowledge and generate answers (decisions) on the basis of the knowledge they comprise.

This division of classes also shows the evolution process of information systems from static to fully dynamic, managing the models in decision support systems (Pietrzykowski et al. 2007). It should also be noted that expert systems currently being developed and artificial intelligence systems are ranked as a subclass of decision support systems.

It is planned that the developed knowledge base will be an element of the system considered as an expert system. With its open formula, cooperation is also possible with the systems that belong to other classes.

## 4 EXPERT SYSTEMS

A typical expert system could be described as structure presented below:
- user interface – a module which is responsible for interaction between the system and the user (navigator). The module input allows user to ask questions and to specify additional information. The module also gathers input data from other sys-

tems. The module output provides the user with answers and explanations;
- knowledge base – data which are characterised by special structure adjusted to store logical rules. The rules are quickly searched for in accordance with given criteria;
- inference mechanism – main part of the system that is responsible for the process of solving the problem. To do so, the mechanism uses knowledge base, logical rules and additional information;
- explanation mechanism – part of the interface that provides the explanation and gives legitimacy for the system answer (decision) (Giarratano & Riley 2004, Jackson 1998).

The most important issues that shall be examined during the process of creating knowledge base of the expert system which offers COLREGS interpretation and gives solutions for aiding navigational decisions are:
- database structure definition;
- data acquisition.

Because the expert system with rule knowledge base is to be considered, specific problems that should be worked out are as follows:
- rules based on COLREGS and local laws or regulations;
- inference mechanisms that are supposed to seek proper rules and formulate conclusions;
- priorities mechanism that is responsible for building proper hierarchy of exact rules and regulations;
- working out the inference mechanism based on good seamanship procedures, that works simultaneously with the mechanism that considers COLREGS rules and local regulations.

## 5 KNOWLEDGE ACQUISITION

The rules stored in the knowledge base are derived from regulations provided in the COLREGS. The regulations were taken under examination that led to a set of logical rules. The connection between regulations and rules was taken into consideration in rules' notation. It is essential during the explanation process of the answer given by system. The operation of extracting data from special rules can be performed in the same way.

In case of extracting any informal knowledge that accounts for the principles of good seamanship, the process is more complex. For example, Jones or Cockroft diagrams which apply to restricted visibility situations can be treated as such source of knowledge. The knowledge itself must be then checked, verified and supplemented by experts - navigators.

## 6 SOLUTION

The knowledge base and the system that are proposed will contain a database divided into three parts that apply to:
- rules from COLLISION REGULATIONS,
- rules that concern the principles of good seamanship,
- additional rules that arise from special regulations. It is important to bear in mind that local rules have higher priority than general rules. Therefore, it is possible that some conflicts will occur and corresponding local and general rules may be contradictory.

Ultimately, the proposed expert system is expected to work as stand-alone module or as be a part of Navigational Decision Support System, developed at the Maritime University of Szczecin.

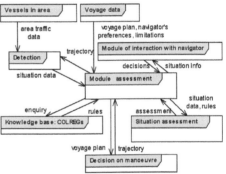

Figure 1. General architecture of navigational decision support system on a seagoing vessel (Pietrzykowski & Uriasz 2009)

The diagram above shows arrangement of the proposed system. It will replace the knowledge base module and take over its function and tasks (Fig.1).

Due to the fact that the knowledge base will be used for building the expert system and consequently it will operate in the decision support system, the following structure of the expert system is taken into consideration (Fig.2).

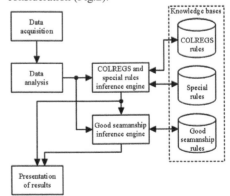

Figure 2. Structure of the proposed expert system.

The functions of the proposed expert system are:
- Data acquisition – the module that acquires data required for the system to work, such as: position, speed, navigational statuses, headings, relative bearings, aspects, weather conditions, etc.;
- Data analysis – the module that analyses acquired data, performs data standardization and calculates formulas and parameters to be used in inference mechanisms;
- COLREGS and special rules inference engine – inference mechanism which performs the interpretation of knowledge contained in the rules stored in knowledge bases; on the basis of data sent from 'Data analysis' module, optimal rules are selected – the rules that fit the examined situation. Then, the selected rules are used in forward chaining process and, consequently, the navigator is informed if the vessel is a give-way or stand-on one. In addition, the rules and special rules which are applicable at the moment will be presented.
- Good seamanship inference engine – inference mechanism running in parallel to the above mentioned one, using a knowledge base containing rules reflecting the principles of good seamanship; diagram of operation is very similar to the previous one, but the output gives a suggestion of conduct in a particular situation; the operation of this module is independent from the previous one, so it is possible to give suggestions for appropriate conduct in spite of an absence of priority such as manoeuvres resulting from the Cockcroft diagram for restricted visibility;
- Knowledge bases – set of knowledge bases containing rules resulting from the extraction of knowledge from COLREGS, local regulations and the rules of good seamanship, supplemented with links to relevant regulations and explanations;
- Presentation of results – this module prepares the information obtained from inference mechanisms for presentation and transfers them to the user interface.

Data needed to carry out the inference are derived from a host computer or directly from the navigation systems of the vessel. As assumed, the proposed system always includes all information available. If some data are required and are not among those transmitted, it may be necessary to refine data, taking place via the master system. Obtained data are processed in order to adapt them to the internal representation of knowledge present in the system. This processing is primarily responsible for standardization and for searching for contradictions (e.g., incompatibility of data obtained from different sources, taking into account malfunctions, measurement error or human factor). In the proposed system, input information is obtained in the form of vector data and it may contain contradictions that must be

eliminate. The processed data are the basis for the inference process.

Inference mechanisms work with dedicated knowledge bases, which contain rules extracted from the COLREGS, local regulations (special rules) and rules of good seamanship, a transcript of the knowledge of expert navigators, supplemented by other sources of non-codified rules for determining the proper conduct in different situations. The method of writing the rules was adapted to the transferred input values obtained from the data analysis module, enabling the user to find them promptly.

For the storage of rules there is a dedicated database with appropriately designed structure. Defined in this structure are the tables for each set of rules and additional data used in the process for requesting and managing the work of the whole expert system. The first step in the process of inference is performed by a query to the appropriate tables, which results in a set of rules satisfying the conditions compatible with the existing navigational situation. Then, the provided rules of inference will be carried forward, resulting in the generation of a conclusion, passed to the module output.

Because there exist different sets of rules for the COLREGS (and local regulations) and the principles of good seamanship, two conclusions are obtained in parallel: on the right way with an indication of applicable regulation and rules, and the proposed procedure in the situation.

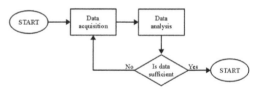

Figure 3. Functional diagram of data acquisition.

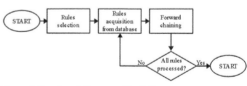

Figure 4. Functional diagram of the inference process.

The system operation is based on the following algorithm shown in Figures 3, 4:
1 Get input from the user interface and navigation equipment;
1 Adapt to the requirements of the input search rules;
2 Find rules matching the analysed situation in the respective knowledge bases;
3 If rules were found in the database with local regulations, they must be taken into account in decision making;

4 If rules were found in COLREGS, determine their hierarchy against the local regulations;
5 Decide on the basis of the rules contained in the databases of local and COLREGS;
6 If emergency arises resulting from the absence of rules and regulations of local COLREGS, acquire again data (clarify) (go to point 1). If the data cannot be refined, send information about the lack of data required to make a decision;
7 If the base of the principles of good seamanship contains rules for the situation, determine suggestions resulting from these rules;
8 Show the results of the expert system work.

Marine navigation equipment and systems on board are the source of input data for running the expert system based on the knowledge base. Where an ambiguous situation is identified, that may result from the inability to establish certain facts, a request will be made to supplement data, for example by a dialogue between system and navigator.

Rules in the knowledge base should be supplemented with additional information, such as the degree of validity of the a rules e.g. rules with the lowest degree of validity are used only to carry out the inference process, rules of higher degree will also appear on the screen as a warning, and a rule having the highest degree will additionally activate a sound alarm.

If the system has an additional knowledge base that contains rules based on local laws, it is possible to automatically take it into account when making a decision. The moment of inclusion or exclusion depends on the geographical position of the vessel obtained from navigational devices (e.g. GPS).

7 EXAMPLES

Below are examples of the proposed expert system functioning. The system receives selected input data and provides a response in the form of conclusions and explanations. The system response and risk of collision which is determined by the decision support system provides the basis for working out the manoeuvre according to COLREGS. The answer is preceded by an analysis of the rules contained in the knowledge base. Examples 1 to 4 are shown on figure 5.

1 Input data: restricted visibility = NO; Distance = 5Nm; Own priority (S1) = 6 (power-driven vessel); Other vessel's priority (S2) = 6; Relative bearing = 005° PS - Port Side; Aspect = 150° SS - Starboard Side; Own speed (V1) > Other vessel's speed (V2)
Conclusion: Give a way
Explanation: conclusion based on rule No. 13 of COLREGS – Overtaking.

2 Input data: restricted visibility = NO; Distance = 3Nm; Own priority = 6; Other vessel's priority =

6; Relative bearing = 100° SS; Aspect = 040° PS; Own speed < Other vessel's speed
Conclusion: Give a way
Explanation: conclusion based on rule No. 15 of COLREGS – Crossing situations.

3 Input data: restricted visibility = NO; Distance = 3Nm; Own priority = 6; Other vessel's priority = 6; Relative bearing = 003° SS; Aspect = 004° PS; Own speed ≈ Other vessel's speed
Conclusion: Head on vessels, Alter your curse to starboard
Explanation: conclusion based on rule No. 14 of COLREGS – Head-on situation.

4 Input data: restricted visibility = YES; Distance = 2Nm; Own priority = 6; Other vessel's priority = 6; Relative bearing = 010° SS; Aspect = 010° PS; Own speed ≈ Other vessel's speed
Conclusion: all actions should be taken to avoid collision – reduce your speed.
Explanation: conclusion based on rule No. 19 of COLREGS (all actions should be taken to avoid collision).

5 Input data: restricted visibility = NO; Distance = 10Nm; Own priority = 6; Other vessel's priority = 6; Own speed ≈ Other vessel's speed
Conclusion: none
Explanation: The distance is considered as safe.

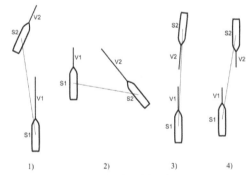

Figure 5. Examples of navigational situations.

Explanatory notes and assumptions applied to the above mentioned examples
– 8Nm is a distance assumed to be safe by the system and no actions need to be taken in navigational situation;
– Vessel's priorities gradation is adopted from simplified "privileged hierarchy" (Rymarz 1995).

8 SUMMARY

The construction of decision support systems can significantly expand the capabilities of navigational decision support systems. Their important element is the knowledge base, which may be a separate structure or part of expert systems. The effectiveness of decision support systems largely depends on the correctness of the knowledge base, its structure and mode of action.

The following assumptions and proposals of implementation have been presented:

1  To use an expert system as a sub module of decision support system;
2  To prepare a knowledge base based on rules;
3  To divide a database to 3 parts containing respectively COLREGS rules, special rules and principles of good seamanship rules to improve efficiency and to allow the knowledge to be updated.

# REFERENCES

COLREG 1972. *Convention on the International Regulations for Preventing Collisions at Sea.* International Maritime Organization

Giarratano J.C. & Riley G.D 2004. *Expert Systems, Principles and Programming*

Jackson P. 1998. *Introduction to Expert Systems*

Pietrzykowski Z., Chomski J., Magaj J. & Bąk A. 2007. *Aims and tasks of the navigational support system on a sea going vessel*, In J. Mikulski (ed.) Advanced in Transport Systems Telematics 2, Publisher Faculty of Transport, Silesian University: 251-259. Katowice

Pietrzykowski Z., Magaj J. & Chomski J. 2009. *A navigational decision support system for sea-going ships*, Measurement Automation and Monitoring

Pietrzykowski Z. & Uriasz J. 2009.Knowledge representation in a ship's navigational decision support system. In Adam Weintrit (ed.), *Marine Navigation and Safety of Sea Transportation*:45-5., Gdynia

Rymarz W. 1995. Podręcznik Międzynarodowego Prawa Drogi Morskiej. Gdynia

Wołejsza P. 2005. *An Algorithm of an Anti-collision Manoeuvre.* Międzynarodowa Konferencja Naukowo-Techniczna Inżynieria Ruchu Morskiego. Świnoujście

# 9. A Method for Assessing a Causation Factor for a Geometrical MDTC Model for Ship-Ship Collision Probability Estimation

J. Montewka*, F. Goerlandt, H. Lammi & P. Kujala
*\*Aalto University, School of Engineering, Finland; Maritime University of Szczecin, Poland*
*Aalto University, School of Engineering, Finland*

ABSTRACT: In this paper a comparative method for assessing a causation factor for a geometrical model for ship-ship collision probability estimation is introduced. The results obtained from the model are compared with the results of an analysis of near-collisions based on recorded AIS data and then with the historical data on maritime accidents in the Gulf of Finland.
The causation factor is obtained for three different meeting types, for a chosen location and prevailing traffic conditions there.

## 1 INTRODUCTION

The MDTC (Minimum Distance To Collision) model for ship-ship collision probability estimation is a geometrical model, with a detailed description given in the following papers: (Montewka et al. 2010), (Montewka et al. 2011). In order to provide the probability of an accident, the model uses the commonly adopted approach, which combines a frequency of ship-ship meeting situations given an assumption of blind navigation, and a causation factor, which quantifies the proportion of cases in which such a meeting ends up as a collision, due to human or technical reasons.

The causation factor is a sensitive part of a model, very much location dependent, thus it is not justified to use the same value for the different models (Gluver and Olsen 1998). Applying a causation probability value derived from a study in another sea area may save some effort, but then the actual conditions are not addressed at all (Hanninen and Kujala 2009).

Two approaches can be recognized in the literature in order to estimate the causation factor. The simplified approach is based on a historical data, where the causation factor is assumed a ratio between the registered number of accidents and the estimated number of collision candidates (Fujii and Siobara 1971), (MacDuff 1974), (Inoue and Kawase 2007).

A second approach is more sophisticated, based on the concept of either event tree (Pedersen 1995), (Martins and Maturana 2010) or Bayesian Networks (DNV 2003), (Hanninen and Kujala 2009). This way of modelling is undoubtedly more time consuming than the first approach, however it allows getting an insight into the chain of events leading to an accident instead of providing just a number.

In order to determine the causation factor for the MDTC model for three different ship-ship encountering types (crossing, head-on and overtaking), we based our study on a modified first approach, which is relatively quick and straightforward thus robust. We perform two stage analysis, which combines the statistical data on maritime accidents and an analysis of near-collisions based on recorded AIS data.

The causation factor is being defined here as a ratio between the modelled number of collision candidates and the actual number of accidents. However the available statistics on maritime accidents are not very detailed, and the type of an accident is not included there. Thus there is a need to find a proxy between a recorded number of accidents and a modelled number of collision candidates (Heinrich et al. 1980), (Inoue and Kawase 2007), (Gucma and Marcjan 2010).

It seems justified to analyze the safety of navigation on the basis of the numbers of both accidents and near-miss situations. Such a combination of analyses may better reflect the collision hazard, as pointed out by (Inoue et al. 2004) and (Inoue and Kawase 2007).

In air transportation there has been a tendency to seek out proxy for aviation safety. One commonly used measure is that of the "air-miss", often called a "near-miss". According to (Button and Drexler 2006) *"a near-miss involves an aircraft intruding upon a predetermined safety zone or envelope around another aircraft"*. The reporting procedures of near-miss in aviation are well founded providing valuable statistics. In the maritime sector similar procedures are missing, thus the near-miss can be detected only by analysis of recorded data and back propagation of recorded events.

Following this idea, this paper proposes also a methodology to evaluate the occurrence of near ship-ship collisions in an open sea area, based on the AIS data. The method for near-collisions analysis presented in this work is rooted in a well-established concept of a ship domain proposed by (Fujii and Tanaka 1971). An overview of the near collision detection method is then given and applied to the summer traffic in the Gulf of Finland.

Finally, we compare the results obtained from the MDTC model, expressed as the number of "collision candidates" with the number of near-collisions and the number of accidents recorded in the chosen area of the Gulf of Finland. This approach allows us to quantify the number of modelled "collision candidates", with blind navigation assumption behind, to the number of cases that ended up as close encounters, where collision evasive actions were taken. Such quantification is carried out for three major types of meeting scenario (crossing, head-on, overtaking). By combining this accurate enough data with an average annual number of accidents that happened (which are random, and almost non predictable), the causation factor for the MDTC model is obtained.

## 2 RESEARCH MODEL

### 2.1 Accident analysis

The annual number of ship-ship collisions in the analyzed location of the Gulf of Finland (the waterways junction between Helsinki and Tallinn) is obtained from HELCOM database, that covers a time period between 1987 and 2007 (Pettersson et al. 2010). During this time, three accidents of this type took place. Two of them happened during summer time, and one was related to the ice conditions, which are out of scope of the analysis presented in this paper.

According to the aforementioned statistics there was, on average, one summer collision per ten years. This assumption is simplified, as the rate of collision occurence is random, as the first collision happened in 1996, second in 2001 and between the years 2001 and 2007 no summer collision happened in the area of investigation. Notwithstanding, we assume that the annual ship-ship collision frequency in the analyzed area equals 0.1.

Unfortunately, the database provided by HELCOM does not contain any information regarding type of ship-ship encounter, at which the accident took place. Thus it is not feasible to compare a modelled number of collision candidates in given en-

counter type (crossing, head-on, overtaking) with an appropriate number of the accidents. At this point the results of near- collisions analysis are utilized and considered a proxy between a model and the recorded accident data.

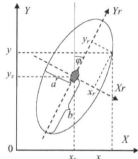

Figure 1: The ship domain applied in the near-collision analysis, with the following axes: a = 1.6LOA, b = 4LOA (Wang et al. 2009)

### 2.2 Near-collisions analysis

The near-collision analysis applied in this paper is based on a concept of a ship domain, which according to definition given by (Goodwin 1975), is the area around the vessel which the navigator would like to keep free of other vessels, for safety reasons.

Since the first introduction of the ship domain concept by (Fujii and Tanaka 1971), various researchers have attempted to quantify the size of this domain. An overwiev of the different proposed domains is given in (Wang et al. 2009). Even though the ship domain is a well established concept, certain problems with the application can be identified as pointed out by (Jingsong et al. 1993). Domains can be classified by their shape: circular, elliptical and polygonal domains. A distinction can also be made between fuzzy domains and crisp domains. Fuzzy domains such as that proposed by (Pietrzykowski 2008) and (Wang 2010) seem preferable in terms of safety analysis of marine traffic, but are at present still under development. Crisp domains use a simple classification of a situation between safe or unsafe, which evidently is a simplification. Moreover, the sizes of the domains proposed in the literature vary quite significantly (Wang et al. 2009).

In this paper, the smallest ship domain found in the literature, by (Fujii and Tanaka 1971), is applied. This is justifiable, since the aim of the method proposed in this paper is finding the most critical encounters between ships. This domain is defined as an ellipse with the major axis along the ship's length (LOA)and the minor axis perpendicular to the ship's beam, as illustrated in Figure 1. The half-length of the major axis is taken as 4LOA while the half-length of the minor axis is taken as 1.6LOA. A number of comments should be made in the use of this domain:

- the domain is symmetric, which implies that the possible influence of the COLREGs is not taken into account;
- another consequence of this symmetry is the fact that passing behind the stern is considered as dangerous as passing in front of the bow;
- in the meeting between ships, the largest ship has the largest domain; this means that for the largest vessel, the situation is classified as dangerous, whereas for the smallest vessel, the situation may still be evaluated as safe;
- the domain is affected by ship length only, neither ship type nor hydrometeorological conditions are included in the analysis.

However the latter can be supported by the recent research, which revealed that the ship domain has a relatively low correlation with the sea state and wind force (Kao et al. 2007).

In this section, a brief description of analysis of AIS data in order to estimate a number of near-collisions in the selected area of the Gulf of Finland is given. Recorded AIS data consists of millions of data points, containing static and dynamic information regarding a ship. In order to analyze the maritime traffic in the GOF, this data need to be grouped into routes. Routes are defined here as a set of trajectories between a departure and arrival harbor as introduced by (Goerlandt and Kujala 2011). The AIS data is first gathered per ship, based on the MMSI number. After sorting this data chronologically, the data per ship is further split up to form individual ship trajectories, using a methodology described by (Aarsther and Moan 2009). These trajectories are then further processed and grouped per route. The sample rate of these vessel positions in the trajectories is about 5 minutes on average. In order to enable a comparison between vessel positions at exactly the same time instant, the trajectory data is artificially enhanced to contain data for each second. The extrapolation for the vessel position is performed using an algorithm suitable for data in the WGS-84 reference frame following (Vincenty 1975). The ship speed is linearly interpolated between known values. It should also be noted that certain vessel types are not taken into account into the analysis, like tugs are left out of the analyzed database. This is done because these vessels are meant to operate in a close vicinity of merchant vessels. The near collision detection algorithm is shown in Figure 2.

The basic idea is to scan the database for events where the ship contour of one vessel (i.e. the ship area in terms of ship length and width) enters the ship domain of another vessel. If the domain is violated, the event is labeled as a near collision and relevant details such as time of occurrence, location, encounter type, ship types and ship flags are stored for further analysis. The near collision detection algorithm is encoded in MATLAB.

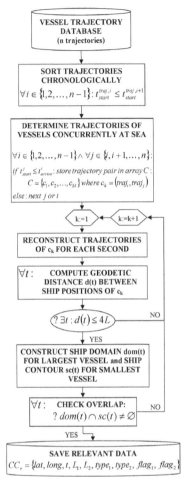

Figure 2: Near collision detection algorithm

The algorithm starts with evaluating whether or not the trajectories of the two considered vessels occur in an overlapping timeframe. If so, the closest distance between vessel positions for contemporary time instances is computed using an algorithm appropriate for geodetic computations according to (Vincenty 1975). If this closest distance between points in trajectories is smaller than the extreme value of the ship domain, the actual vessel contour in terms of length and width are constructed for the smaller vessel and the ship domain is constructed for the larger vessel, for each second. Concurrent ship domain and a vessel contour are evaluated to overlap or not. If there is an overlap of a ship domain, the relevant situational data is stored. If there is no overlap, the next case is investigated.

In the analysis of the locations of the near collisions, a distinction is made between three different encounter situations, as defined in the Collision Regulations by (Organization 2002). Thus crossing, head-on and overtaking are considered. Having the data for the whole Gulf of Finland, we focus on a selected area, which is a crossing of waterways between Helsinki and Tallinn. The area is bounded by the following meridians: 024.5deg E and 025deg E and the parralels: 59deg N - 60deg N.

The results obtained in the course of the analysis, for the time period analyzed (01.04.2007-30.10.2007) are depicted graphically in Figure 3.

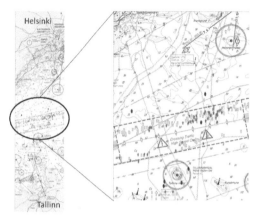

Figure 3: Results of the near collision analysis

The annual numbers of near collisions, based on the obtained data, ordered according to the traffic scenario are shown in Table 1.

Table 1: The annual number of near-collision events in the waterways crossing in the Gulf of Finland.

| Ships meeting | Annual number of near-collisions |
|---|---|
| Crossing | 95.0 |
| Head-on | 14.0 |
| Overtaking | 252.0 |
| Overtaking adjusted | 54.5 |

In the yearly perspective the elliptical Fujii domain leads to 14 ship domain violations for head-on encounters and 95 for crossing encounters. However, 252 cases are identified for overtaking encounters. This is due to the fact that the Fujii domain does not take the regulation of traffic in terms of traffic separation schemes into account. In order to get more meaningful results, a heuristic solution for this is applied, by requiring that the number of domain violations for overtaking is equal to the average number of critical encounters for head-on and crossing (labeled "Overtaking adjusted" in Table 1). To this effect, the Fujii domain is evaluated with a reduced width of $1.25L\_max$ for overtaking encounters (as opposed to the original $1.6L\_max$), where $L\_max$ is the length of the largest vessel in the encounter.

## 2.3 Collision probability modelling

The MDTC model, which is a geometrical model, estimates a probability of collision between two ships based on a well founded formula (Kristiansen 2004):

$$P = N_A P_C \qquad (1)$$

where $N_A$ is the number of collision candidates, often named a geometrical probability of a collision course and $P_C$ is the causation probability, also called the probability of failing to avoid a collision when on a collision course. A ship on a collision course is called a collision candidate, which may end up as a collision as a result of technical failure or human error. The causation probability quantifies the proportion of cases in which a collision candidate ends up as a collision.

As a number of collision candidates NA depends on a number of factors, which are described in the following part of this chapter, the input data should be carefully chosen and interpreted before an analysis is carried out. The input values are location dependent, and within a specific location they are very often also time dependent, for instance:
- an intensity of traffic in the given area (if scheduled traffic is observed over the given area, the intensity of ships will change in the course of the day),
- a frequency of occurrence of given ship type in the given area (in general it can be correlated with scheduled traffic, in certain hours more ships of given type can be expected than in an- other time spans).

It is also important to observe a correlation between ship's main particulars and ship type if stochastic modeling is adopted.

MDTC model applied in this study distinguishes between three types of ships encounters, these are: crossing, overtaking and head-on. The probability of having an accident in case of vessels crossing each other course, is calculated by means of the following formula (Endoh 1982), (Montewka et al. 2010):

$$N_{cros\sin g} = \sum_{ij} \frac{E'[V_{ij}]\lambda_i\lambda_j MDTC}{V_i V_j \sin\alpha} \qquad (2)$$

where $E'[V_{ij}]$ denotes the expected relative velocity of all pairs of vessels of types $i$ and $j$, $\lambda$ denotes the intensity of the vessels of given type entering the given waterway, $V$ is the velocity of the vessels according to type, and $\alpha$ is the angle of intersection between the courses of vessels in groups $i$ and $j$.

In case of parallel meetings, namely overtaking and head-on meetings, the common formula is used, and the difference is in a value of intersection angle $\alpha$. In case of overtaking $\alpha <10$deg and in case of reciprocal courses 175deg$< \alpha <$185deg.

$$N_{parallel} = T_0 P_0 P_{time} \qquad (3)$$

where $T_0$ is the overtaking rate (the number of vessels which will overtake another while on parallel courses, irrespective of the passing distance), $P_0$ is the probability that the vessels come close to each other and $P_{time}$ is the probability that these two ships being close to each other will meet in a certain time period. The latter also reduces the theoretical possibility of ship colliding themselves and is estimated for scheduled traffic between Helsinki and Tallin. This probability is not taken into account in case of E-W traffic, which is more random in nature. The overtaking rate is obtained by means of the following equation (Endoh 1982), (Montewka et al. 2010):

$$T_0 = \frac{N^2}{2L} E'[V_{ij}] \qquad (4)$$

where $N$ is the expected number of vessels in the waterway on parallel courses, $L$ is the length of waterway, and $E'(V_{ij})$ denotes the expected relative velocity of all pairs of vessels of types $i$ and $j$. The expected relative velocity between two vessels is determined as follows:

$$E[V_{ij}] = \sqrt{\left(V_i^2 + V_j^2 - 2V_i V_j \cos\alpha\right)} \qquad (5)$$

where $V_i$ is the velocity of a vessel of given group, $\alpha$ means the angle of intersection, which is defined as the difference between the courses of vessels in groups $i$ and $j$.

The probability that the vessels come to a distance, that results in a collision ($P_0$) is simply estimated as follows:

$$P_0 = P\left(d < \frac{B_i + B_j}{2}\right) \qquad (6)$$

where $d$ is the distance between two ships while overtaking and $B$ is the breadth of a vessel of a given class $i$ and $j$. In order to obtain the results as close as possible to the results of near-collisions analysis, the same criteria have to be used. Thus the critical distance for ships on parallel courses is adopted from the near-collisions algorithm, and equals 1.25LOA.

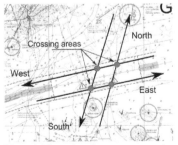

Figure 4: The analyzed waterways crossing
Source: (Montewka et al. 2010)

In the course of our analysis we used the AIS data, which covered a period of seven months of the year 2007, in which the Gulf of Finland remained ice free. The analysis presented in this paper considers specific location in the Gulf of Finland which is waterways junction between Helsinki and Tallinn. These water- ways experience dense RoPax traffic between Finland and Estonia as well as dense traffic in Traffic Separation Scheme (TSS) heading East and West.

The ship data-base contains the following data: MMSI number of ship, time stamp, ship type, ship length, ship breadth, ship speed, ship course, ship position (latitude and longitude according to WGS84 reference system). These particulars are used in order to calculate the probability of ship collisions in the analyzed waterways.

The area under examination and main traffic streams composition is shown in Figure 4. Four traffic streams are analyzed, according to traffic pattern observed in the area: North (N), South (S), East (E) and West (W). The N-S streams consist mostly of scheduled RoPax ships sailing between Helsinki and Tallinn while the E-W streams are composed of all other kinds of ships.

Maritime traffic in the area is assumed to be a stochastic process, and is modelled by means of random sampling and Monte-Carlo methodology. The initial traffic database is decomposed into four smaller databases, according to the four main traffic streams (Figure 4). Then each stream is modelled separately, taking into account the non uniform distribution of ships in time over each stream. The histograms of parameters used in the course of marine traffic analysis are presented in Figures 5 and 6.

The MDTC value, which acts as an input for the equation 2, is drawn from the appropriate chart (Figures: 7, 8, 9). The charts were obtained in the course of an analysis with the use of a model of ship motion given the maneuvering pattern and a ship type (Montewka et al. 2011). The maneuvering pattern, which decides if both of the ships involved in collision situation perform collision evasive actions or only one of them, is chosen randomly with the same probability of occurrence for each of them ($p = 0.5$). Such an assumption may sound simplified, however there is not enough evidence in the literature to disregard it. In case where the maneuvering pattern one is chosen, the algorithm checks if there is a tanker involved, if so then an appropriate MDTC value only for tankers is chosen.

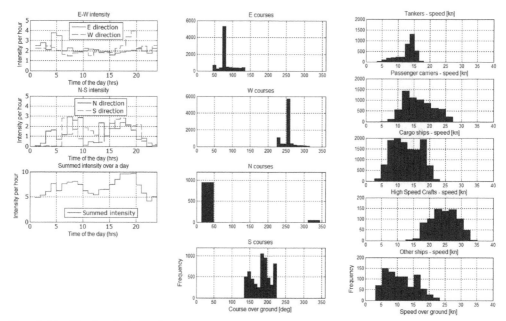

Figure 5: Intensities of marine traffic streams

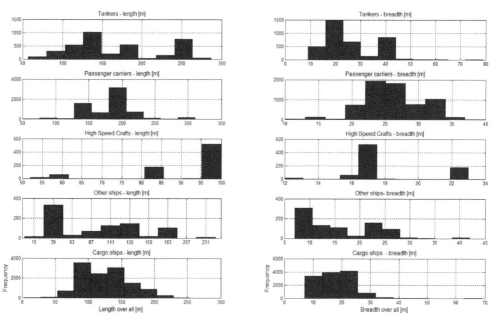

Figure 6: Histograms of main particulars of ships over analyzed area.

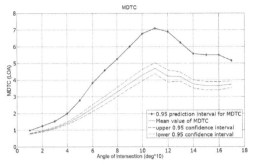

Figure 7: The obtained MDTC chart for the maneuvering pattern No 1 (Montewka et al. 2011)

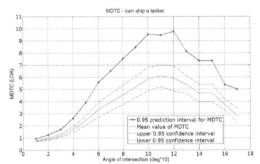

Figure 8: The obtained MDTC chart for tankers - the maneuvering pattern No 1 (Montewka et al. 2011)

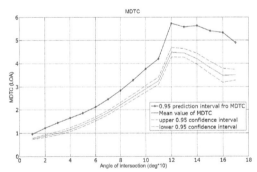

Figure 9: The obtained MDTC chart for the maneuvering pattern No 2 (Montewka et al. 2011)

The annual number of collision candidates obtained with hte use of MDTC model is presented in Table 2. The results are divided into three meeting scenarios (crossing, head-on and overtaking). Within these scenarios there are different sub-scenarios which represents meetings of ships sailing in various streams.

Table 2: The annual number of collision candidates in the waterways crossing in the Gulf of Finland, obtained by means of MDTC model (the avarage values).

| Ships meeting | Annual number of collision candidates |
|---|---|
| Crossing | 5538 |
| Head-on N-S | 1.0 |
| Head-on E-W | 57.0 |
| Head-on All | 58.0 |
| Overtaking N-N | 164 |
| Overtaking S-S | 156 |
| Overtaking E-E | 28 |
| Overtaking W-W | 34 |
| Overtaking All | 382 |

## 3 RESULTS

In the course of presented analyses we obtain a data regarding near-collisions, number of accidents and modelled number of collision candidates for the specific location in the Gulf of Finland.

The aim of this reasearch is to develope a causation factor for the MDTC model by means of a comparative study. The values of the causation factor ($P_C$) are strongly location dependent, as the original studies regarding this parameters have been conducted in the specific locations (eg. straits in Japan, the Dover Strait) it is difficult to assess how the results obtained there can be transferable to other sea areas. The $P_C$ value is also highly dependent on a geometrical model used for the probability of ship accident estimation, thus transferring the same value between different models seems not justified from the scientific point of view.

In our approach we estimate the causation factor that is related to the MDTC model, based on the following formula:

$$P_C(m) = SHF_m \frac{N_A}{\sum_m N_{near-coll}} \qquad (7)$$

$$SHF_m = \frac{N_{near-coll(m)}}{N_{coll-cand(m)}} \qquad (8)$$

where *SHF* is a *ship handling factor* defined for each type of meeting *m* individually (m = [*head-on, overtaking, crossing*]), for the specific value of S H F see Table 3, $N_{near-coll}$ is a number of observed near-collisions, $N_{coll-cand}$ is a number of modelled collision candidates and $N_A$ is a number of recorded accidents. As a result the causation factors for three types of ship/ship encounter were estimated (Table 4).

Ship handling factors presented in Table 3 governs a ship handling process, showing a difference between blind navigation model and real traffic for different encounter types.

Table 3: The ship handling factor (SHF) for three types of ship-ship encounter, for a specific location in the Gulf of Finland.

| Type of meeting | The SHF for given events |
|---|---|
| Crossing | 1.7 * 10e - 2 |
| Head-on | 2.4 * 10e - 1 |
| Overtaking | 1.4 * 10e - 1 |

In Table 4 the values of the causation factor for the MDTC model are gathered. However further research which would cover the greater sea area leading towards a better definition of the causation factor should be carried out.

The numbers for the causation factor proposed here consider a specific geometrical model (MDTC), ordered traffic with waterways crossing, continous surveillance from VTS stations, presence of Traffic Separation Schemes and an intense RoPax cross traffic. The proposed causation factors make a distinction between type of ship-ship encounter. The model is applicable only for the "summer traffic", which means, that presence of ice is not considered.

Table 4: The causation factor for the MDTC model for three types of ship-ship encounter.

| Type of meeting | The causation factor |
|---|---|
| Crossing | 1.04 * 10e - 5 |
| Head-on | 1.46 * 10e - 4 |
| Overtaking | 0.85 * 10e - 4 |

The general relations between analyzed types of event (modelled number of collision candidates, observed number of near-collisions and recorded number of accidents) for the analyzed location are depicted in Figure 10.

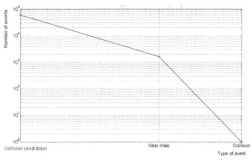

Figure 10: The general relations between each type of event

## 4 CONCLUSIONS

This paper addresses a problem of defining the causation factor for a given geometrical model. We propose a straightforward methodology, which is based on recorded near-collisions (obtained in the course of AIS data analysis) and actual collisions (obtained from HELCOM accidents database). The method estabilishes the ratios between the recorded number

of accidents, the recorded number of the near-collisions and the modelled number of the collision candidates. Knowing these values, it is possible to define a causation factor that constitutes a link between a geometric model for ship-ship collision frequency estimation and a number of accidents due to the given parameters of marine traffic and surroundings.

Making a comparative study we defined the causation factors for the MDTC model, for three ship-ship encounter types. The estimated values of causation factors for the selected area of the Gulf of Finland and given types of vessels sailing there are of the following orders of magnitude: 10e - 5 for ships crossing and 10e - 4 for ships meeting each other on parallel courses.

Although the methodology behind this analysis is straightforward, the results are promising, however there is a need for more extensive analysis, that would cover a larger sea area.

## ACKNOWLEDGMENT

The authors appreciate the financial contributions of the following entities: the EU, Baltic Sea Region (this research was founded by EfficienSea project), Merenkulun säätiö from Helsinki, the city of Kotka and the Finnish Ministry of Employment and the Economy.

## REFERENCES

Aarsther, Karl, G. and T. Moan (2009). Estimating navigation patterns from AIS. The Journal of Navigation 62(04), 587–607.

Button, K. and J. Drexler (2006). Are measures of air-misses a useful guide to air transport safety policy? Journal of Air Transport Management 12(4), 168–174.

DNV (2003). Formal safety assessment - large passanger ships, annex ii: Risk assesment - large passenger ships - navigation. Technical report.

Endoh, S. (1982). Aircraft collision models. Master Thesis. Massachusetts Institute of Technology. MSc thesis, Massachusetts Institute of Technology.

Fujii, Y. and R. Siobara (1971). The analysis of traffic accidents. The Journal of Navigation 24(4), 534–543.

Fujii, Y. and K. Tanaka (1971). Traffic capacity. The Journal of Navigation 24, 543–552.

Gluver, H. and D. Olsen (1998). Ship collision analysis. Taylor & Francis.

Goerlandt, F. and P. Kujala (2011). Traffic simulation based ship collision probability modeling. Reliability Engineering & System Safety 96(1), 91–107.

Goodwin, E. M. (1975). A statistical study of ship domains. The Journal of Navigation 28(03), 328–344.

Gucma, L. and K. Marcjan (2010). The incident based system of navigational safety management of coastal areas. In P. Gelder, L. Gucma, and D. Proske (Eds.), 8th International Probabilistic Workshop. Maritime University, Szczecin.

Hanninen, M. and P. Kujala (2009). The effects of causation probability on the ship collision statistics in the gulf of finland. In A. Wentrit (Ed.), Marine Navigation and Safety of Sea Transportation, London, pp. 267–272. Taylor and Francis.

Heinrich, H., D. Petersen, and N. Roos (1980). Industrial accident prevention (5th ed.). New York: McGraw-Hill.

Inoue, K. and M. Kawase (2007). Innovative probabilistic prediction of accident occurrence. In A. Weintrit (Ed.), Marine navigation and safety of sea transportation, London, pp. 31–34. Taylor & Francis.

Inoue, K., H. Seta, M. Kawase, Y. Masaru, H. Daichi, U. Hideo, H. Kohei, and M. Kenji (2004). Assessment model of ship handling safety by noting unsafe situation as an index. Journal of the Kansai Society of Naval Architects (241), 205–210.

Jingsong, Z., W. Zhaolin, and W. Fengchen (1993). Comments on ship domains. The Journal of Navigation 46(03), 422–436.

Kao, S.-L., K.-T. Lee, K.-Y. Chang, and M.-D. Ko (2007). A fuzzy logic method for collision avoidance in vessel traffic service. The Journal of Navigation 60(01), 17–31.

Kristiansen, S. (2004). Maritime Transportation: Safety Management and Risk Analysis. Butterworth-Heinemann.

MacDuff, T. (1974). The probability of vessels collisions. Ocean Industry, 144–148.

Martins, M. and M. Maturana (2010). Human error contribution in collision and grounding of oil tankers. Risk Analysis 30(4), 674–698.

Montewka, J., F. Goerlandt, and P. Kujala (2011). A new definition of a collision zone for a geometrical model for ship collision probability estimation. In A. Weintrit (Ed.), 9th INTERNATIONAL NAVIGATIONAL SYMPOSIUM ON MARINE NAVIGATION AND SAFETY OF SEA TRANSPORTATION, Gdynia. Gdynia Maritime University.

Montewka, J., T. Hinz, P. Kujala, and J. Matusiak (2010). Probability modelling of vessel collisions. Reliability Engineering & System Safety 95, 573–589.

Organization, I. M. (2002). COLREG: Convention On The International Regulations For Preventing Collisions At Sea (1st ed.). London: Sterling Book House.

Pedersen, P. T. (1995). Collision and grounding mechanics. Copenhagen, pp. 125–157. The Danish Society of Naval Architects and Marine Engineers.

Pettersson, H., T. Hammarklint, and D. Schrader (2010, October). Wave climate in the baltic sea 2008. HELCOM Indicator Fact Sheets 2009. Online.

Pietrzykowski, Z. (2008). Ship's fuzzy domain - a criterion for navigational safety in narrow fairways. The Journal of Navigation 51, 499–514.

Vincenty, T. (1975). Direct and inverse solutions of geodesics on the ellipsoid with application of nested equations. Survey Review 176, 88–93.

Wang, N. (2010). An intelligent spatial collision risk based on the quaternion ship domain. The Journal of Navigation 63(04), 733–749.

Wang, N., X. Meng, Q. Xu, and W. Zuwen (2009, October). A unified analytical framework for ship domains. The Journal of Navigation 62(4), 643–65

# 10. The Sensitivity of Safe Ship Control in Restricted Visibility at Sea

J. Lisowski
*Gdynia Maritime University, Electrical Engineering Faculty, Department of Ship Automation, Gdynia, Poland*

ABSTRACT: The structure of safe ship control in collision situations and computer support programmes on base information from the ARPA anti-collision radar system has been presented. The paper describes the sensitivity of safe ship control to inaccurate data from the ARPA system and to process control parameters alterations. Sensitivity characteristics of the multi-stage positional non-cooperative and cooperative game and kinematics optimization control algorithms on an examples of a navigational situations in restricted visibility at sea are determined.

## 1 SAFE SHIP CONTROL

### 1.1 *Structure of control system*

The challenge in research for effective methods to prevent collisions has become important with the increasing size, speed and number of ships participating in sea carriage. An obvious contribution in increasing safety of shipping has been application of the ARPA (Automatic Radar Plotting Aids) anti-collision system (Fig. 1).

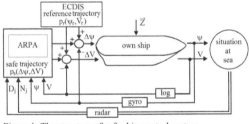

Figure 1. The structure of safe ship control system.

### 1.2 *Information of the state process*

The ARPA system enables to track automatically at least 20 encountered j objects as is shown on Figure 2, determination of their movement parameters (speed $V_j$ , course $\psi_j$) and elements of approach to the own ship ($D^j_{min} = DCPA_j$ - Distance of the Closest Point of Approach, $T^j_{min} = TCPA_j$ - Time to the Closest Point of Approach) and also the assessment of the collision risk $r_j$ (Bist 2000, Bole 2006).

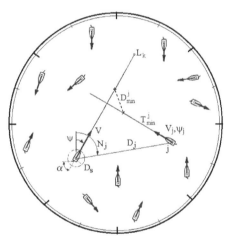

Figure 2. Navigational situation passing of the own ship with j met ship moving with $V_j$ speed and $\psi_j$ course.

The risk value is defined by referring the current situation of approach, described by parameters $D^j_{min}$ and $T^j_{min}$, to the assumed evaluation of the situation as safe, determined by a safe distance of approach $D_s$ and a safe time $T_s$ – which are necessary to execute a collision avoiding manoeuvre with consideration of distance $D_j$ to j-th met ship (Cahill 2002).

The functional scope of a standard ARPA system ends with the trial manoeuvre altering the course $\pm \Delta \psi$ or the ship's speed $\pm \Delta V$ selected by the navigator (Cockcroft & Lameijer 2006, Gluver & Olsen 1998).

### 1.3 Computer support of navigator

The problem of selecting such a manoeuvre is very difficult as the process of control is very complex since it is dynamic, non-linear, multi-dimensional, non-stationary and game making in its nature. In practice, methods of selecting a manoeuvre assume a form of appropriate steering algorithms supporting navigator decision in a collision situation. Algorithms are programmed into the memory of a Programmable Logic Controller PLC (Fig. 3) (Lisowski 2008).

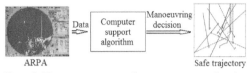

ARPA                                          Safe trajectory

Figure 3. The system structure of computer support of navigator manoeuvring decision in collision situation.

## 2 COMPUTER PROGRAMMES OF NAVIGATOR SUPPORT

### 2.1 Base model of process

The most general description of the own ship passing the j number of other encountered ships is the model of a differential game of j number of moving control objects (Fig. 4).

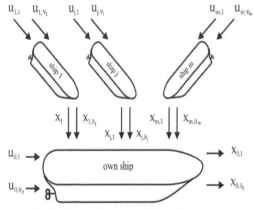

Figure 4. Block diagram of the basic differential game model of safe ship control process.

The properties of control process are described by the state equation:

$$\dot{x}_i = f_i \left[ \left( x_{0,\vartheta_0}, ...., x_{m,\vartheta_m} \right), \left( u_{0,v_0}, ...., u_{m,v_m} \right), t \right] \qquad (1)$$

where:

$\vec{x}_{0,\vartheta_0}(t)$ - $\vartheta_0$ dimensional vector of process state of own ship determined in time $t \in [t_0, t_k]$,

$\vec{x}_{j,\vartheta_j}(t)$ - $\vartheta_j$ dimensional vector of the process state for j-th ship,

$\vec{u}_{0,v_0}(t)$ - $v_0$ dimensional control vector of own ship,

$\vec{u}_{j,v_j}(t)$ - $v_j$ dimensional control vector of j-th ship (Isaacs 1965, Lisowski 2010, Engwerda 2005).

The constraints of the control and the state of the process are connected with the basic condition for the safe passing of the ships at a safe distance $D_s$ in compliance with COLREG Rules, generally in the following form:

$$g_j(x_{j,\vartheta_j}, u_{j,v_j}) = \left( D_s - D_{\min}^j \right) \le 0 \qquad (2)$$

For the class of non-coalition games, often used in the control techniques, the most beneficial conduct of the own ship as a player with j-th ship is the minimization of her goal function in the form of the payments – the integral payment and the final one:

$$I_{0,j} = \int_{t_0}^{t_k} [x_{0,\vartheta_0}(t)]^2 dt + r_j(t_k) + d(t_k) \rightarrow \min \qquad (3)$$

The integral payment represents loss of way by the ship while passing the encountered ships and the final payment determines the final risk of collision $r_j(t_k)$ relative to the j-th ship and the final deflection of the ship $d(t_k)$ from the reference trajectory (Fig. 5) (Modares 2006, Nisan et al. 2007).

### 2.2 Programme of multi-stage positional non-cooperative game MSPNCG

The optimal steering of the own ship $u_0^*(t)$, equivalented for the current position p(t) to the optimal positional steering $u_0^*(p)$. The sets of acceptable strategies $U_j^0[p(t_k)]$ are determined for the encountered ships relative to the own ship and initial sets $U_0^{jw}[p(t_k)]$ of acceptable strategies of the own ship relative to each one of the encountered ship. The pair of vectors $u_j^m$ and $u_0^j$ relative to each j-th ship is determined and then the optimal positional strategy for the own ship $u_0^*(p)$ from the condition (4).

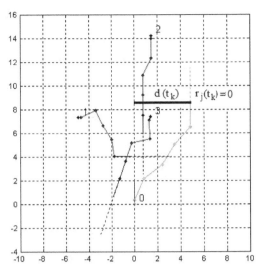

Figure 5. The final risk of collision $r_j(t_k)$ relative and the final deflection $d(t_k)$ from the reference trajectory in situation passing of three met ships.

$$I_0^* = \min_{\substack{u_0 \in \bigcap\limits_{j=1}^{m} U_0^j}} \max_{u_j^m \in U_j} \min_{u_0^j \in U_0^j} \int_{t_0}^{t_{L_k}} u_0(t)\,dt = S_0^*(x_0, L_k) \qquad (4)$$

The function $S_0$ refers to the continuous function of the manoeuvring goal of the own ship, characterising the distance of the ship at the initial moment $t_0$ to the nearest turning point $L_k$ on the reference $p_r(t_k)$ route of the voyage (Millington & Funge 2009, Osborne 2004).

The optimal control of the own ship is calculated at each discrete stage of the ship's movement by applying the Simplex method to solve the problem of the triple linear programming, assuming the relationship (4) as the goal function and the control constraints (2).

Using the function of *lp – linear programming* from the Optimization Toolbox Matlab, the positional multi-stage game non-cooperative manoeuvring MSPNCG program has been designed for the determination of the own ship safe trajectory in a collision situation (Lisowski 2010).

### 2.3 *Programme of multi-stage positional cooperative game MSPCG*

The quality index of control (4) for cooperative game has the form:

$$I_0^* = \min_{\substack{u_0 \in \bigcap\limits_{j=1}^{m} U_0^j}} \min_{u_j^m \in U_j} \min_{u_0^j \in U_0^j} \int_{t_0}^{t_{L_k}} u_0(t)\,dt = S_0^*(x_0, L_k) \qquad (5)$$

### 2.4 *Programme of non-game kinematic optimization NGKO*

Goal function (4) for kinematics optimization has the form:

$$I_0^* = \min_{\substack{u_0 \in \bigcap\limits_{j=1}^{m} U_0^j}} \int_{t_0}^{t_{L_k}} u_0(t)\,dt = S_0^*(x_0, L_k) \qquad (6)$$

## 3 THE SENSITIVITY OF SAFE SHIP CONTROL

### 3.1 *Definition of safe control sensitivity*

The investigation of sensitivity of game control fetch for sensitivity analysis of the game final payment (3) measured with the relative final deviation of $d(t_k)=d_k$ safe game trajectory from the reference trajectory, as sensitivity of the quality first-order.

Taking into consideration the practical application of the game control algorithm for the own ship in a collision situation it is recommended to perform the analysis of sensitivity of a safe control with regard to the accuracy degree of the information received from the anti-collision ARPA radar system on the current approach situation, from one side and also with regard to the changes in kinematical and dynamic parameters of the control process (Lisowski 2009, Straffin 2001).

Admissible average errors, that can be contributed by sensors of anti-collision system can have following values for:
- radar,
  - bearing: $\pm 0,22^{\circ}$,
  - form of cluster: $\pm 0,05^{\circ}$,
  - form of impulse: $\pm 20$ m,
  - margin of antenna drive: $\pm 0,5^{\circ}$,
  - sampling of bearing: $\pm 0,01^{\circ}$,
  - sampling of distance: $\pm 0,01$ nm,
- gyrocompas: $\pm 0,5^{\circ}$,
- log: $\pm 0,5$ kn,
- GPS: $\pm 15$ m.

The algebraic sum of all errors, influent on picturing of the navigational situation, cannot exceed for absolute values $\pm 5\%$ or for angular values $\pm 3^{\circ}$.

### 3.2 *The sensitivity of safe ship control to inaccuracy of information from ARPA system*

Let $X_{0,j}$ represent such a set of state process control information on the navigational situation that:

$$X_{0,j} = \{V, \psi, V_j, \psi_j, D_j, N_j\} \qquad (7)$$

Let then $X_{0,j}^{ARPA}$ represent a set of information from ARPA system containing extreme errors of measurement and processing parameters:

$$X_{0,j}^{ARPA} = \{V \pm \delta V, \psi \pm \delta\psi, V_j \pm \delta V_j, \psi_j \pm \delta\psi_j,$$
$$D_j \pm \delta D_j, N_j \pm \delta N_j\} \qquad (8)$$

Relative measure of sensitivity of the final payment in the game $s_{inf}$ as a final deviation of the ship's safe trajectory $d_k$ from the reference trajectory will be:

$$s_{inf} = (X_{0,j}^{ARPA}, X_{0,j}) = \frac{d_k^{ARPA}(X_{o,j}^{ARPA})}{d_k(X_{0,j})} \qquad (9)$$

$$s_{inf} = \{s^V, s^\psi, s^{V_j}, s^{\psi_j}, s^{D_j}, s^{N_j}\} \qquad (10)$$

### 3.3 Sensitivity of safe ship control to process parameters alterations

Let $X_{param}$ represents a set of parameters of the state process control:

$$X_{param} = \{t_m, D_s, \Delta t_k, \Delta V\} \qquad (11)$$

Let then $X'_{param}$ represents a set of information containing extreme errors of measurement and processing parameters:

$$X'_{param} = \{t_m \pm \delta t_m, D_s \pm \delta D_s, t_k \pm \delta t_k, \Delta V \pm \delta\Delta V\} \qquad (12)$$

Relative measure of sensitivity of the final payment in the game as a final deflection of the ship's safe trajectory $d_k$ from the assumed trajectory will be:

$$s_{dyn} = (X'_{param}, X_{param}) = \frac{d'_k(X'_{param})}{d_k(X_{param})} \qquad (13)$$

$$s_{dyn} = \{s^{t_m}, s^{D_s}, s^{t_k}, s^{\Delta V}\} \qquad (14)$$

where:
$t_m$ - advance time of the manoeuvre with respect to the dynamic properties of the own ship,
$t_k$ - duration of one stage of the ship's trajectory,
$D_s$ – safe distance,
$\Delta V$ - reduction of the own ship's speed for a deflection from the course greater than $30°$ (Baba & Jain 2001).

## 4 SENSITIVITY CHARACTERISTICS OF SAFE SHIP CONTROL IN RESTRICTED VISIBILITY AT SEA

Computer simulation of MSPNCG, MSPCG and NGKO algorithms, as a computer software support-

ing the navigator manoeuvring decision, were carried out on an example of a real navigational situations of passing j=3, 12 and 20 encountered ships. The situations were registered in Kattegat Strait on board r/v HORYZONT II, a research and training vessel of the Gdynia Maritime University, on the radar screen of the ARPA anti-collision system Raytheon (Figs 6-7).

Figure 6. The place of identification of navigational situations in Kattegat Strait.

Figure 7. The research-training ship of Gdynia Maritime University r/v HORYZONT II.

### 4.1 Navigational situation for j=3 met ships

Computer simulation of MSPNCG, MSPCG and NGKO programmes was carried out in Matlab/Simulink software on an example of the real navigational situation of passing j=3 encountered ships in Kattegat Strait in restricted visibility when $D_s$=2 nm and were determined sensitivity characteristics for the alterations of the values $X_{0,j}$ and $X_{param}$ within ±6% or ±3° (Figs 8-14).

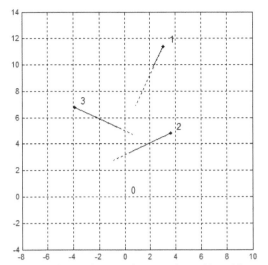

Figure 8. The 12 minute speed vectors of own ship and 3 encountered ships in situation in Kattegat Strait.

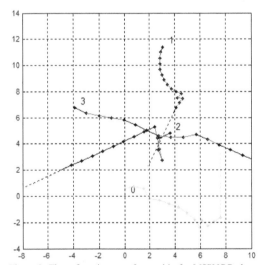

Figure 9. The safe trajectory of own ship for MSPNCG algorithm in restricted visibility $D_s=2$ nm in situation of passing $j=3$ encountered ships, $r(t_k)=0$, $d(t_k)=7.60$ nm.

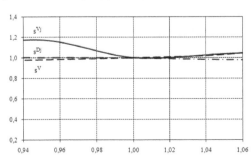

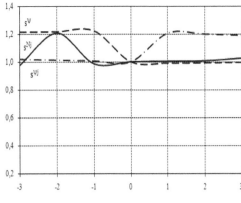

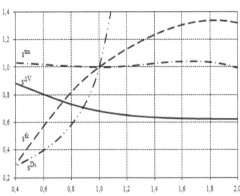

Figure 10. Sensitivity characteristics of safe ship control according to MSPNCG programme on an example of the navigational situation $j=3$ in the Kattegat Strait.

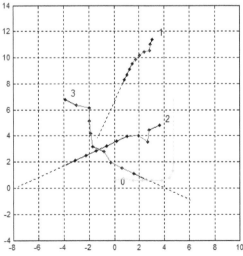

Figure 11. The safe trajectory of own ship for MSPCG algorithm in restricted visibility $D_s=2$ nm in situation of passing $j=3$ encountered ships, $r(t_k)=0$, $d(t_k)=4.71$ nm.

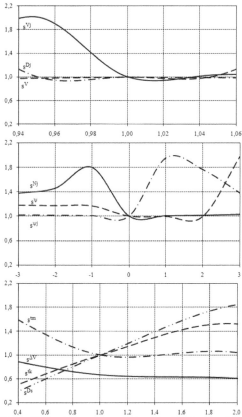

Figure 12. Sensitivity characteristics of safe ship control according to MSPCG programme on an example of the navigational situation $j=3$ in the Kattegat Strait.

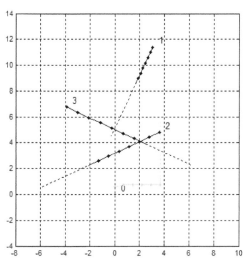

Figure 13. The safe trajectory of own ship for NGKO algorithm in restricted visibility $D_s=2$ nm in situation of passing $j=3$ encountered ships, $r(t_k)=0$, $d(t_k)=3.70$ nm.

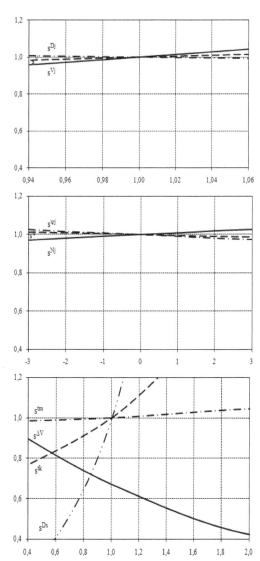

Figure 14. Sensitivity characteristics of safe ship control according to NGKO programme on an example of the navigational situation $j=3$ in the Kattegat Strait.

### 4.2 Navigational situation for $j=12$ met ships

Computer simulation of MSPNCG, MSPCG and NGKO programmes was carried out in Matlab/Simulink software on an example of the real navigational situation of passing $j=12$ encountered ships in Kattegat Strait in restricted visibility when $D_s=2$ nm and were determined sensitivity characteristics for the alterations of the values $X_{0j}$ and $X_{param}$ within $\pm6\%$ or $\pm3^\circ$ (Figs 15-21).

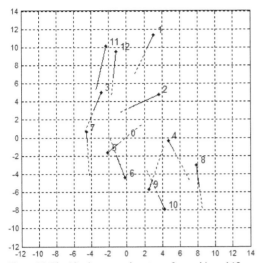

Figure 15. The 12 minute speed vectors of own ship and 12 encountered ships in situation in Kattegat Strait.

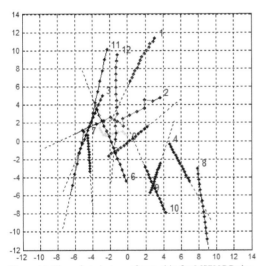

Figure 16. The safe trajectory of own ship for MSPNCG algorithm in restricted visibility $D_s=2$ nm in situation of passing $j=12$ encountered ships, $r(t_k)=0$, $d(t_k)=3.20$ nm.

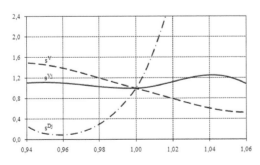

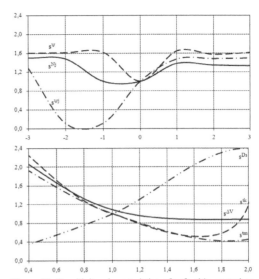

Figure 17. Sensitivity characteristics of safe ship control according to MSPNCG programme on an example of the navigational situation $j=12$ in the Kattegat Strait.

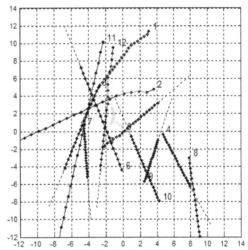

Figure 18. The safe trajectory of own ship for MSPCG algorithm in restricted visibility $D_s=2$ nm in situation of passing $j=12$ encountered ships, $r(t_k)=0$, $d(t_k)=1.40$ nm.

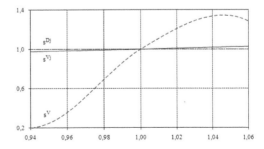

81

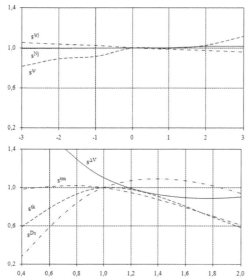

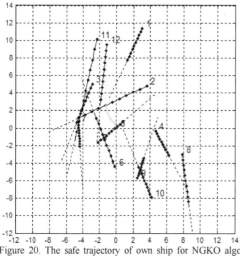

Figure 19. Sensitivity characteristics of safe ship control according to MSPCG programme on an example of the navigational situation $j=12$ in the Kattegat Strait.

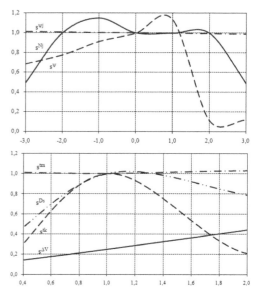

Figure 21. Sensitivity characteristics of safe ship control according to NGKO programme on an example of the navigational situation $j=12$ in the Kattegat Strait.

### 4.3 Navigational situation for j=20 met ships

Computer simulation of MSPNCG, MSPCG and NGKO programmes was carried out in MATLAB/SIMULINK software on an example of the real navigational situation of passing $j=20$ encountered ships in Kattegat Strait in restricted visibility when $D_s=2$ $nm$ and were determined sensitivity characteristics for the alterations of the values $X_{0,j}$ and $X_{param}$ within $\pm 6\%$ or $\pm 3^{\circ}$ (Figs 22-28).

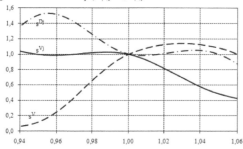

Figure 20. The safe trajectory of own ship for NGKO algorithm in restricted visibility $D_s=2$ $nm$ in situation of passing $j=12$ encountered ships, $r(t_k)=0$, $d(t_k)=1.23$ $nm$.

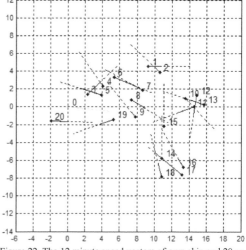

Figure 22. The 12 minute speed vectors of own ship and 20 encountered ships in situation in Kattegat Strait.

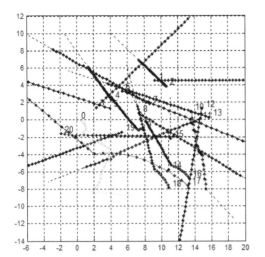

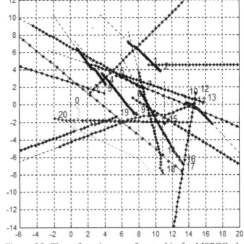

Figure 23. The safe trajectory of own ship for MSPNCG algorithm in restricted visibility $D_s=2$ *nm* in situation of passing $j=20$ encountered ships, $r(t_k)=0$, $d(t_k)=8.06$ *nm*.

Figure 25. The safe trajectory of own ship for MSPCG algorithm in restricted visibility $D_s=2$ *nm* in situation of passing $j=20$ encountered ships, $r(t_k)=0$, $d(t_k)=6.64$ *nm*.

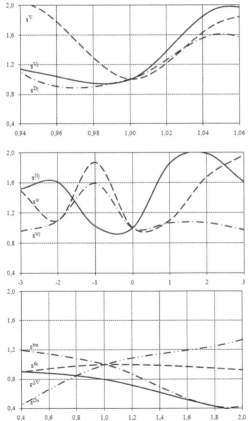

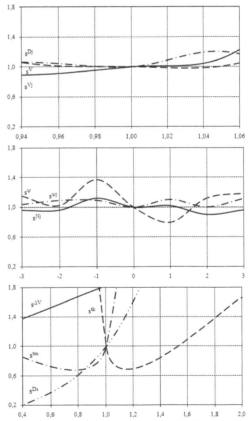

Figure 24. Sensitivity characteristics of safe ship control according to MSPNCG programme on an example of the navigational situation $j=20$ in the Kattegat Strait.

Figure 26. Sensitivity characteristics of safe ship control according to MSPCG programme on an example of the navigational situation $j=20$ in the Kattegat Strait.

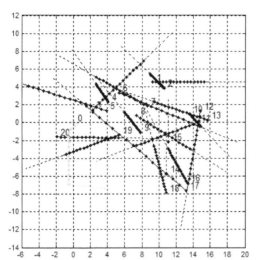

Figure 27. The safe trajectory of own ship for NGKO algorithm in restricted visibility $D_s=2$ nm in situation of passing $j=20$ encountered ships, $r(t_k)=0$, $d(t_k)=6.94$ nm.

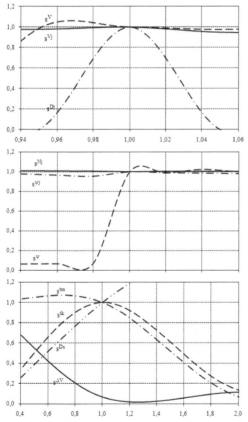

Figure 28. Sensitivity characteristics of safe ship control according to NGKO programme on an example of the navigational situation $j=20$ in the Kattegat Strait.

## 5 CONCLUSIONS

The sensitivity of the final game payment:
- is least relative to the sampling period of the trajectory and advance time manoeuvre,
- most is relative to changes of the own and met ships speed and course,
- it grows with the degree of playing character of the control process and with the quantity of admissible strategies.

The less sensitivity of safe ship control in collision situations is represented by NGKO algorithm and highest by MSPNCG algorithm.

The considered control algorithms are, in a certain sense, formal models of the thinking process of navigator steering of the ship motion and making manoeuvring decisions.

Therefore they may be applied in the construction new model of ARPA system containing the computer supporting of navigator decisions.

## REFERENCES

Baba, N. & Jain, L.C. 2001. *Computational intelligence in games*. New York: Physica-Verlag.
Bist, D.S. 2000. *Safety and security at sea*. Oxford-New Delhi: Butter Heinemann,
Bole, A., Dineley, B. & Wall, A. 2006. *Radar and ARPA manual*. Amsterdam-Tokyo: Elsevier.
Cahill, R.A. 2002. *Collisions and thair causes*. London: The Nautical Institute.
Cockcroft, A.N. & Lameijer, J.N.F. 2006. *The collision avoidance rules*. Amsterdam-Tokyo: Elsevier.
Engwerda, J.C. 2005. *LQ dynamic optimization and differential games*. West Sussex: John Wiley & Sons.
Fadali, M.S. & Visioli, A. 2009. *Digital control engineering*. Amsterdam-Tokyo: Elsevier.
Gluver, H. & Olsen, D. 1998. *Ship collision analysis*. Rotterdam-Brookfield: A.A. Balkema.
Isaacs, R. 1965. *Differential games*. New York: John Wiley & Sons.
Lisowski, J. 2008. Computer support of navigator manoeuvring decision in congested water. *Polish Journal of Environmental Studies*, Vol. 17, No. 5A:1-9.
Lisowski, J. 2009. *Sensitivity of safe game ship control on base information from ARPA radar*. Chapter in monograph "Radar Technology": 61-86, Croatia: Intech.
Lisowski, J. 2010. Optimization decision support system for safe ship control. Chapter in monograph "Risk Analysis VII Simulation and Hazard Mitigation": 259-272, Southampton-Boston: WIT Press.
Millington, I. & Funge, J. 2009. Artificial intelligence for games. Amsterdam-Tokyo: Elsevier.
Modarres, M. 2006. Risk analysis in engineering. Boca Raton: Taylor & Francis Group.
Nisan, N., Roughgarden, T., Tardos, E. & Vazirani, V.V. 2007. *Algorithmic game theory*. New York: Cambridge University Press.
Osborne, M.J. 2004. *An introduction to game theory*. New York: Oxford University Press.
Straffin, P.D. 2001. *Game theory and strategy*. Warszawa: Scholar (in polish).

# 11. Experimental Research on Evolutionary Path Planning Algorithm with Fitness Function Scaling for Collision Scenarios

P. Kolendo, R. Śmierzchalski & B. Jaworski
*Gdansk University of Technology, Gdansk, Poland*

ABSTRACT: This article presents typical ship collision scenarios, simulated using the evolutionary path planning system and analyses the impact of the fitness function scaling on the quality of the solution. The function scaling decreases the selective pressure, which facilitates leaving the local optimum in the calculation process and further exploration of the solution space. The performed investigations have proved that the use of scaling in the evolutionary path planning method makes it possible to preserve the diversity of solutions by a larger number of generations in the exploration phase, what could result in finding better solution at the end. The problem of avoiding collisions well fitted the algorithm in question, as it easily incorporates dynamic objects (moving ships) into its simulations, however the use scaling with this particular problem has proven to be redundant.

## 1 INTRODUCTION

The problem of path planning occurs in numerous technical applications, such as, motion planning for mobile robots [12], ship weather routing in ocean sailing, or safety path planning for a ship in a collision situation at sea [5]. The problem is defined in the following way: having given a moving object and the description of the environment, plan the path for object motion between the beginning and end location which avoids all constraints and satisfies certain optimization criteria. The problem can be divided into two basic tasks: the off-line task, in which we look for the path of the object in the unchanging environment, and the on-line task, in which the object moves in the environment that meets the variability and uncertainty restrictions. The on-line mode of the path planning relates to the control of the moving object in the non-stationary environment, in which parts of some obstacles reveal certain dynamics.

The main goal of the present paper is to present collision scenarios simulation results acquired using evolutionary path planner and to analyse how the fitness function scaling impacts the solution [1,2,14]. Particular instance of the path planning problem as the navigation problem of avoiding collision at sea [5, 6] is considered. By taking into account certain boundaries of the manoeuvring region, along with the navigation obstacles and other moving ships, we reduce the problem to the dynamic optimization task with static and dynamic constrains. We consider this

an adaptive evolutionary task of estimating the ship path in the unsteady environment. The research was performed using Evolutionary Planner/Navigator (υEP/N++) system [7, 8, 9] which takes into account specific nature of the process of avoiding collisions, by using different types of static and moving constraints to model the real environment of moving targets and their dynamic characteristics.

## 2 EVOLUTIONARY ALGORITHMS

Evolutionary Algorithms (EA) are optimization methods that try to mimic evolutionary path in order to find the best solution for a specific problem. Each member of a generation - a set of potential solutions - is being rated against a fitness function to determine member's individual adaptation rate - the quality of the solution. Best fits are being then selected to prepare a new generation. Also, additionally, there is a small chance of offspring's mutation that helps to keep the population differentiated. This process is repeated until an optimal solution is found or the maximum, presumed number of generations is reached. To find out more about EA please refer to position [13] of the bibliography.

## 3 PLANNER INTRODUCTION

When determining the safe trajectory for the so-called *own ship*, we look for a trajectory that compromises the cost of necessary deviation from a given route, or from the optimum route leading to a destination point, and the safety of passing all static and dynamic obstacles, here referred to as *strange ships* (or *targets*). In this paper the following terminology is used: the term *own ship* means the ship, for which the trajectory is to be generated, and *strange ship* or *target* mean other ships in the environment, i.e. the objects which are to be avoided. All trajectories which meet the safety conditions reducing the risk of collision to a satisfactory level constitute a set of feasible trajectories. The safety conditions are, as a rule, defined by the operator based on the speed ratio between the ships involved in the passing manoeuvre, the actual visibility, weather conditions, navigation area, manoeuvrability of the ship, etc. The most straightforward way of determining the safe trajectory seems to be the use of an additional automatic device – a decision supporting system being an extension of the conventional Automatic Radar Plotting Aids (ARPA) system.

Other constraints resulting from formal regulations (e.g. traffic restricted zones, fairways, etc) are assumed stationary and are defined by polygons – in a similar manner to that used in creating the electronic maps. When sailing in the stationary environment, the own ship meets other sailing strange ships/targets (some of which constitute a collision threat).

It is assumed that the dangerous target [6] is each target that has appeared in the area of observation and can cross the estimated course of the own ship at a dangerous distance. The actual values of this distance depend on the assumed time horizon. Usually, the distances of 5-8 nautical miles in front of the bow, and 2-4 nautical miles behind the stern of the ship are assumed as the limits for safe passing. In the evolutionary task, the targets threatening with a collision are interpreted as the moving dangerous areas having shapes and speeds corresponding to the targets determined by the ARPA system.

The path $S$ is safe (i.e., it belongs to the set of safe paths) if any line segment of $S$ stays within the limits of the environment $E$, does not cross any static constraint and at the times $t$ determined by the current locations of the own ship does not come in contact with the moving representing the targets. The paths which cross the restricted areas generated by the static and dynamic constrains are considered unsafe, or dangerous paths.

The safety conditions are met when the trajectory does not cross the fixed navigational constraints, nor the moving areas of danger. The actual value of the safety cost function is evaluated as the maximum value defining the quality of the turning points with respect to their distance from the constraints.

The υEP/N++ is a system based on evolutionary algorithm [1] incorporating part of the problem maritime path planning specific knowledge into its structures. The evolutionary approach provides many benefit such as real or close to real time operations, complex search and high level of adjustment possibilities. Due to the unique design of the chromosome structure and genetic operators the υEP/N++ does not need a discretised map for search, which is usually required by other planners. Instead, the υEP/N++ "searches" the original and continuous environment by generating paths with the aid of various evolutionary operators. The objects in the environment can be defined as collections of straight-line "walls". This representation refers both to the known objects as well as to partial information of the unknown objects obtained from sensing. As a result, there is little difference for the υEP/N++ between the off-line planning and the on-line navigation. In fact, the υEP/N++ realises the off-line planning and the on-line navigation using the same evolutionary algorithm and chromosome structure.

A crucial step in the development of the evolutionary trajectory planning systems was made by introducing the dynamic parameters: time and moving constraints. In the evolutionary algorithm used for trajectory planning eight genetic operators were used, which were: soft mutation, mutation, adding a gene, swapping gene locations, crossing, smoothing, deleting a gene, and individual repair [8, 9]. The level of adaptation of the trajectory to the environment determines the total cost of the trajectory, which includes both the safety cost and that connected with the economy of the ship motion along the trajectory of concern.

The current version of planner is also updated with the possibility of using fitness function scaling, which can essentially improve the quality of the results. Scaling is used to increase or suppress the diversity of the population, by controlling the selection pressure. It helps to maintain diversity of individuals at the initial phase of the computation, or to find a final solution at the end of it. The process of computation can be improved in the two abovementioned ways by using different scaling functions at proper times and changing the scaling parameters.

The scaling schemes which can be applied in the EA are: linear scaling, power law scaling, sigma truncation scaling, transform ranking scaling, ranked and exponential scaling [1, 2, 3, 10, 11]

## 4  SIMULATION ENVIROMENT

The operations of the evolutionary path planning algorithm υEP/N++ [8] was examined for multiple situations, one of them depicted as an example in Fig. 1. The simulation is usually performed in the environment populated with static and dynamic constraints, however tests presented in this paper consider only dynamic objects. The static constraints are represented by black polygons, while the dynamic objects (characterised by their own course and speed) were marked by grey hexagons. The figures below show only the dynamic objects (targets), that reveal a potential point of collision [10] with the own ship. The positions of the dynamic objects are displayed for the best route, which is bolded in the figures. In all here reported experiments the population size was 30, i.e. the evolutionary system processed 30 paths.

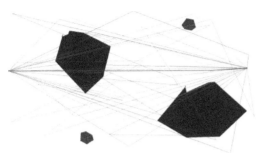

Figure 1. Example Simulation with the initial population.

As previous research show [16], scaling of the fitness function should allow later convergence of the results, thus presenting better solutions. Our tests were to check if this is the case also with the most common collision scenarios. Experiments were performed multiple times and the results represent the mean of what the simulations have shown. Due to the nature of EA, each program run is unique and it happened several times that a run without scaling produced results similar to those of the mean of the scaled ones (and the other way round), although those were marginal and depended mostly on the disruption of paths in the initial population. However it is worth noting that the planner is always able to find a good, feasible solution.

The runs performed for scaled cases were using power scaling with a power factor of 2. The program was set to the maximum of 400 generations, however during the tests generations were observed step by step (each generation was observed separately) in order to notice if scaling is working as expected and if the moment of final convergence is picked up precisely. The moment of achieving the final solution is one of the most crucial differences between a scaled and non-scaled EA. To show this clearly, each simulation was performed both for a scaled and non-scaled runs which allowed to form proper remarks.

## 5 SIMULATIONS

The simulations were performed for three most common collision situations and for two more complex. In each of them the own ship is represented by its own trajectory (as its size is negligible compared to the environment and target's safety zone). Both own ship and target have their starting positions equally far from the Point of the Potential Collision (*PPC*) and move at the same speed of 10 knots. The three scenarios differ from each other by the targets' trajectories. It is important to underline that our planner plots a path only for the own ship and the strange ship course is fixed and cannot be changed. $\upsilon$EP/N++ task is thereby to plot a new path that would avoid collision and reduce the costs of the course change to minimum while keeping the whole procedure save. The course of own ship is always $0^0$, while the targets' course is $240^0$ in situation 1 (Figure 2a), $300^0$ in situation 2 (Figure 2b) and $120^0$ in the last scenario (Figure 2c). Also two additional, advanced simulations were performed, which were constructed based on material in [15]. The example results, representing the mean of the observed runs for both scaled and non-scaled attempts are shown on Figures 4 to 8. Shortcut Gen. refers to number of the generation of the run shown in a segment. On those figures we can see the process of path plotting from the earliest generations (which quality is far from the ultimate one) through the final ones, observing how the $\upsilon$EP/N++ tries to calculate an optimal route. The figures show that once an good feasible path is found, the algorithm utilises genetic operators in order to shorten and smoothen the plotted course. The desired course is also often tangent to the target's safe area, as this correlates with the goals listed.

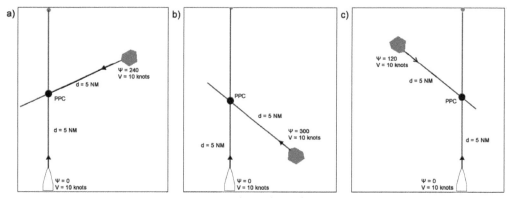

Figure 2. Basic Simulation Situations – target ship on a) $240^0$ b) $300^0$ c) $120^0$ course

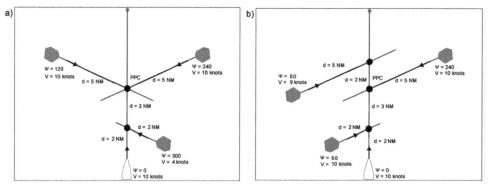

Figure 3. Advanced Simulation Situations.

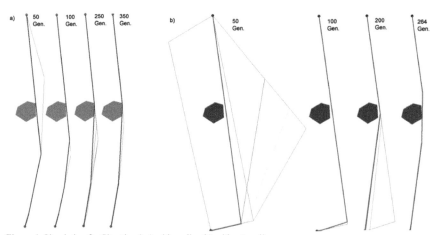

Figure 4. Simulation for Situation 1 a) with scaling b) without scaling

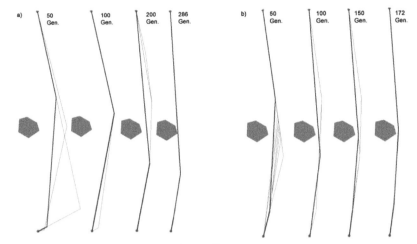

Figure 5. Simulation for Situation 2 a) with scaling b) without scaling

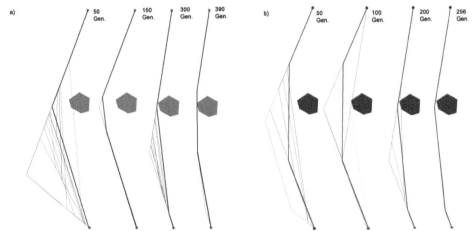

Figure 6. Simulation for Situation 3 a) with scaling b) without scaling

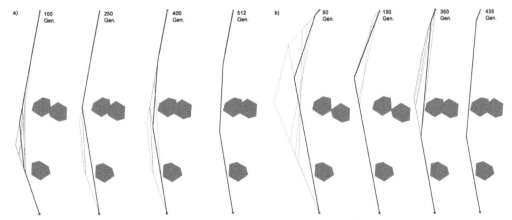

Figure 7. Simulation for Advanced Situation from Figure 3a a) with scaling b) without scaling

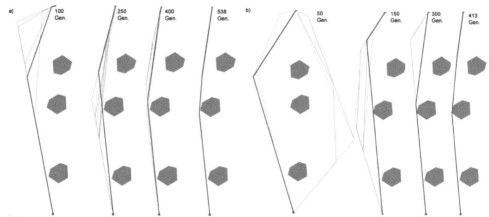

Figure 8. Simulation for Advanced Situation from Figure 3b a) with scaling b) without scalin

## 6 RESULTS AND CONCLUSIONS

The simulation results above proven that υEP/N++ is able to efficiently find a feasible and acceptable path both in case the scaling is applied and when it is not present, just as shown in detail in [14]. As one can clearly see from figures 2,3 and 4 applying scaling extended the time of the final solution convergence and allowed new paths to form during the generation run. However, the end result was similar, and both versions (with and without scaling) of the algorithm provided comparably good results. However it is also worth to underline that when scaling was applied, the paths calculated by the algorithm became smoother (required course corrections of smaller value), thus more economical as best seen on Fig. 5. As the presented examples are rather non-complex, it is only logical to deduce that applying scaling for this particular problem proven non-necessary, although the computation time extension was barely noticeable. As scaling is a great tool to widen the area of search, it has little effect when faced with a noncomplex challenge, at least for tasks typical for υEP/N++. However, as scaling application did smooth the paths, one can't ignore the improvement noticed, even for a problem of such a small magnitude. Scaling will perform much better in tasks where a path through a hugely populated area has to be plotted, where it's effect will be much better noticeable.

This example research only worked with the power scaling, but as further work is done, where different scaling methods are tested and compared with one another, it could be worth to extend the scaling rating subroutine to be able to run different scaling options.

This paper shows how important it is to set proper scaling strategy that would cope well with the task ahead. Although the experiments presented here were using only two different strategies, their different affect was apparent. One can easily notice that as

it is important to select efficient genetic operators, it is as important to match the right scaling scheme. υEP/N++ is equipped with procedures that grade genetic operators and as the run goes, it chooses and utilizes the best one of them. As further research goes, a similar routine can be devices for scaling, so that the algorithm can turn it on, when it seems it can better the results and abandon it, when it becomes redundant. One of the features of Genetic Algorithms is its great scalability, however the parameters have to always be well adjusted to the problem, even to a particular case that is being research. The algorithm has to be able to dynamically adjust to the problem's requirements if it's to be used in the industrial scale.

## ACKNOWLEDGEMENTS

The authors thank the Polish Ministry of Science and Higher Education for funding this research under grant no. N514 472039.

## BIBLIOGRAPHY

1. Goldberg D.E., *Genetic Algorithms in Search, Optimization, and Machine Learning*, Addison-Wesley Longman Publishing Co., Inc. Boston, MA, USA, 1989
2. Farzad A. Sadjali, *Comparison of fitness scaling functions in genetic algorithms with applications to optical processing*, , Proc. of SPIE Vol. 5557, pp. 358
3. Hopgood A., Mierzejewska A., *Transform Ranking: a New Method of Fitness Scaling in Genetic Algorithms*, Research and development in intelligent systems XXV: proceedings of AI-2008, pp.349-355.
4. Michalewicz Z., *Genetic Algorithms + Data Structures = Evolution Programs*. Spriger-Verlang, 3rd edition, 1996.
5. Śmierzchalski R., *Trajectory planning for ship in collision situations at sea by evolutionary computation*, In Proceedings of the IFAC MCMC'97, Brijuni, Croatia, 1997.

6. Śmierzchalski R., *Ships' domains as collision risk at sea in the evolutionary method of trajectory planning*, Computer Information and Applications Vol II, 2004, pp. 117 – 125.

7. Śmierzchalski R. and Michalewicz, Z., *Adaptive Modeling of a Ship Trajectory in Collision Situations at Sea*, In Procced-ings of the 2nd IEEE World Congress on Computational Intelligence, ICEC'98, Alaska, USA 1998, pp. 364 - 369.

8. Śmierzchalski R. and Michalewicz, Z., *Modeling of a Ship Trajectory in Collision Situations at Sea by Evolutionary Algorithm*, IEEE Transaction on Evolutionary Computation, Vol.4, No.3, 2000, pp.227-241.

9. Śmierzchalski R., Michalewicz, Z., *Path Planning in Dynamic Environments*, chapter in "Innovations in Machine Intelligence and Robot Perception", Springer-Verlag, 2005. pp.135-154.

10. Tidor B., Michael de la Maza, *Boltzmann Weighted Selection Improves Performance of Genetic Algorithms*, MIT, Artificial Intelligence Laboratory, December 1991, pp. 1-18.

11. Wall Mathew, GAlib: A C++ Library of Genetic Algorithm Components, MIT 1996

12. Yap, C.-K., *Algorithmic Motion Planning, In Advances in Robotics*, Vol.1: Algorithmic and Geometric Aspects of Robotics}, J.T. Schwartz and C.-K. Yap Ed., , Lawrence Erlbaum Associates, 1987, pp.95 - 143.

13. Eiben E.A., Smith J.E.: *Introduction to evolutionary computing*, Springer 2003

14. P. Kolendo, R. Śmierzchalski, B. Jaworski: *Scaling Fitness Function in Evolutionary Path Planning Method*, 20th IEEE International Symposium on Industrial Electronics IEEE-ISIE 2011

15. Young-Il Lee, Yong-Gi Kim: *A Fuzzy Collision Avoidance Technique for Intelligent Ship.*

# 12. A New Definition of a Collision Zone For a Geometrical Model For Ship-Ship Collision Probability Estimation

J. Montewka ‡, F. Goerlandt & P. Kujala
*‡Aalto University, School of Engineering, Finland; Maritime University of Szczecin, Poland*
*Aalto University, School of Engineering, Finland*

ABSTRACT: In this paper, a study on a newly developed geometrical model for ship-ship collisions probability estimation is conducted. Most of the models that are used for ship-ship collision consider a collision between two ships a physical contact between them. The model discussed in this paper defines the collision criterion in a novel way. A critical distance between two meeting ships at which such meeting situation can be considered a collision is calculated with the use of a ship motion model. This critical distance is named the minimum distance to collision (MDTC). Numerous factors affect the MDTC value: a ship type, an angle of intersection of ships' courses, a relative bearing between encountering ships and a maneuvering pattern. They are discussed in the paper.

## 1 LITERATURE REVIEW

A number of models for ship-ship collision probability estimation can be found in the literature. They can be divided into two major groups namely: static and dynamic models. The static models are simpler and less time consuming for computation, however their accuracy can be questioned. The dynamic models are more complex, in principle they need more input data than static models, but their results are comprehensive. In this section the short overview of existing models will be made, and our contribution to the existing knowledge will be put forward.

### 1.1 Static models

The most known approaches were introduced by (Fujii et al. 1970) and (MacDuff 1974). Models of this kind have been commonly used in recent decades and won the popularity among researches mainly due to simplicity and robustness. However they have also some drawbacks, lack of ship dynamics or assumption regarding a collision between two ships. A collision is defined as a meeting of two ships in a distance named the "collision diameter", which means almost the physical contact. Such an assumption may lead to an understanding that in any ship-ship encounter at a distance greater than the "collision diameter" these ships are able to avoid a collision, which in most cases is not true. Despite the drawbacks the model was adopted by (Pedersen 1995), and with minor modifications was used to de-termine the safety of navigation in many European waters: (Otto et al. 2002), (Sfartsstyrelsen 2008). Hence in Europe it is mostly known as Pedersen model. Another method for the frequency of collision estimation, making an assumption regarding uncorrelated traffic, was outlined by (Fowler and Sorgrad 2000). A critical situation is assumed to occur when ships come to close quarters to a distance of 0.5 Nm of each other, which is constant regardless of a meeting scenario. A model for encounter probability estimation proposed by (Kaneko 2002) defines a critical area of an optional form of a closed boundary, around a ship which violation means collision. Kaneko in his model recognizes two shapes of the critical area: rectangular and circular, but again the size of the area is fixed. A series of papers utilizing the ship domain approach to ship-fixed object collision assessment was published also by (Gluver and Olsen 1998) and (Pedersen 2002).

However, none of the model listed above takes ship dynamics into consideration.

### 1.2 Dynamic models

Another group of models utilize marine traffic simulations. A group of researches led by Merrick proposed a risk analysis methodology for maritime traffic in coastal areas based on system simulation (Merrick et al. 2002), (Merrick et al. 2003). Maritime traffic is simulated in the time domain based on routes obtained from expert opinion and vessel arrival records. Finally, these were combined with the simulation output in order to carry out a risk analysis (van Dorp and Merrick 2009).

Another probabilistic model for the assessment of navigational accidents in an open sea area was outlined by (Gucma and Przywarty 2007). The method makes use of a simplified model of maritime traffic, which is simulated in the time domain. A recent model, introduced by (Goerlandt and Kujala 2011) is based on an extensive time-domain simulation of maritime traffic in a given area. Vessel movements are modelled based on data obtained from a detailed study of route-dependent vessel statistics. The collision candidates are detected by a collision detection algorithm which assesses the spatio-temporal propagation of the simulated vessels in the studied area.

Markov, semi-Markov and Random Field theory based models for maritime traffic safety estimation were introduced recently (Smalko and Smolarek 2009), (Smolarek and Guze 2009), (Smolarek 2010), (Guze and Smolarek 2010). However the main assumption of the models proposed is that traffic flow is stationary, which is not applicable to areas with scheduled traffic. Recently a geometrical model for estimation the probability of ship collisions while overtaking were introduced by (Lizakowski 2010), his model considers human factor and the fairway and ship dimensions. However all these models are advanced mathematically they do not take into account ship dynamics nor human factors.

A multicomponent model for an inland ship safety estimation was presented by (Galor 2010). However each of the proposed model's component is essential, the model itself constitutes rather an introduction to the further quantitative analysis of the problem.

For the first time the idea of ship manoeuvrability implementation into a collision assessment model was presented by (Curtis 1986). However, this model was limited to one ship type, which was a very large crude carrier (VLCC), and only overtaking and head-on situations were considered.

### 1.3 Authors' contribution

A new criterion for ship-ship collision probability estimation and a new model have been introduced by (Montewka et al. 2010). The model considers ship maneuverability and traffic parameters; the new collision criterion is named the Minimum Distance To Collision (MDTC). MDTC is a critical distance between two ships being on collision courses, at which they must perform collision evasive actions, in order to pass safely. The MDTC is estimated by means of ship motion model and series of experiment for various ship meeting scenarios.

This paper is a continuation of our previous research, it consists of the detailed analysis of the MDTC values for a wide ranges of input variables (they are defined in the following Chapters) and two

patterns of performing collision evasive action. The maneuvering patter one means that own ship is performing a collision evasive action and the other ship is not acting, in the maneuvering patter two both ships are involved in avoiding a collision. Performance of turning circle is considered a collision evasive action.

## 2 INTRODUCTION TO MDTC MODEL

The MDTC model introduced in a previous work of (Montewka et al. 2010) and developed further in this paper, is based on an initial assumption, that two ships collide if the distance between them becomes less than a certain value, named a MDTC. This MDTC value is not a fixed number, but it is calculated dynamically for each type of vessel and encounter individually. Thus it changes with the situation. The main factors affecting the MDTC value are: the vessels maneuverability, the angle of intersection labelled α in Figure1a, the relative bearing from one vessel to the other labelled β in Figure1b and a pattern of evasive maneuvers (one vessel swinging or both). In the previous study, a simplified methodology was applied, which assumed that two vessels met at a constant relative bearing while proceeding with their service speeds. Presented study considers a wide range of relative bearings, varying from 10 do 80 degrees (counting from the own ship's bow) and takes into account two different engine settings for each ship type, therefore providing more detailed results.

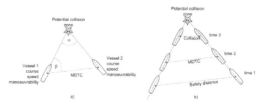

Figure 1: A definition of MDTC and major factors affecting it
Source: (Montewka et al. 2010)

## 3 RESEARCH MODEL

The theory of the model and preliminary research aiming to define the "collision zone" were presented by (Montewka et al. 2010). In this paper the results of studies with respect to different ship types and ship speeds and varying meeting angles are shown. However only planar motion of a ship is taken into account and assumption regarding ship navigating through deep water is made. We also assume, that the prevailing weather conditions do not deteriorate significantly the maneuverability of ships sailing in the analyzed part of the Gulf of Finland. In order to validate it, we simulated a maneuver of turning circle to starboard side, for the chosen ship type, which was a RoPax (for ship particulars see Table 1), for two different wave conditions (no wave, and an average wave height for the Gulf of Finland). According to (Pettersson et al. 2010) and (Raamet et al. 2010) the average monthly weave height recorded in the analyzed area (sea between Helsinki and Tallin) does not exceed 2 meters, and as a such was adopted for the simulation. For this purpose the Laidyn ship motion model was adopted (Matusiak 2007).

The results allowed us to keep our assumptions, as a difference between the trajectories of a ship in two different heights of a wave seems to be negligible for the purposes of our research (Figure 2).

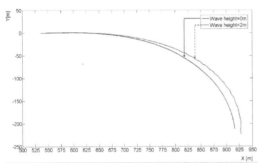

Figure 2: Turning circles of RoPax performed for two different wave heights

## 3.1 Ships considered

In the course of our analysis we are considering four major ship types: a passenger ship, a containers carrier, a RoPax and a tanker. In each scenario, ships are assumed to proceed with two different engine settings (except for a passenger vessel which is assumed to sail always at a maximum speed) which result in forty two encountering scenarios, as depicted in Figure 4. The following abbreviations are used: 'FA' is full ahead and 'HA' means half ahead. The 'FA' abbreviation corresponds to a mean speed of a ship of given type as obtained from recorded AIS data. The abbreviation 'HA' does not correspond to an actual engine setting, it rather reflects a spread of recorded speed values for a given class of ships in the analyzed area. The value of 'HA' for given ship type was calculated by subtracting the standard deviation from the mean value for a given type of ship. The main particulars of the analyzed vessels are listed in Table 1.

Table 1: Ships particulars.

| Ship type | LOA [m] | B [m] | T [m] | v [kn] |
|---|---|---|---|---|
| Container carrier | 150.0 | 27.2 | 8.5 | 20;17 |
| RoPax | 158.6 | 25.0 | 6.1 | 20;18 |
| Tanker | 139.0 | 21.0 | 9.0 | 14;11 |
| Passenger | 185.0 | 27.7 | 6.5 | 25 |

## 3.2 Encountering scenarios

Each of an encountering scenario is run for seventeen different crossing angles ($\alpha$), varying from 010 to 170 degrees with 10 degrees increment. Where 010 means almost overtaking (vessel B on a course of 350deg) and 170 stands for almost head-on meeting (Vessel B on a course of 190deg), as depicted in Figure 3. The situation shown there considers own ship seeing another at 45 degrees relative bearing. In the course of the experiment, each crossing angle is calculated for a range of relative bearings, from 10 to 80 degrees, counting from the own ship's bow.

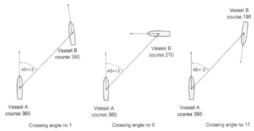

Figure 3: Relative positions of vessels, with three chosen crossing angles, before they start to maneuver, (Montewka et al.2010)

For each ship-ship encounter at a given intersection angle ($\alpha$) and at a given relative bearing ($\beta$), one MDTC value is obtained. As specified in a block diagram depicted in Figure 4, in total 5712 MDTC values are obtained. Then for each intersection angle ($\alpha$) the maximum MDTC value among eight (as there are eight relative bearings considered) is drawn. Also the relative bearing which is the most inconvenient from a collision evasive point of view, and which requires the most space to make an action is indicated. For further statistical analysis 714 out of all 5712 MDTC values are selected for each maneuvering pattern.

## 3.3 Maneuvering patterns

In case of a maneuvering pattern number one, the own ship is performing an evasive action, by turning circle, and another ship is following her initial course. In case of maneuvering pattern number two, two ships are performing turning circles in order to avoid collisions. The following simplifications in the presented methodology are done:
- in case of evasive pattern where two vessels perform turning circles, they both start their maneuvers at the same time;
- ships are turning away from each other, which implies course alteration away from each other to avoid collision and to shorten the time at close quarters (such assumption meets requirements of the COLREG, which states, that ships must avoid

altering courses towards each other if in close quarters);
- the settings of ships' engines and rudders are constant during maneuvers;
- the influence of weather conditions is omitted;
- the hydromechanical ship-ship interactions are omitted.

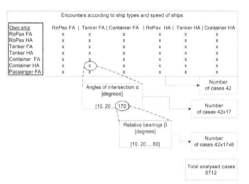

Figure 4: Research model

## 3.4 *MDTC estimation*

In order to calculate the value of MDTC for a given pair of vessels, an iterative algorithm is used, as depicted in Figure 6. The basic assumption is that the two ships collided at a time instant t0. Then starting from this time the reverse iterative algorithm is applied. It uses backward calculation method in a space-time domain. Two trajectories of two ships are drawn and the consecutive positions of ship's centre of gravity are plotted every second ($dt$=1s). If corresponding ships' contours following the trajectories have at least one common point, indicating that they both collided, the algorithm increases the initial distance between these two ships by constant value of 0.1LOA$_{average}$. The trajectories are redrawn starting from the new initial positions of the ships. This process is repeated until the two contours of ships have no overlaps at any time instant for a given relative bearing.

New initial positions are defined by moving ship B from ship A away. For a given meeting scenario (a given angle of intersection α and a given relative bearing β) the ships are moved away along a line of a given relative bearing (β line). For the simplicity of calculations it is assumed, that own ship holds her initial position, while the other ship is moved away along the β line.

In the situation where two trajectories have no common points and the contours of the ships do not over-lap, the initial position of vessel B is recorded (as the initial position of own ship A was always (0,0)), and the distance between these two positions is calculated and stored. This distance, is named MDTC for a given relative bearing. As each meeting

scenario is analyzed for a range of relative bearings (from 10 to 80 degrees), the procedure presented is repeated for all relative bearings, yielding eight values of MDTC for each angle of intersection α. Finally, the maximum value of MDTC among these eight is drawn. This maximum value is considered a MDTC value for a given angle of intersection. This procedure is repeated for all angle of intersection, then for each maneuvering patterns. Thus the MDTC charts are obtained.

In order to determine the MDTC charts, all routines are encoded in MATLAB. As a polygonal region, which could represent a ship contour an ellipse is chosen. To determine, whether two contours of the ships (represented as ellipses) overlap, the following MATLAB function is applied (MathWorks 2010):

$$\mathbf{IN} = inpolygon(X,Y,xv,yv), \qquad (1)$$

it returns a matrix **IN** of the same size as X and Y. Each element of (**IN**) is assigned the value 1 or 0 depending on whether the point $(X(p,q),Y(p,q))$ is inside the polygonal region whose vertices are specified by the vectors $xv$ and $yv$.

For the sake of computation effectiveness each ellipse is transformed into discrete form and the number of points that represent the ellipse is 24. The ellipse's axes are defined in the following way:

$$a = 0.5LOA$$

$$b = 0.5B, \qquad (2)$$

where $a$ denotes a major axis, $b$ is a minor axis, $LOA$ means length overall of a ship and $B$ is a ship's breadth.

A MDTC value for a given encounter implies a safe passage of two vessels, which corresponds to a situation where these two vessels approximated by the ellipses, will always be separable and will not touch each other at any time step of a collision evasive action. A graphical interpretation of above is depicted in Figure 5, where both ships are at the closest distance in the time step 81sec., however they are still separable. A block diagram showing an algorithm applied in the study to estimate a MDTC chart for a given meeting scenario is depicted in Figure 6.

## 4 DATA ANALYSIS

The data obtained in the course of MDTC calculations (see Figure 6) considers different ship types, different engine settings and two different maneuvering patterns, as stated in Figure 4. In the next step the statistical analysis of the obtained data is performed.

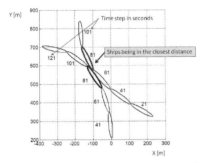

Figure 5: Ships as ellipses and interpretation of a non-contact passage

In Figure 7 data sets concerning MDTC assuming a maneuvering pattern number one (own ship involved in collision evasive action), according to a ship type, are presented. Whereas the data depicted in Figure 8 shows appropriate values for MDTC, according to a ship type for maneuvering pattern number two (both vessels are involved).

To determine whether the differences which can be noted visually are significant from the statistical point of view, the appropriate statistical tests are performed. The following are hypothesized:

- $H_0$: the obtained values of MDTC, for a range of intersection angles α, are drawn from the same population (or equivalently, from different populations with the same distribution), thus MDTC do not depend on a ship type.
- $H_1$: the medians of analyzed variables are not all equal, thus the MDTC values do not originate from the same population, and they are a ship type dependent.

In order to validate these hypotheses we perform a nonparametric Kruskal-Wallis test which compares samples from two or more groups, as the obtained data do not follow a normal distribution. In the case presented here we analyze 42 different encounters, each consisting of 17 crossing angles, as depicted in Figure 4. We form a 42-by-17 matrix, where each column of the matrix represent an independent sample containing 42 mutually independent observations, and a number of columns is equivalent to a number of crossing angels α. The function that Kruskal-Wallis test is based on compares the medians of the samples in a matrix, and returns the p-value for the null hypothesis.

In the course of the analysis the obtained p-value vary for two maneuvering patterns concerned. In the case where both vessels make a turn (the maneuvering pattern No 2) the p-value yields 0.9988. This shall not cast any doubt on the null hypothesis, and suggests that all sample medians come from the same population.

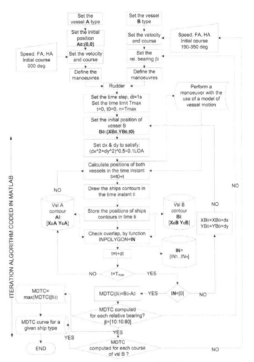

Figure 6: Block diagram for MDTC calculation

However, the results obtained for the maneuvering pattern No 1, where the own ship performs collision evasive action only, are more scattered therefore not so straightforward in inference. The results of the statistical tests concerning both maneuvering patterns are gathered in Table 2. Analyzing a full range of intersection angles, for maneuvering pattern No 1, there is no evidence for not rejecting the null hypothesis. This can lead to thinking that at least one data set originates from a different population that the other data sets.

Table 2: Results of statistical tests ordered by the angle of intersection

| Maneuvering pattern | Segment [deg] | α value | p value | Hypothesis rejected |
|---|---|---|---|---|
| No 1 | 10-170 | 0.05 | 0.002 | $H_0$ |
| No 2 | 10 − 170 | 0.05 | 0.9988 | $H_1$ |

However dividing the intersection angles range into segments, and analyzing them separately, makes it feasible to defend the null hypothesis.

In order to made a cross check of the results obtained in this analysis, in the next step we order a data set according to a ship type, and run the Kruskal-Wallis tests on smaller samples. The results obtained are presented in Table 3.

Table 3: Results of statistical tests according to a ship type - the maneuvering pattern No 1.

| Ship type | Segment [deg] | α value | p value | Hypothesis rejected |
|---|---|---|---|---|
| Container | 10-170 | 0.05 | 0.4424 | $H_1$ |
| RoPax | 10-170 | 0.05 | 0.9891 | $H_1$ |
| Tanker | 10-170 | 0.05 | 0.1237 | $H_1$ |
| Passenger | 10-170 | 0.05 | 0.9307 | $H_1$ |

Table 4: Results of statistical tests according to a ship type and the angle of intersection - the maneuvering pattern No 1.

| Ship type | Segment [deg] | α value | p value | Hypothesis rejected |
|---|---|---|---|---|
| Tanker | 10-120 | 0.05 | 0.4411 | $H_1$ |
| Tanker | 120-140 | 0.05 | 0.6671 | $H_1$ |
| Tanker | 140-170 | 0.05 | 0.8306 | $H_1$ |

Having variable "Tanker" excluded the p-value for the null hypothesis for the maneuvering pattern No 1 becomes higher than adopted level α.

Table 5: Results of statistical tests according to the angle of intersection, the maneuvering pattern No 1, tankers excluded.

| Maneuvering pattern | Segment [deg] | α value | p value | Hypothesis rejected |
|---|---|---|---|---|
| No 1 | 10-170 | 0.05 | 0.20 | $H_1$ |
| No 1 | 10-120 | 0.05 | 0.98 | $H_1$ |
| No 1 | 120-140 | 0.05 | 0.70 | $H_1$ |
| No 1 | 140-170 | 0.05 | 0.39 | $H_1$ |

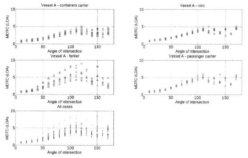

Figure 7: MDTC values for a given ship type - maneuvering pattern No 1 (own vessel performing an evasive action)

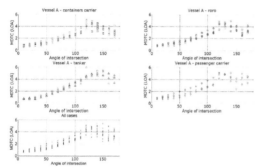

Figure 8: MDTC values for a given ship type - maneuvering pattern No 2 (both vessels performing collision evasive maneuvers)

It can be noticed that two cases ("RoPax" and "Passenger") defend the null hypothesis for a full range of intersection angles. In a case of "Container" the null hypothesis can not be rejected, but the $p - value$ obtained is not as high as in the previous cases, however still acceptable. In a case of "Tanker", although the $p - value$ obtained is greater than the adopted level α, however the null hypothesis can be rejected as early as a confidence level α is 0.13. This obviously can cast some doubts on the null hypothesis. In order to have an insight into a data set regarding variable "Tanker" we divide a range of intersection angles into three segments, and run again the statistical test. The results obtained are presented in Table 4.

The following conclusions can be made:
- for the maneuvering pattern No 1, if the variable "Tanker" is excluded, the statistical tests prove, that the data for other three variables ("Container", "RoPax" and "Passenger") are drawn from the same population;
- for maneuvering pattern No 1, the MDTC values for tankers are obtained in the course of a separate analysis;
- for maneuvering pattern No 2, there are strong evidences, that all data come from the same population, thus MDTC is not a ship type depended variable.

5  RESULTS

The obtained MDTC values are categorized according to an intersection angle α, and we make attempts to define distributions of MDTC values for each angle α and for given maneuvering patterns. Because of the limited survey sample, and data scatter none of commonly known distributions, neither continuous nor discreet, fit the data. Thus further analysis is conducted using one of the sampling methodology, namely a non parametric bootstrap procedure.

For each maneuvering pattern, the following values are estimated: a mean and a standard deviation of a MDTC for a given angle of intersection (α). In order to obtain these parameters, the following procedure is adopted (Vose 2008):
- to collect the data set of $n$ samples $\{x_1...x_n\}$ in our case $n=42$
- to create $B$ bootstrap samples $\{x_1^*...x_n^*\}$ where each $x_i^*$ is a random sample with replacement from $\{x_1...x_n\}$, in our case $B = 10^6$;
- to estimate, for each bootstrap sample $\{x_1^*...x_n^*\}$, the required statistics $\hat{s}$. The distribution of these

$B$ estimates of $s$ represents the bootstrap estimate of uncertainty about the true value of $s$.

The outcome of the bootstrap analysis are as follows:
– the mean MDTC values, for each intersection angle $\alpha$;
– the 0.95 confidence intervals around the mean values (represented as dotted lines in a Figure 6).

Then using the upper confidence interval of the mean value, the 0.95 prediction interval is calculated. The upper prediction band obtained is shown as a solid line with diamonds. The results obtained, for two maneuvering patterns are depicted in the following Figures: 9, 10, 11.

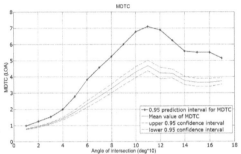

Figure 9: The obtained MDTC chart for the maneuvering pattern No 1

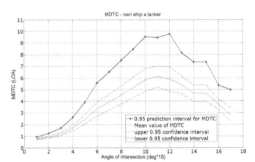

Figure 10: The obtained MDTC chart for tankers - the maneuvering pattern No 1

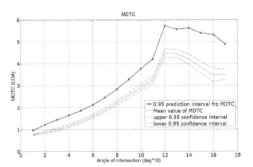

Figure 11: The obtained MDTC chart for the maneuvering pattern No 2

## 6 CONCLUSIONS

This paper addresses a chosen aspects of marine traffic safety modelling. This means a novel method for definition a collision-zone for ship-ship meetings. This parameter named MDTC is an input for a model that estimates the probability of ship-ship collision. It takes into account ship dynamics, traffic patterns and indirectly the human actions (the maneuvering patterns).

In the course of the analysis presented in this paper three different charts representing three types of collision zones were obtained. The statistical analysis shows that the dimension of a collision zone depends mostly on a maneuvering pattern. In case where both ships perform collision evasive actions, one chart describes all types of ships analyzed. However in case where only one ship performs a collision evasive maneuver, two charts are obtained, where one considers tankers and another the remaining ship types.

The experiment leading to MDTC chart estimation is based on a ship planar motion model, and the assumptions concerning deep water and lack of external forces and hydromechanical ship-ship interactions are made. However the size of the vessels under consideration allows the statement, that the sea conditions prevailing in the Baltic Sea, and especially in the Gulf of Finland do not affect the results significantly.

Another important factor affecting the actual number of modelled accidents is a causation factor. This topic is not addressed by research presented in this paper.

## ACKNOWLEDGMENT

The authors appreciate the financial contributions of the following entities: the EU, Baltic Sea Region (this research was founded by the EfficienSea project), the Merenkulun säätiö from Helsinki, the city of Kotka and the Finnish Ministry of Employment and the Economy.

## REFERENCES

Curtis, R. (1986). A ship collision model for overtaking. The Journal of Navigation 37(04), 397–406.

Fowler, T. G. and E. Sorgrad (2000). Modeling ship transportation risk. Risk Analysis 20(2), 225–244.

Fujii, Y., H. Yamanouchi, and N. Mizuki (1970). On the fundamentals of marine traffic control. part 1 probabilities of colliion and evasive actions. Electronic Navigation Research Institute Papers 2, 1–16.

Galor, W. (2010). The model of risk determination in sea-river navigation. Journal of Konbin 14-15(1), 177–186.

Gluver, H. and D. Olsen (1998). Ship collision analysis. Taylor & Francis.

Goerlandt, F. and P. Kujala (2011). Traffic simulation based ship collision probability modeling. Reliability Engineering & System Safety 96(1), 91–107.

Gucma, L. and M. Przywarty (2007). The model of oil spill due to ship collisions in southern baltic area. In A. Weintrit (Ed.), Marine navigation and safety of sea transportation, London, pp. 593–597. Taylor & Francis.

Guze, S. and L. Smolarek (2010). Markov model of the ship's navigational safety on the open water area. In International Scientific Conference Transport of 21st century. Warsaw University of Technology.

Kaneko, F. (2002). Methods for probabilistic safety assessments of ships. Journal of Marine Science and Technology 7, 1–16.

Lizakowski, P. (2010). The probability of collision during vessel overtaking. Journal of Konbin 14-15(1), 91–99.

MacDuff, T. (1974). The probability of vessels collisions. Ocean Industry, 144–148.

MathWorks, T. (2010, November). Matlab. online: http://www.mathworks.com.

Matusiak, J. (2007). On certain types of ship responses disclosed by the two-stage approach to ship dynamics. Archives of Civil and Mechanical Engineering 7(4), 151–166.

Merrick, J. R. W., J. R. van Dorp, J. P. Blackford, G. L. Shaw, J. Harrald, and T. A. Mazzuchi (2003). A traffic density analysis of proposed ferry service expansion in san francisco bay using a maritime simulation model. Reliability Engineering & System Safety 81(2), 119–132.

Merrick, J. R. W., J. R. van Dorp, T. Mazzuchi, J. R. Harrald, J. E. Spahn, and M. Grabowski (2002). The prince william sound risk assessment. INTERFACES 32(6), 25–40.

Montewka, J., T. Hinz, P. Kujala, and J. Matusiak (2010). Probability modelling of vessel collisions. Reliability Engineering & System Safety 95, 573–589.

Otto, S., P. T. Pedersen, M. Samuelides, and P. C. Sames (2002). Elements of risk analysis for collision and grounding of a roro passenger ferry. Marine Structures 15(4-5), 461–474.

Pedersen, P. T. (1995). Collision and grounding mechanics. Copenhagen, pp. 125–157. The Danish Society of Naval Architects and Marine Engineers.

Pedersen, P. T. (2002). Collision risk for fixed offshore structures close to high-density shipping lanes. Proceedings of the Insitution of Mechanical Engineers, Part M: Journal of Engineering for the Maritime Environment 216(1), 29–44.

Pettersson, H., T. Hammarklint, and D. Schrader (2010, October). Wave climate in the baltic sea 2008. HELCOM Indicator Fact Sheets 2009. Online.

Raamet, A., T. Soomere, and I. Zaitseva-Parnaste (2010). Variations in extreme wave heights and wave directions in the north-eastern baltic sea. In Proceedings of the Estonian Academy of Sciences, Tallinn, pp. 182–192. Estonian Academy of Sciences: Estonian Academy of Sciences. Available online at www.eap.ee/proceedings.

Sfartsstyrelsen (2008). Risk analysis of sea traffic in the area around bornholm. Technical report, COWI, Kongens Lyngby.

Smalko, Z. and L. Smolarek (2009). Estimate of collision threat for ships routes crossing. In L. Gucma (Ed.), Proceedings of XIIIth International Scientific and Technical Conference on Marine Traffic Engineering, pp. 195–199. Maritime University of Szczecin.

Smolarek, L. (2010). Dimensioning the navigational safety in maritime transport. Journal of Konbin 14-15(1), 271–280.

Smolarek, L. and S. Guze (2009). Application of cellular automata theory methods to assess the risk to the ship routes. In L. Gucma (Ed.), Proceedings of XIIIth International Scientific and Technical Conference on Marine Traffic Engineering, pp. 200–204. Maritime University of Szczecin.

van Dorp, J. R. and J. R. W. Merrick (2009). On a risk management analysis of oil spill risk using maritime transportation system simulation. Annals of Operations Research.

Vose, D. (2008). Risk analysis: a quantitative guide. John Wiley and Sons.

# 13. Uncertainty in Analytical Collision Dynamics Model Due to Assumptions in Dynamic Parameters

K. Ståhlberg, F. Goerlandt, J. Montewka* & P. Kujala
*Aalto University, School of Engineering, Department of Applied Mechanics, Marine Technology*
*P.O. Box 15300, FI-00076 AALTO, Espoo, Finland*
*\* Aalto University, School of Engineering, Marine Technology, Espoo, Finland*
*Maritime University of Szczecin, Institute of Marine Traffic Engineering, Poland*

ABSTRACT: The collision dynamics model is a vital part in maritime risk analysis. Different models have been introduced since Minorsky first presented collision dynamics model. Lately, increased computing capacity has led to development of more sophisticated models. Although the dynamics of ship collisions have been studied and understanding on the affecting factors is increased, there are many assumptions required to complete the analysis. The uncertainty in the dynamic parameters due to assumptions is not often considered.
In this paper a case study is conducted to show how input models for dynamic parameters affect the results of collision energy calculations and thus probability of an oil spill. The released deformation energy in collision is estimated by the means of the analytical collision dynamics model Zhang presented in his PhD thesis. The case study concerns the sea area between Helsinki and Tallinn where a crossing of two densely trafficked waterways is located. Actual traffic data is utilized to obtain realistic encounter scenarios by means of Monte Carlo simulation. Applicability of the compared assumptions is discussed based on the findings of the case study.

## 1 INTRODUCTION

Ship-ship collisions are rare events that potentially might have disastrous impact on the environment, human life and economics. To find effective risk mitigating measures the risk must be reliably assessed. Proper assessment of the ship-ship collision risk requires understanding on the complicated chain of events. Simplifying assumptions on certain parameters are necessary as the research in this field is not comprehensive. Especially, the important link between the encounter of the colliding vessels and the actual moment of impact contain obvious uncertainties.

In this paper a case study is conducted to compare models found in literature for dynamic parameters in collision scenario. The case study concerns collisions in which the struck vessel is an oil tanker. The traffic is simulated by means of a Monte Carlo simulation based on AIS data to obtain realistic encounter scenarios for the analyzed area. The assumptions are then applied to encounter scenario to obtain the complete impact scenario. The deformation energy released in the collision is calculated by analytic method (Zhang 1999) and the damage extents are estimated with simple formula to normalize the results of deformation energy calculations. The effects of assumptions for dynamic parameters to collision risk are discussed.

## 2 COLLISION RISK EVALUATION

### 2.1 *Concept of risk*

Risk is a product of probability p and consequences c and is expressed with (Kujala et al, 2010)

$$R = \sum p_i \cdot c_i \tag{1}$$

where i denotes certain chain of events or scenario.

### 2.2 *Tanker Collisions*

In case of ship-ship collisions scenario is a function of vast number of static and dynamic parameters. The parameters used in this study are listed in Table 1.

Table 1. Collision parameters used in this study.

| | Description | Unit | Type |
|---|---|---|---|
| M | Mass | [kg] | Static |
| L | Length | [m] | Static |
| B | Width | [m] | Static |
| $m_x$ | Added mass coefficient, surge motion | [-] | Static |
| $m_y$ | Added mass coefficient, sway motion | [-] | Static |
| j | Added mass coefficient, rotation around centre of gravity | [-] | Static |
| R | Radius of ship mass inertia around centre of gravity | [m] | Static |

| $V_x$ | Surge speed | [m/s] | Dynamic |
|---|---|---|---|
| $V_y$ | Sway speed | [m/s] | Dynamic |
| x | x-position of centre of gravity | [m] | Static |
| y | y-position of centre of gravity | [m] | Static |
| $x_c$ | x-position of impact point, in coordinate system ship A | [m] | Dynamic |
| $y_c$ | y-position of impact point, in coordinate system ship A | [m] | Dynamic |
| α | collision angle | [rad] | Dynamic |
| $\mu_0$ | coefficient of friction | [-] | Static |
| e | coefficient of restitution | [-] | Static |

The static parameters can be derived from AIS data, statistics and theory of ship design. Modeling of the dynamic parameters is often based on statistics of the collisions.

Ship-ship collision risk evaluation schematic is outlined in Figure 1 for the case of an oil tanker being struck vessel.

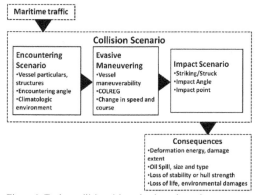

Figure 1. Tanker collision risk evaluation schematic

The first step of the risk analysis is modeling the traffic in the analyzed area. Modeling may be done via simulation of individual vessel movements as proposed by Merrick et al. (2003), van Dorp et al. (2009), Uluşçu et al. (2009) and Goerlandt & Kujala (2010) or alternatively by simulating the traffic flows as proposed by Pedersen (1995, 2010) or Montewka et al (2010). The encounter scenarios are obtained as a result of the traffic simulation. The impact scenarios may be then obtained with the models discussed in detail in Section 3.3.

Second part of the risk analysis is the evaluation of the consequences which begins with the estimating the released deformation energy that is absorbed by the vessel structures. Collision dynamics models to calculate the deformation energy can be divided into two groups, time domain and analytical (Wang et al 2000), based on applied calculation method. Analytical closed form methods have been proposed by Minorsky (1959), Vaughan (1977), Hutchison (1986), Hanhirova (1995), and Zhang (1999). Models based on time domain calculations are proposed by Chen (2000) and Tabri et al. (2009). In analytical models the external dynamics and internal mechan-

ics are uncoupled while in time domain methods these are coupled.

## 3 COMPARISON METHODS

### 3.1 *Traffic simulation and encounter scenarios*

The traffic simulation is described here shortly as the simulation itself is not crucial regarding the comparison of impact models. The simulation is described in detail in (Ståhlberg, 2010)

The traffic in the analyzed area is obtained from AIS data. The data contains traffic information from the month of July 2006 in the sea area between Helsinki and Tallinn where densely trafficked waterways cross. In Figure 2 the analyzed area and the data points are presented. The four main waterways in the crossing area are named after compass quarters in form of "from-to" as shown in Figure 2. The considered waterway combinations and resulting encounter types are listed with reference numbers in Table 2.

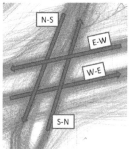

Figure 2. Plot of AIS data points in analyzed area

The AIS data is filtered to distinguish the traffic between waterways and ship types. The numbers of passages through the analyzed area per ship type are listed in Table 3. The Monte Carlo simulation flowchart starting from the filtered AIS data is shown in Figure 3. The result of the simulation is the encounter situations based on the traffic data.

Table 2. The considered waterway combinations and resulting encounter types with respective reference numbers.

| Ref number | Route | Encounter type |
|---|---|---|
| 1 | N-S, E-W | Crossing |
| 2 | N-S, W-E | Crossing |
| 3 | S-N, E-W | Crossing |
| 4 | S-N, W-E | Crossing |
| 5 | W-E, E-W | Head-on |
| 6 | E-W, W-E | Head-on |
| 7 | E-W, E-W | Overtaking |
| 8 | W-E, W-E | Overtaking |

Table 3. Number of passages per ship type and route.

| Ship | Route | | | |
|---|---|---|---|---|
| Type | N-S | S-N | E-W | W-E |
| HSC | 741 | 740 | 0 | 0 |
| PAX | 253 | 254 | 26 | 14 |
| Cargo | 5 | 4 | 768 | 742 |
| Tanker | 0 | 0 | 218 | 215 |
| Other | 3 | 3 | 36 | 35 |

* HSC = High Speed Craft, PAX = Passenger vessel, Cargo = Cargo vessel

The Monte Carlo simulation to create encounter scenarios is run 10000 times for those combinations of main waterways in which the tanker may be struck vessel. In the utilized data set tankers were recorded sailing only on "E-W" and "W-E" waterways. In this study the probability of a vessel involved in collision is weighted with the number of voyages in the area.

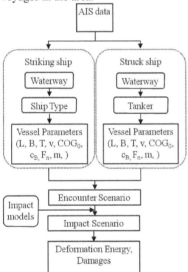

Figure 3. Flowchart of Monte Carlo simulation

### 3.2 Impact scenario simulation

With the encountering vessels' characteristics known the impact scenarios are simulated here by applying the compared models for the dynamic parameters. The models may be considered to be the "evasive maneuvering" model shown in Figure 1.

The compared assumptions are presented in Figures 4-7 and the distribution parameters are compiled into Table 4.

In "Blind Navigator" –model there are no maneuvering actions taken to avoid the collision and thus the speeds and courses are unchanged from the encounter scenario. The collision location is assumed to be uniformly distributed along the struck vessel's length. This model is used by Van Dorp & Merrick (2009) and COWI(2008). Based on the analysis of

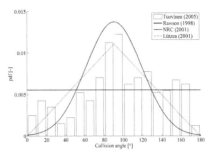

Figure 4. Input distributions for collision angle, Lützen: initial angle 90°, Brown (2002) quasi-equivalent to NRC (2001)

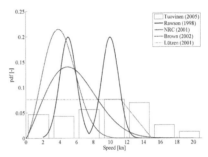

Figure 5. Input distributions for striking ship speed, Lützen with initial speed of 15 kn

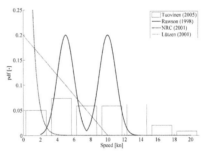

Figure 6. Input distributions for struck ship speed, Lützen with initial speed of 10 kn, Brown (2002) quasi-equivalent to NRC (2001)

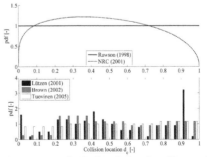

Figure 7. Input distributions for location of impact along struck ship's length, 0 = aft, 1 = fore

Table 4. Overview of impact scenario models.

| Impact Model | Collision Angle, β [deg] | $V_A$ [kn] | $V_B$ [kn] | Impact Point, d [x/L] |
|---|---|---|---|---|
| Blind Navi | $β=β_0$ | $V_A=V_{A0}$ | $V_B=V_{B0}$ | U(0,1) |
| Rawson (1998) | U(0,180) | bi-normal N(5,1) N(10,1) Truncated {2, 14} | idem to $V_A$ | U(0,1) |
| NRC (2001) | N(90,29) | W(6.5,2.2) | E(0.584) | B(1.25,1.45) {0, 1} |
| Lützen (2001) | $T(0,β_0,180)$ | $U(0,0.75V_{A0})$ $T(0.75V_{A0},V_{A0})$ | $T(0,V_{B0})$ | Empirical See FIG 7 |
| Brown (2002) | N(90,29) | W(4.7,2.5) | E(0.584) | Empirical See FIG 7 |
| Tuovinen (2005) | Empirical See FIG 4 | Empirical See FIG 5 | Empirical See FIG 6 | Empirical See FIG 7 |

* Distributions are marked as follows, U=Uniform(min, max) N=Normal($μ$, $σ$), T=Triangular(min, triangle tip, max), E=Exponential($λ$), B=Beta($α$, $β$, min, max), W=Weibull(k, $λ$)

collisions in (Cahill, 2002) and (Buzek & Holdert, 1990) it seems extremely rare that neither vessel takes any action.

Lützen's (2001) set of assumptions implies that the struck vessel is more prone to speed reduction than the striking vessel while the impact angle is tri-angularly distributed between 0° and 180° with the tip of the distribution at the encounter angle. The longitudinal impact location is given by empirical distribution. Although there is no explanation how the distributions for collision angle and velocities are derived these are included into the comparison be-cause of the existing relation between encounter and impact scenarios.

Rawson et al (1998) model is based on statistics of the grounding accidents with assumption that the collision speed being similarly distributed as grounding speed. Velocities of the colliding vessel are distributed according to a double normal distri-bution in which the averages are described to repre-sent the service speed, i.e. no speed reduction, and half of service speed. The same speed distribution is used for both striking and struck vessel. Collision angle and collision location are uniformly distribut-ed between 0°…180° and along the struck vessel's length respectively.

Tuovinen (2006) compiled statistics from over 500 collisions. Statistics have been used here as pre-sented originally, in form of empirical distributions.

Brown (2002) and NRC (2001) give quite similar distributions. Brown gives lower velocity for the striking vessel. These models both assume that strik-ing vessel has higher velocity than struck at the mo-ment of impact. It is noteworthy that these two mod-els suggest much lower collision speeds than other models. Collision angle is normally distributed around right angle. In NRC model the collision loca-tion is beta distributed so that midship section is rammed at higher probability than the fore and aft of

the vessel while Brown suggests empirical distribu-tion.

Overall, the distributions Lützen suggested are the only ones taking the encounter into account in any way and other models give same distributions for dynamic parameters irrespective of encounter scenario. None of these models indicate how to de-termine which vessel is striking and which is struck. It is assumed here that the probabilities of vessel be-ing striking or struck are equal for all models as no other probabilities were suggested in these models. The compared models do not have the possibility of initial sway nor yaw speeds, which in case of ma-neuvering is unlikely.

It can be seen in the Figures 4-7 that models, with exception of Brown and NRC, give distinctively dif-ferent distributions for the dynamic parameters. Considering the struck vessel speed being lower in all the models expect Rawson it appears likely that the collision statistics from which the distributions are derived include collisions in which the struck vessel is in anchorage or in berth. Tuovinen's (2005) statistics include approximately 6% of such cases and 41% of open seas collisions. Brown (2002) states that the share is significant as in about 60% of collisions struck vessel speed is zero.

### 3.3 Deformation energy calculations

Zhang presented in his PhD thesis (Zhang, 1999) a simplified calculation method for released deformation energy in ship-ship collision. Zhang's method is based on rigid body mechanics and conservation of momentum. The method is derived based on the dynamics of two rods colliding on a frictionless surface and has three degrees of freedom. The hydrodynamic effects are considered as constant added masses. Both vessels may have initially forward and sway speeds. During the collision the rotational movements are considered as small and are neglected. After the collision both vessels are allowed to have rotational velocity. Figure 8 illustrates the impact scenario and defines the used co-ordinate systems. The formulation is not presented here due to its lengthiness.

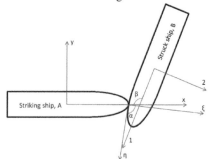

Figure 8. Impact scenario and the co-ordinate systems

### 3.4 *Damage calculation*

The method of damage calculation used here is presented in Goerlandt et.al. (2011). The focus in case of a tanker being a struck vessel is on the possibility that cargo oil is spilled. That requires penetration of one or more oil cargo tanks. Thus the penetration depth must exceed the double side width added with the dislocation of the inner shell when a rupture occurs. Additionally, the collision location along stuck tanker hull must be within the boundaries of the cargo tanks. Smailys & Česnauskis (2006) suggested following limits for cargo area for tankers operating in the Baltic Sea.

$$0.2L \le d \le 0.94L \qquad (2)$$

where L is vessel length and d is distance of impact point from amidships along the centerline.

For the purposes of this study the simple criterion for oil cargo tank penetration is used and is expressed as critical energy, $E_{cr}$, with

$$E_{cr} = \begin{cases} 12.5, DWT < 5000 \\ 12.5 \left(1 + \dfrac{\left(W_{ds} - 1\right)\left(\dfrac{DWT}{1000} - 5\right)}{35}\right), 5000 \le DWT \le 40000 \\ \dfrac{DWT}{2000} + 5, DWT > 40000 \end{cases} \qquad (3)$$

where $W_{ds}$ is double side width given in meters in ABS (2010) classification rules by:

$$W_{ds} = \begin{cases} 1, DWT < 10000 \\ 0.5 + \dfrac{DWT}{20000}, 10000 \le DWT \le 30000 \\ 2, DWT > 30000 \end{cases} \qquad (4)$$

This criterion is obtained from a simple linear regression in the example cases discussed in (Zhang 1999, Lützen 2001, HSE 2000). It is further assumed that the effect of striking vessel bow geometry is negligible and that the energy absorbed by the striking vessel is taken into account in $E_{cr}$. Even though the evaluation of the critical energy is based on a very simplified model and better alternatives are available in the literature (Brown 2002, Ehlers 2008), this criterion is withheld due to its simplicity. Application of the simple criterion of (Eq. 3) affects all impact scenario models in a similar way, such that the conclusions are still valid. The actual value of $E_{cr}$ is in this respect not essential as it is only used as a reference to better present the differences in impact models. In this study the collision consequences analysis is limited to evaluating if the deformation energy in direction normal to the struck vessel hull, $E_\xi$, exceeds $E_{cr}$ that is required to breach a cargo tank, while neither the actual structural damages nor the amount of oil spilled are not considered.

## 4 SIMULATION RESULTS AND DISCUSSION

In this section, the results of the Monte Carlo simulations for the relative velocity, collision energy and hull breach probability are given for the impact scenario models.

### 4.1 *Relative velocity*

The relative velocity $V_{rela}$ is considered as the velocity that the bow of the striking vessel is approaching the collision point at the struck vessel side. In vector form it is given with:

$$\vec{V}_{rela} = \vec{V}_A - \vec{V}_B \qquad (5)$$

The released deformation energy is highly depending on the $V_{rela}$ at the moment of impact. Relative velocities obtained from simulation for "Blind navigator" and Lützen model in selected encounter situations are presented in Figure 9. The other four models give similar results for $V_{rela}$ irrespective of the encounter situation and thus results are presented only for waterway combination 1.

The "Blind Navigator" model is giving much higher values of $V_{rela}$, apart from head-on encounter, than other models as expected. There are two peaks in the result distributions of "Blind navigator" for crossing encounter situations. The lower peak represents passenger vessel cases and higher peak High Speed Crafts as striking vessel.

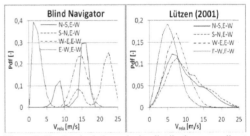

Figure 9. Simulated relative velocity distributions according to impact models in which encounter is considered

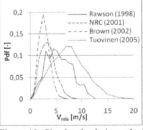

Figure 10. Simulated relative velocity distributions according to impact models in which encounter is not considered

The angle between N-S and W-E traffic flows is approximately 120° while between N-S and E-W traffic the angle is 60°. The effect of angle on relative velocity can be seen by comparing "Blind Navigator" results in Figure 9, the larger angle results in higher $V_{rela}$. The Lützen model appears to be relatively insensitive to variation of the encounter angle as only slight difference can be observed. This is due to the reduction of the struck vessel speed. The Lützen model gives the impact speed of the struck vessel to be on average ⅓ of the initial velocity.

The models that are derived from statistics by Rawson, NRC, Brown and Tuovinen give much more diverse results for $V_{rela}$ than may be anticipated as the available accident data is limited and one would expect that the statistics would be practically based on the same data. It should be noted that these four model result in similar distributions for all encounter scenarios. Thus while the $V_{rela}$ is lower in case of crossing encounter it is higher in case of overtaking compared to "Blind navigator" and Lützen models.

## 4.2 Deformation energy

In here only the transversal deformation energy $E_\xi$ is considered because it represents the deformation energy in direction of penetration depth. The simulation results for $E_\xi$ in each simulated encounter are normalized by dividing it with respective critical energy $E_{cr}$. In Figures 11-13 the cumulative distributions for normalized deformation energy $E_{\xi N}$ for each impact scenario model are presented for selected waterway combinations.

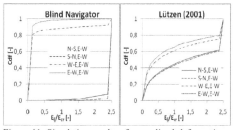

Figure 11. Simulation results of normalized deformation energy for "Blind navigator" and Lützen (2001) impact models.

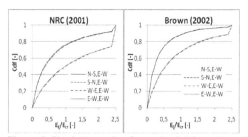

Figure 12. Simulation results of normalized deformation energy for "Blind navigator" and Lützen (2001) impact models.

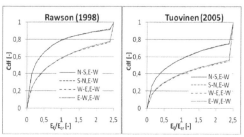

Figure 13. Simulated relative velocity distributions according to impact models in which encounter is considered

In "Blind Navigator" and Lützen models $V_{rela}$ and impact angle are dominating factors in normalized $E_\xi$ as seen in figure 11 when comparing results of crossing encounters with head-on and overtaking encounters. For head-on encounters normalized $E_\xi$ is little higher than for overtaking but much lower than in crossing encountering. This is because even if $V_{rela}$ is high the deformation energy is mostly released in η-direction along the struck vessel side.

The vessels sailing on W-E and E-W waterways are often on round trip to Gulf of Finland and thus the vessels are recorded in most cases on both waterways. Furthermore the loading condition was assumed to be fully laden for all vessels. For these reasons the vessel mass distributions are equivalent.

The same applies for N-S and S-N waterways except that the vessels are sailing between Helsinki and Tallinn. Additionally, the vessel masses on latter waterway pair are much lower than that of the prior. The differences in the vessel masses are resulting in differences between waterway combinations in the Figures 12, 13 as the distributions of $V_{rela}$ are equivalent for all encounter scenarios in these models.

In figures 14, 15 the normalized cumulative distributions are compiled into same graph for crossing encounter and head-on encounter respectively with $E_{cr}$ marked with vertical line.

From Figures 14, 15 similar observations as from Figures 11-13 can be made. The four models derived from statistics each result in higher $E_{\xi N}$ in head-on encounter than crossing while the opposite occurs for the "Blind navigator" and Lützen models. The same is valid for overtaking as was shown in Figures 12, 13.

## 4.3 Probability of oil cargo tank penetration

The oil cargo tank is penetrated when $E_{\xi N}$ is greater than 1 and the impact location is within tank limits given by Equation 2. The number of simulated impact scenarios in which the impact location is outside tank limits are listed in Table 5.

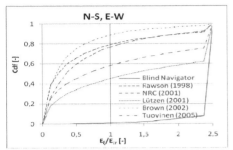

Figure 14. Simulated transversal deformation energy relative to critical energy in crossing encounter.

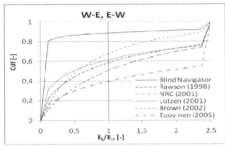

Figure 15. Simulated transversal deformation energy relative to critical energy in head-on encounter.

Table 5. Number of simulated collision scenarios of total 10000 simulations in which collision location is outside cargo tank limits given by Eq 2.

| Blind Navi | Rawson (1998) | NCR (2001) | Lützen (2001) | Brown (2002) | Tuovinen (2005) |
|---|---|---|---|---|---|
| 2626 | 2602 | 1700 | 1772 | 1223 | 2449 |

In Table 6 the numbers of simulated collisions resulting in an oil spill per impact model are presented. The same is visualized in Figure 13.

Table 6. Number of simulated collision scenarios in which oil cargo tank penetration occurs of total 10000 simulations.

| Ref No* | Blind Navi | Rawson (1998) | NCR (2001) | Lützen (2001) | Brown (2002) | Tuovinen (2005) |
|---|---|---|---|---|---|---|
| 1 | 7283 | 1581 | 1889 | 4695 | 926 | 3153 |
| 2 | 7379 | 1612 | 1955 | 4671 | 930 | 3002 |
| 3 | 7335 | 1648 | 1930 | 4743 | 982 | 3169 |
| 4 | 7321 | 1629 | 1934 | 4557 | 977 | 3033 |
| 5 | 604 | 3230 | 4146 | 3232 | 2901 | 4705 |
| 6 | 563 | 3098 | 4089 | 3054 | 2794 | 4550 |
| 7 | 105 | 3121 | 4107 | 2842 | 2738 | 4551 |
| 8 | 43 | 3192 | 4142 | 2940 | 2841 | 4591 |

* See Table 2 for explanation of Reference numbers

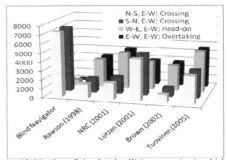

Figure 13. Number of simulated collision scenarios in which oil cargo tank penetration occurs of total 10000 simulations

Taking the collision location into account does change the results but very little. The differences between the models remain obvious. The collision following crossing encounter results in an oil spill in three out of four cases according to "Blind navigator" model. Brown's model suggest that oil spill would occur only once in ten collisions.

## 5 CONCLUSIONS

In this paper a number of proposed models for impact scenarios from literature have been applied to the output of a maritime traffic simulation model to create impact scenarios. The released deformation energy is calculated with an analytical collision dynamics model for each impact scenario. Based on the obtained deformation energy the cargo tank penetration probability is estimated. The simulation results for relative velocity, transversal deformation energy and oil cargo tank penetration are compared between different impact scenario models.

The results of this case study indicate that the models give a widely varying average hull breach probability. In particular, the uncertainty on cargo tank breach probabilities dependence of initial encountering is significant, which is an important factor in the analysis of oil spill risk in specific location i.e. crossing or merging of waterways.

The distributions of collision energy for models based on statistics depend almost solely on the striking vessel mass instead of the actual encounter scenario. In the statistics that the models are based on there are no collisions where a high speed craft is involved. Further it is reasonable to assume that these statistics include collisions, in which the struck vessel is in anchorage, leading to underestimation in struck vessel speed at the moment of impact in collisions occurring at open seas.

None of the statistics is broken up for cases in different sea areas nor is the encountering related with the collision. These lacking in data are partly due to the limited number of collision cases available, lack of transparency and unsatisfactory reporting standards.

It is very likely that the statistical models are grossly underestimating the effect of encounter speed for both vessels in the area concerned in this case study. This leads to the conclusion that the understanding of the conditions of ship collision in a risk modeling framework is very limited at present.

The proposed models for impact scenarios are moreover burdened with some inherent conceptual limitations. The most significant limitation is the unsatisfactory modeling of evasive maneuvering, which links the initial encounter situation to the impact scenario. The results clearly indicates that especially the parameters which navigators have a possibility to affect in evasive maneuvering, i.e. vessel speed and collision angle, play a determining role in the evaluation of the consequences. Further research on this matter is needed.

## ACKNOWLEDGMENT

The authors appreciate the financial contributions of the following entities: the EU, Baltic Sea Region (this study was partly founded by EfficienSea project), Merenkulun säätiö from Helsinki, the city of Kotka and the Finnish Ministry of Employment and the Economy.

## REFERENCES

American Bureau of Shipping (ABS). 2010. Rules for Building and Classing Steel Vessels. *American Bureau of Shipping.* Houston, USA.

Brown A.J. 2002. Collision scenarios and probabilistic collision damage. *Marine Structures,* 15(4-5):335-364.

Buzek, F.J. & Holdert H.M.C. (1990). Collision Cases Judgements and Diagrams, *Lloyd's of London Press Ltd.*

Cahill, R.A. 2002. Collisions and their causes, third ed. *The Nautical Institute,* London.

Chen, D. 2000. Simplified Collision Model (SIMCOL). M.Sc. thesis. *Virginia Tech,* Blacksburg, USA.

COWI for the Danish Maritime Authority. (2008). Risk Analysis of Sea Traffic in the Area around Bornholm, *COWI report no.* P-65775-002, January 2008.

Ehlers, S. Broekhuijsen, J. Alos H.S. Biehl F. Tabri K. 2008. Simulating the collision response of ship side structures: A failure criteria benchmark study. *International Shipbuilding Progress,* 55:127-144.

Goerlandt, F. & Kujala, P. 2010. Traffic simulation based collision probability modeling. *Reliability Engineering and System Safety,* doi:10.1016/j.ress.2010.09.003

Goerlandt F, Ståhlberg K, Kujala P. 2011. Comparative study of input models for collision risk evaluation. *Ocean Engineering – manuscript under review.*

Hanhirova, K. 1995. External Collision Model, Safety of Passenger/RoRo Vessels, *Helsinki University of Technology, Ship Laboratory*

Health and Safety Executive (HSE). 2000. Collision resistance of ship-shaped structures to side impact, *Health and Safety Executive,* London, United Kingdom, ISBN 0-7176-1997-4

Hutchison, B.L. 1986. Barge Collisions, Rammings and Groundings – an Engineering Assessment of the Potential for Damage to Radioactive Material Transport Casks, Technical Report SAND85-7165 TTC-05212

Kujala, P. Hänninen, M. Arola, T. Ylitalo, J. 2009. Analysis of the marine traffic safety in the Gulf of Finland. *Reliability Engineering and System Safety,* 94(8):1349-1357.

Lützen, M. 2001. Ship collision damage. *PhD thesis, Technical University of Denmark.*

Merrick, J.R.W. van Dorp, J.R. Harrald, J. Mazzuchi, T. Spahn, J. Grabowski, M. 2003. A systems approach to managing oil transportation risk in Prince William Sound. *Systems Engineering,* 3(3):128-142.

Minorsky, V.U. (1959), An Analysis of Ship Collisions with Reference to Protection of Nuclear Power Plants, *Journal of Ship Research,* October 1959.

Montewka, J. Hinz, T. Kujala, P. Matusiak, J. 2010. Probability modeling of vessel collision. *Reliability Engineering and System Safety,* 95(5):573-589.

National Research Council (NRC), 2001. Environmental Performance of Tanker Designs in Collision and Grounding, *Special Report 259, The National Academies Press.*

Pedersen, P.T. 1995. Collision and grounding mechanics. *The Danish society of Naval Architects and Marine Engineers* 125-157.

Pedersen, P.T. 2010. Review and application of ship collision and grounding analysis procedures. *Marine Structures,* 23(3):241-262.

Rawson, C. Crake, K. Brown, A. 1998. Assessing the environmental performance of tankers in accidental grounding and collision, *SNAME Transactions* 106:41-58.

Smailys, V. & Česnauskis, M. 2006. Estimation of expected cargo oil outflow from tanker involved in casualty. *Transport – 2006,* vol 21, No 4, p. 293-300.

Ståhlberg, K. 2010. Estimating deformation energy in ship-ship collisions with stochastic modeling, *M.Sc. Thesis, Aalto University, School of Science and Technology,* Espoo, Finland

Tabri, K. Varsta, P. Matusiak, J. 2009. Numerical and experimental motion simulations of non-symmetric ship collisions. *Journal of Marine Science and Technology,* 15(1):87-101.

Tuovinen, J. 2005. Statistical analysis of ship collisions. *M.Sc. thesis. Helsinki University of Technology,* Espoo, Finland.

Ulusçu, Ö.S. Özbaş, B. Altiok, T. Or, İ. 2009. Risk analysis of the vessel traffic in the strait of Istanbul. *Risk Analysis,* 29(10):1454-1472.

van Dorp, J.R. & Merrick, J.R.W. 2009. On a risk management analysis of oil spill risk using maritime transportation system simulation. *Annals of Operations Research,* doi: 10.1007/s10479-009-0678-1.

Vaughan, H. (1977). Damage to Ships Due to Collision and Grounding, *DNV Technical Report* No. 77-345.

Wang, G. Spencer, J. Chen, Y. (2001). Assessment of a Ship's Performance in Accidents, *Journal of Marine Structures,* 15:313-333.

Zhang, S. 1999. The mechanics of ship collisions. *PhD thesis, Technical University of Denmark.*

# 14. Applied Research of Route Similarity Analysis Based on Association Rules

Zhe Xiang, Ru-ru Liu, Qin-you Hu & Chao-jian Shi

*Merchant Marine College, Shanghai Maritime University, China*

ABSTRACT: In recent years, with the development of information technology, businesses have accumulated a lot of useful historical data, as the shipping industry does. These data can be found deposited a large number of "knowledge", for example, Shipping records for historical information, Ship-Port relations information, Ship-ship relations information, Port & shipping route relations, Shipping route information. It can provide intellectual support to shipping informatization development.

Association rules in data mining technology is one of important technologies. The technology, based on statistical methods, can mine the associated and implied "knowledge" from data warehouse ,which has a large number of accumulated data. Apart from this, the technology can also play an important role in the prediction. In this paper, based on FP-growth algorithm, we improve it forming Relevent ships routes.

From the prevalent perspective of data mining, deal with the corresponding vessels' dynamic information, acquired from the AIS, such as data collection, data statistics. On this basis, get the ship-port relation and ship-ship relation after a certain level of data analysis, processing, handing. Furthermore, this paper use the numerous historical ship-port relation and ship-ship relation to build a mathematical model on the ship-port and ship-ship relation. And use the improved association algorithm, FP-growth algorithm, to acquire the strong association rules between ship-port and ship-ship, and eventually mine the similarity of the ship route.

Main points of this paper as follows:

Collect ,count and check the data, which is from ship dynamic information;

Establish the mathematical model between ship-port and ship-ship relation;

Improve the algorithm;

Analyse the similarity of ship route more accurately using the improved algorithm.

## 1 INTRODUCTION

In recent years, data mining has caused great concern to the industry. The so-called mining technology is to present large amounts of data by digging into useful information and knowledge, which applies to business management, production control, market analysis, engineering design and scientific exploration and other fields. This is the data mining technology on information processing in navigation, the use of data mining association rules FP-TREE algorithm, mining the large number of ships AIS information, access to the ship's route similarity analysis.

### 1.1 *FP-TREE algorithm [1]*

Data mining technology is the result of the development of information technology, since the 60 years since the last century, databases, and information technology has systematically evolved from the original document processing to complex and powerful database system, and now the technology is quite mature. Data mining is the definitions of raw data from a large number of extracted or "dig out" the useful information into knowledge.

In the knowledge model of data mining, association rules model is the more important one. The concept of association rules by the Agrawal, Imielinski, Swami suggested that the data in a simple but very useful rule. Association rules models are descriptive models, association rules discovery algorithm is unsupervised learning.

FP-TREE [1][2][3] is one of association rule mining algorithm, which uses divide and conquer strategy, after the first pass scan, the frequency of the database into a set of compressed frequent pattern tree (FP-TREE), while retaining the associated information, then FP-TREE library to differentiate into a number of conditions, the length of each library, and a frequency of 1 set of related libraries of these conditions were re-excavation.

## 1.2 AIS information [4]

Automatic Identification System (AIS) is the current advanced ship-aided navigation equipment; the International Maritime Organization has adopted the mandatory installation of AIS requirements. AIS can automatically send the ship a static continuous, dynamic and voyage information, security, short message, but can also automatically receive the information sent around the ship, and exchange information with the coast station.

AIS information includes static information such as name, call sign, MMSI, ship type, ship size and other information, and dynamic information such as vessel position, ground speed, navigational status, draft, destination, estimated time of arrival and other information, but also with the voyage information and security-related text messages.

## 1.3 AIS information on data mining [5][6]

In this paper, using association rules such as FP-TREE algorithms, we deal with the corresponding vessels' dynamic information, such as estimated time of arrival, port of destination and departure information, acquired from the AIS, such as data collection, data statistics. On this basis, get the ship-port relation and ship-ship relation after a certain level of data analysis, processing, handing to build a mathematical model on the ship-port and ship-ship relation to acquire the strong association rules between ship-port and ship-ship, and eventually mine the similarity of the ship route.

## 2 THE PROPOSAL OF THIS PAPER

At first, we used to provide users dynamic-related information services with AIS ship data, such as real-time latitude and longitude of the ship, speed, arrival time, etc. Also including the port border querying. But we are lack of a more intelligent data mining reports service, for example, previous leaves harbor of the ships, the ships' similar route. These are the users concerned, and it is very useful.

Therefore, in order to provide such services, we need to mine existing data to find out the regular pattern between ships and their anchored port and also we need to find out ships' similar routes(the similar route among ships).The conclusion we mine from AIS data can provide information on the effect of enhanced information services . We collected ships infomation through the www.manyships.com dynamic data, and similarly, the Chinese port border has also been part of we collect. It is easy for us to find the ship's ports and similar routes and provide us a lot of supports for data ming.

The most important task of this paper is: data mining, to identify routes of ships similar to the (first find was anchored in this port).

## 3 PREPARATION FOR DATA MING

### 3.1 Data collection

This article uses real-time AIS data from www.manyships.com. AIS database established a table for each ships, with the AIS data updating continuously the database update this table as table 1.

The number of table is Real-time. So, the first step we need to do is to scheduler an interval job to collect data using database snapshots[4]. With this methods,we can collect one week or one month even more AIS data[5] . To say that, for accurcy results,job's interval can not be too long.

### 3.2 Data processing and Summary

After collecting the data,we must immediately calculate whether the ship is in the ports and also calculate its last port .

Finally we proceed the AIS data for each ships. These data included the ship's number(mmsi), name, port of arrival, arrival times, the last port, the last departure time, callsign, type, so as follow. These data be prepared for data mining as table 2.

Each port has its own id, so 'fid' in the picture means the port's id , 'prevfid' means the last port's id. With these data we can be ming AIS data.

Table 1 original datasets snapshot

| mmsi | speed | lon | lat | course | heading | updatetime |
|---|---|---|---|---|---|---|
| 565101000 | .100000014901160 | 7290.17280000000 | 2260.93120000000 | 275.20000000000 | 177.00000000000 | 2010-12-04 08:10:07 |
| 413552790 | 8.60000038146973 | 7313.72160000000 | 2427.64500000000 | 219.00000000000 | 238.00000000000 | 2010-12-03 07:08:02 |
| 413427340 | 8.39999961853026 | 7317.12000000000 | 1831.39360000000 | 264.70000000000 | 17.00000000000 | 2011-01-07 07:24:19 |
| 477383000 | .000000000000000 | 7151.78320000000 | 1933.76100000000 | 317.20000000000 | 306.00000000000 | 2011-01-07 07:24:21 |
| 412206710 | 9.69999980926514 | 7186.13920000000 | 1551.12660000000 | 208.10000000000 | -1.00000000000 | 2010-12-04 08:12:18 |
| 412047720 | .000000000000000 | 7295.70800000000 | 1882.02120000000 | 204.10000000000 | -1.00000000000 | 2011-01-07 07:24:58 |
| 412047210 | .200000002980232 | 7325.06880000000 | 1864.13240000000 | 123.10000000000 | 292.00000000000 | 2011-01-07 07:21:50 |

Table 2 datasets snapshot after procceed

| mmsi | leavetime | fid | fid_name | prevfid | prevn... | prevleavetime | name | callsign | type |
|---|---|---|---|---|---|---|---|---|---|
| 412051550 | 2010-08-30 12:46:5? | 1 | <MEMO> | 47 | <MEMO> | 2010-08-28 08:28:0? | YONG CHI | BRYU | 货轮 |
| 412053050 | 2010-08-26 14:29:2? | 1 | <MEMO> | 5 | <MEMO> | 2010-08-21 20:29:01 | WAN QING SHA | BSPN | 疏浚或水下作业船 |
| 412064000 | 2010-08-26 23:29:0? | 1 | <MEMO> | 207 | <MEMO> | 2010-08-23 05:49:2? | AN GUANG JIANG | BOAU | 货轮 |
| 412070000 | 2010-08-21 23:27:0? | 1 | <MEMO> | 42 | <MEMO> | 2010-08-20 05:55:3? | XING HE | BOLO | 货轮 |
| 412070630 | 2010-08-28 15:08:0? | 1 | <MEMO> | 201 | <MEMO> | 2010-08-27 08:29:4? | HANG CE 501 | 未知 | 其他类型船 |
| 412081690 | 2010-08-27 21:13:0? | 1 | <MEMO> | 204 | <MEMO> | 2010-08-23 20:33:1? | CHANG HANG SHA | BUVU | 货轮 |
| 412187000 | 2010-08-24 12:23:2? | 1 | <MEMO> | 4 | <MEMO> | 2010-08-23 10:07:1? | XIN BIN CHENG | BVFB5 | 货轮 |
| 412205920 | 2010-08-31 22:52:0? | 1 | <MEMO> | 5 | <MEMO> | 2010-08-27 04:29:2? | XIN YUN FENG | BAOZ | 货轮 |
| 412207640 | 2010-08-22 23:27:31 | 1 | <MEMO> | 202 | <MEMO> | 2010-08-22 04:21:5? | GANG TONG HAI 9 | BANA | 未知 |
| 412222000 | 2010-08-28 01:54:2? | 1 | <MEMO> | 2 | <MEMO> | 2010-08-19 20:57:5? | TUO HAI | BOHF | 货轮 |
| 412258000 | 2010-08-29 01:17:3? | 1 | <MEMO> | 31 | <MEMO> | 2010-08-25 05:05:0? | YING CHUN HAI | BIAE3 | 未知 |
| 412272000 | 2010-08-29 23:58:4? | 1 | <MEMO> | 207 | <MEMO> | 2010-08-20 18:13:01 | FENG AN SHAN | BOST | 货轮 |
| 412351000 | 2010-08-30 14:54:2? | 1 | <MEMO> | 5 | <MEMO> | 2010-08-25 07:30:1? | ZI YUN FENG | BHVV | 未知 |

## 4 MINING THE SIMILAR PATH AND THE RELEVENT PORT

In order to retrieve the relative path between the ship, we use association rule discovery methods to determine the correlation between the ports and the ships.

The central idea of this method is: Calculate the ship reaches port summary statistics. Use summary information as a transaction set; The port each ship reached (or leave) is the association rules item. Each transaction is a single arrangement of the ship reaches port. Aim of the algorithm is digging out the relevant port, and getting the ships' similar paths.

### 4.1 Algorithm is defined as follows[1][3]:

(1). $I = \{I_1, I_2 ...... I_m\}$ is item set.

(2). Let $D$ transaction sets, each transaction $T$ is $T \subseteq I$.

(3). Each transaction has a unique identifier TID. In addition, add a field to indicate the transaction from the ship (the number of ship).

For an example of Transaction sets is as table 3:

Table 3 example of Transaction sets

| TID | PORTID | Mmsi |
|---|---|---|
| 1 | 200 , 47 , 48 , 203 , 56 , 207 , 88 , 1 | 412402820 |
| 2 | 47 , 21 , 48 , 200 , 66 , 71 | 412403000 |
| 3 | 21 , 200 , 42 , 59 , 88 , 1 | 412403660 |
| 4 | 21 , 48 , 49 , 88 , 71 , 6 | 412410010 |
| 5 | 47 ,200 , 48 , 62 , 66 , 51 , 71 , 1 | 412429760 |

Each port has an unique identity number ,so as ships. In this paper, Example uses the portid and mmsi to instead of the specific ports and ships.

### 4.2 Algorithm Description:

We intend to make a FP-tree for the transaction sets, the tree node is a transaction set, namely the port (except root node). Edge of the FP-tree is ships' collection, on behalf of the ship go across the two ports. We use 'Mmsi' instead of ship's name.(mmsi is the unique identity of the ships in database.)

The structure of Node data as table 4 :

Table 4 The structure of Node

| Parent | Portid | Support | Child |
|---|---|---|---|

The structure of Node data contains the number of ports, its support count, its child node, its parent node.

The structure of Edge data is as table 5:

Table 5 The structure of Edge

| PortID1 | Mmsi[] | mmsi | Support[] | PortID2 |
|---|---|---|---|---|

Edge data structure as shown above, Mmsi [] expressed as mmsi array, a collection of both ships Mmsi number. Similarly, the corresponding number of ships mmsi also have the corresponding support count. PortID1 and PortID2 mean this edge belongs to which two port nodes.

### 4.3 Steps of the algorithm

According to the minimum support, we use the transaction sets generated a FP-tree by FP-growth algorithm.

1. [1][2] scan transaction set D, in order to acquire all the frequent items contained in D ,we named the collection of these items F,and also caculate their respective support. Frequent items in descending order according to their support, the results recorded as L(table 6):

2. [1][2] Create the root of FP-tree T, marks "null"; Then, for each transaction TID following:

According to the order of L to select and sort the frequent items TID. Let the sorted frequent item list as [x|P], x is the first frequent item, and P is the remaining frequent items; Then call IN-SERT_TREE([x|P],T).The INSERT_TREE([x|P], T) follows the process of implementation: If T have their children named N and meet the N. Por-tid = x. Portid, increase the N's Support 1; else create a new node N, its count is set to 1, link to its parent node T and through the node chain structure link to the node with the same Portid. If P is not empty, recursively call the IN-SERT_TREE (P, N).

In addition, whether the addition of new nodes, each edge must be coupled with corresponding Mmsi TID number. if repeated, Mmsi Support count increased by 1.

Re-scanned in the transaction set, a complete with side information on FP-tree built.

Using the example 3.1 , let the minimum support MIN_SUPPORT = 3, the 1-frequent itemsets is as follows:

1 - frequent item sets as table 6 :

Table 6 1 - frequent

| item | count |
|------|-------|
| 200 | 4 |
| 48 | 4 |
| 47 | 3 |
| 21 | 3 |
| 88 | 3 |
| 71 | 3 |
| 1 | 3 |

And we also acquire the FP-tree as figure 1:

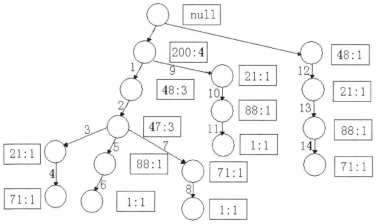

Figure 1 generated tree

The information of edge as table 7:

Table 7 information of edge after calculated

| Edge ID | PortID1 | Mmsi[] | mmsiSupport[] | PortID2 |
|---------|---------|--------|---------------|---------|
| 1 | 200 | 412402820, 412403000, 412429760 | 4,4,2 | 48 |
| 2 | 48 | 412402820, 412403000, 412429760 | 3,3,1 | 47 |
| 3 | 47 | 412403000 | 3 | 21 |
| 4 | 21 | 412403000 | 1 | 71 |
| 5 | 47 | 412402820 | 1 | 88 |
| 6 | 88 | 412402820 | 1 | 1 |
| 7 | 47 | 412403000 | 1 | 71 |
| 8 | 71 | 412402820 | 1 | 1 |
| 9 | 200 | 412403660 | 1 | 21 |
| 10 | 21 | 412403660 | 1 | 88 |
| 11 | 88 | 412403660 | 1 | 1 |
| 12 | 48 | 412410010 | 1 | 21 |
| 13 | 21 | 412410010 | 1 | 88 |
| 14 | 88 | 412410010 | 1 | 71 |

## 4.4 Mining Rules

We has generated a FP - tree, according to frequent 1 - itemsets, we can mine the frequent pattern as ta-ble 8:

Table 8 frequent pattern of Ports

| id | frequent pattern of Ports |
|----|---------------------------|
| 1 | 200-1 |
| 2 | 200-47 |
| 3 | 48-47 |
| 4 | 200-48-47 |
| 5 | 200-48 |
| 6 | 48-71 |

Meanwhile, according to the minimum support (min_support) branch cut, we also can get the ship-related items as table 9:

Table 9 frequent pattern of ships

| Id | frequent pattern of ships |
|----|---------------------------|
| 1  | 412402820-412403000       |

This result shows that a total of 6 groups of ports are associated. What's more,the MMSI number which is 412402820 and 412403000 of the two ships has some similar shipping routes.

Thus, by calculating the ship - port summary information,we not only acquire the Correlation of the ports but also the the Correlation of the ships' routes.

## 5 CONCLUSIONS AND SUMMARY

In this paper,based on the AIS data analysis and processing, we use improved FP-growth algorithm,the algorithm of data mining association rules to analyze the ship arrival information, and build FP-tree, by scanning the transaction sets, creating FP-tree root sub-set of structural conditions of frequent library, and mining the sub-set to get frenquent pattern.Through all these methods,we aggregate and mine the information of ship and port, then can find out the relevant port and ships with the similar routes.

Use of the collected AIS data which is more than one week to run above algorithm method ,we get 42 groups of related ports and 105 ships.(if you want to get more accurate results, you need to capture more of the AIS data) This shows the feasibility of the algorithm. In addition, compared to other association rule algorithm, this algorithm only need to scan the database twice.

## 6 DEFICIENCY AND OUTLOOK

In this paper,we use improved FP-tree Algorithm to dig out the similar route and relevant port,but deficiency as follow:
1. FP-tree need lots of Server's memory.
2. how to set minimum support and confidence is not in-depth reserch.

Thus,our further research will focus on improving the efficiency and application. In addition, data mining application in the navigation area is also the direction of future research

## REFERENCE

[1] MA Xu-hui, Zhang A-hong. Association Rules Generated By the FP Tree Depth-First Algorithm. Computer Knowledge and Technology, 2010, 13: 058
[2] HUI Liang, QIAN Xue-zhong. Non-check mining algorithm of maximum frequent patterns in association rules based on FP-tree. Journal of Computer Applications, 2010, 07: 064
[3] YAN Wei, BAI Wen-yang, ZHANG Yan. Hiding Association Rules with FP-Tree Based Transaction Dataset Reconstrucion. Department of Computer Science and Technology, 2008
[4] JI Xian-biao, SHAO Zhe-ping, PAN Jia-cai, et al. Development of distributed data collection system of AIS information and key techniques[J].Journal of Shanghai Maritime University, 2007, 28(3): 28-31
[5] Lee C Y. An algorithm for path connections and its applications. IRE Trans Electronic Computers, 1961, EC-10: 346-365
[6] ZHANG Wen, TANG Xi-jin, YOSHIDA Taketoshi. AIS: An approach to Web information processing based on Web text mining. Systems Engineering-Theory & Practice, 2010, 01:015

# 15. An Analysis the Accident Between M/V Ocean Asia and M/V SITC Qingdao in Hanam Canal (Haiphong Port)

Vinh Nguyen Cong
*Vietnam Maritime University*

ABSTRACT: The paper presents a special case of the ships accident. The accident was happened without any contact between ships. The cause of the accident is the hydrodynamic interaction between ship in a shallow canal and the analysis is consulted to determine the percent of fault of the accident.

## 1 SUMMARY OF THE ACCIDENT

On the 23rd September 2008, the M/V OCEAN ASIA with engine-trouble is towed through the Ha Nam canal into the port of Haiphong at the speed of 3 knots. At buoy No. 21, the pilot of M/V OCEAN ASIA saw three vessels intended to overtake him so he required the other three vessels stop these actions because of narrow area. But they are continued overtaking M/V Ocean Asia. First one, a tanker overtaken M/V Ocean Asia at buoy No. 21; then M/V Far East Cheer had overtaken M/V Ocean Asia on the portside. The last one, M/V SITC Qingdao proceeded to overtake M/V Ocean Asia.

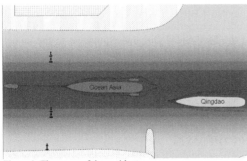

Figure 1. The scene of the accident

At this moment, M/V Ocean Asia was in the center line of the canal. The M/V SITC Qingdao sailed at the speed of 12.7knots. When the distance between two vessels is 0.5 NM, M/V SITC Qingdao reduced her speed and her speed is 9 knots when she was passing M/V Ocean Asia. The distance between side to side of the vessels is 10 – 15m when they were passing.

At the moment, when the stern of the M/V SITC Qingdao had passed the bow of M/V Ocean Asia,

the M/V Ocean Asia quickly crashed into the shore, stranded on the right bank of the canal.

The particulars of these vessels are as follow:
- M/V Ocean Asia:
    o Call sign: 3EMN4
    o Flag: Panama
    o IMO number: 7712353
    o L.O.A: 158.85 m
    o Breadth: 23 m
    o Air draft: 41.8 m
    o Draft F/M/A: 6.3 m /6.4 m /6.6 m
    o GT 10835
    o DWT: 13992 MT
    o Cargo: 345 containers equal 5738 MT
- M/V SITC Qingdao:
    o Call sign: V2BO3
    o Flag: Antigua & Barbuda
    o IMO number: 9207560
    o L.O.A: 144.83 m
    o Breadth: 22.4 m
    o Air draft: 41.5 m
    o Draft F/A: 7.0 m /7.2 m
    o GT 9413
    o DWT: 12649 MT
    o Cargo: 5900 MT

The canal's depth and breadth are 8 m and 80 m respectively. It is one way canal and taking over is prohibited here. The M/V Ocean Asia was towed by three tugboats: M/V Da Tuong towed at the bow by towing line 150 m in length; M/V Transvina and M/V Marina 18 supported at the stern (Fig. 1). At the moment of the accident, there was not current in the canal; water is still.

In this accident, there was no contacting between vessels so there was no evidence on the hulls. To blame on M/V SITC Qingdao, it is necessary to improve that the M/V Ocean Asia came aground due to the influence by overtaking of M/V SITC Qingdao.

## 2 RESEARCH BY SCALED MODELS

To research by scaled models, it is necessary to have condition of the experiment as same as real situation. The formula of the relation between a real vessel speeds and a scaled model vessel is as follows:

$$V_{model} = V_{realvessel} \times \sqrt{k} \qquad (1)$$

where, k is the dimensional ratio of real vessel and model.

In 2003, a research about "Hydrodynamic Interaction between Moving and Stationary Ship in a Shallow Canal" were carried out by Stefan Kyulevcheliev, Svetlozar Georgiev, Ship and Industrial Hydrodynamics Department, Bulgarian Ship Hydrodynamics Centre, Varna, Bulgaria and Ivan Ivanov Shipbuilding Department, Technical University of Varna, Bulgaria. They use two scaled models with the parameters are show in the Table 1.

Table 1. Parameter of the models

|  | Moving hull | Stationary hull |
|---|---|---|
| Length between perpendiculars [m] | 4.000 | 4.400 |
| Length at waterline [m] | 3.881 | 4.400 |
| Beam [m] | 0.456 | 0.456 |
| Draft [m] | 0.080 | 0.080 |
| Block coefficient | 0.928 | 0.870 |
| Displacement [m³] | 0.131 | 0.139 |

The experimental set-up with the accepted coordinate system for the forces and moments are shown schematically in Figure 3 below. The relative position of the two models has been determined by detecting a marker at the bow of the moving model with a laser fitted at the stationary model.

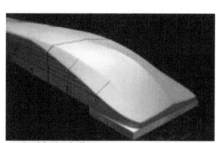

a) Moving hull

b) Stationary hull

Figure 2. The models were used in the experiments

Both the models have been free to sink and trim, but fixed in all other degrees of freedom. The moving model has run along the centerline of the tank with a width of 2.6 m.

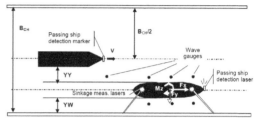

Figure 3. Experimental set-up

The measuring unit, mounted on the stationary hull, has comprised two strain gauges for the longitudinal force, allowing in this way deduction of the yawing moment, and one strain gauge for the lateral force. Two lasers detecting the distance to fixed horizontal plane attached to the canal wall have measured the sinkage of the hull.

Experiment was carried out with the sets of canal depth, model speed, and distance between two vessels when overtaking others (Stefan 2001). In this paper, the results of experiments which have conditions similar to conditions of the accident of the M/V Qingdao and M/V Ocean Asia will be presented.

The Figure 4 and Figure 5 show the experimental results when the moving model was overtaking stationary model at the speed of 0.875 m/s and ratio of canal depth and draft (H/T), respectively, 1.5, 2 and 2.5.

The length between perpendiculars of SITC Qingdao LBP is 134m. The length of the moving model in experiment is 4m. So the speed of 9 knots of the M/V SITC Qingdao equivalent to the model speed as follows (Cohen 1983):

$$V_{model} = 9 \times \sqrt{\frac{4}{134}} = 1.56 \ (knots) = 0.800(m/s) \qquad (2)$$

The speed of moving model in the experiment was 0.875m/s, so it can be considered as same as the speed of SITC Qingdao when she was overtaking M/V Ocean Asia.

About the depth of the canal, the rate of draft and depth of both vessels draft SITC Qingdao and Ocean Asia were only about H/T = 1.1 ÷ 1.14. From Figure 3 to Figure 5, forces and torque acts on the stationery are depended in inversely proportional to the value of H/T. This leads to the amplitude of forces, moment acts on the M/V Ocean Asia will be greater than the value represented on a graph of the case H/T = 1.5.

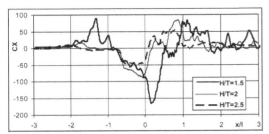

Figure 4. Hydrodynamic load CX at equal speed and different water depths

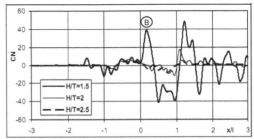

Figure 6. Hydrodynamic load CN at equal speed and different water depths

On the Figure 4, when the distance from SITC Qingdao to the Ocean Asia was about two lengths of a vessel (x/l=-1.5) the M/V Ocean Asia was pushed a bit forward. Since bow SITC Qingdao access to the stern of M/V Ocean Asia (x/l = -1), the M/V Ocean Asia was pulled back until the stern of M/V SITC Quingdao is passed the bow of M/V Ocean Asia (x/l = 1). During this period, the force got the maximum value as the bow of M/V SITC Qingdao passing the bow of M/V Ocean Asia.

The transverse force acting on the hull M/V Ocean Asia was variable as follows: When the distance of two vessels is half of vessel's length (x/l = -0.5, Fig. 5), the M/V Ocean Asia was pushed toward the shore. Then, from x/l=0 to x/l=1, M/V Ocean Asia was pulled to M/V SITC Quingdao. This pulling force was maximum value at the position A on the Figure 5.

The transverse force and moment acting on the M/V Ocean Asia peaked almost at the same time as the bow of SITC Qingdao passing the bow of Ocean Asia (points A and B in Figure 5 and Figure 6). Under the influence of pulling force and moment, the M/V Ocean Asia changed her course toward the right bank of the canal.

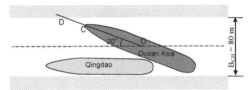

Figure 7 The distance from Ocean Asia to the shore

According to accident document, at the position M/V Ocean Asia run on ground, its heading was about from 40° to 45° compared to axial canal. Assume that when SITC Qingdao passed M/V Ocean Asia, the heading of Ocean Asia was deflected away from the direction of canal 20° (Fig. 7). The distance to shore from the bow of Ocean Asia is:

$$CD = OD - OC \quad (3)$$

where:

$$OD = \frac{B_{CH}/2}{\sin 20°} = \frac{40}{\sin 20°} = 117\,(m) \quad (4)$$

$$OC = \frac{1}{2} \times LOA_{Ocean\ Asia} = \frac{1}{2} \times 158 = 79\,(m) \quad (5)$$

The distance from Ocean Asia to the shore is:

$$CD = 117 - 79 = 38\,(m) \quad (6)$$

The M/V Ocean Asia is towed at speed of 3 knots. It takes 24.6 seconds to pass over the distance CD = 38 meters. This is a very short period of time, it not allows the M/V Ocean Asia and tugs to have any actions to against this effect.

Moreover, the tugboat M/V Da Tuong towed by the towing line 150 m in length, so it is not available to adjust the course of M/V Ocean Asia. Two other tugboats at the stern of the M/V Ocean Asia also

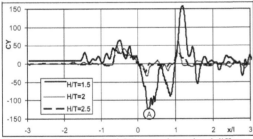

Figure 5. Hydrodynamic load CY at equal speed and different water depths

At the moment, when the stern of M/V Quingdao passing the bow of M/V Ocean Asia (x/l=1), the transverse force was change the direction quickly and the M/V Ocean Asia was pushed strongly to the shore.

The Figure 6 shows the moment effects on the hull of M/V Ocean Asia. Before x/l=0, the moment changed the direction frequently. It courses swaying to the M/V Ocean Asia. When the bow of SITC Qingdao was passing the bow of Ocean Asia, the moment rises quickly (point B in Fig. 6). It turns the bow of Ocean Asia strongly toward to the right canal.

could not reduce the changing course suddenly of M/V Ocean Asia because the M/V SITC Qingdao was passing too close to M/V Ocean Asia, there was no room for tugboat maneuvering. In additional, these two tugboats were at the stern of M/V Ocean Asia while the hydrodynamic force acted on the vessel's bow. That's why in this situation, the three tugboats could not against the changing course of M/V Ocean Asia which caused the accident.

## 3 EFFECT OF NARROW CANAL

In the above experiments, the overtaken vessel was near the shore. In the case of the accident, the M/V Ocean Asia was towed at the center line of the canal and the M/V SITC Qingdao ran very near the shore so the forces and moment effect on M/V Ocean Asia are not only as the analysis above, there are additional impacts.

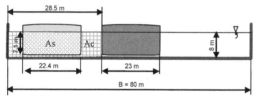

Figure 8 Cross section of the canal when the M/V SITC Qingdao was over taking M/V Ocean Asia

At the cross section describing on the Figure 8, the maximum allowed speed of the SITC Quindao is just 1.7 knots while it ran at the speed of 9 knots. At this speed, it creates much of chaos water flow and has lost control.

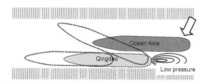

Figure 9 Effect due to narrow canal

When M/V SITC Qingdao passing M/V Ocean Asia, the flow of water between vessels was strong chaotic. A big volume of water at the bow of SITC Qingdao could not transfer to the stern because of narrow canal. It creates a low pressure area behind the M/V SITC Qingdao. To fill up the water to this low pressure area, large volume water on the starboard stern of M/V Ocean Asia had moved and it causes the heading of M/V Ocean Asia changes suddenly toward to the shore (Fig. 8).

## 4 CONCLUSIONS

Thus, when the M/V SITC Qingdao overtook M/V Ocean Asia, they simultaneously affected by:
- Forces and moment as analysis in the section 2, and
- Low pressure area behind M/V SITC Qingdao due to narrow canal as analysis in the section 3.

Both of effects have peak at the same time so they create powerful effect, making the M/V Ocean Asia sudden shift to the right bank of the canal.

According the above analysis, the cause of the accident between M/V SITC Qingdao and M/V Ocean Asia are as follows:
- M/V SITC Qingdao had violated regulations on safety, running at large speed and overtaking at the narrow area, where it is prohibited by the Port Authority.
- The pilots on the vessels did not have the necessary coordination of their work. The pilot on the M/V SITC Qingdao was not obeyed the rule and regulation of the canal which prohibits overtaking and limits the speed at this section of the canal. If M/V SITC Qingdao sailed at lower speed and M/V Ocean Asia kept out of the way of the M/V SITC Qingdao, they accident could not happen.
- M/V Ocean Asia did not be applied any safety measures. The anchor of the vessel did not use as it should be. M/V Ocean Asia is towed vessel but it not means that M/V Ocean Asia has no responsibilities in this case.
- M/V Ocean Asia was towed so the ability to maneuver is limited, especially the ability to use a rudder to keep track. This makes the risk of incidents such as when she is overtaken by others.

M/V Ocean Asia ran aground due to these above reasons. According to the analysis, it can be said that the accident of M/V Ocean Asia was caused by the overtaking of M/V SITC Qingdao.

## NOMENCLATURE

| | |
|---|---|
| B | Beam of ship(s) |
| $B_{CH}$ | Canal width |
| CN | $= 1000\ Mz\ /\ (0.5\ \rho\ \Box VM^2\ L_S^2\ T)$ |
| CX | $= 1000\ Fx\ /\ (0.5\ \rho\ VM^2\ L_S\ T)$ |
| CY | $= 1000\ Fy\ /\ (0.5\ \rho\ VM^2\ LS\ T)$ |
| FH | Froude number (based on water depth) |
| Fx | Longitudinal force |
| Fy | Lateral force |
| H | Water depth |
| hw | Wave elevation |
| $L_M$ | Length of moving ship |
| $L_S$ | Length of stationary ship |
| l | $= 0.5\ (L_M + L_S)$ |
| Mz | Yawing moment |
| T | Draft of ship(s) |
| $V_M$ | Speed of moving model |

x/l    Non-dimensional stagger between the middle sections of the ships

YW    Spacing between stationary ship side and canal wall

YY    Spacing between ship sides

$\rho$    Mass density of water

## REFERENCES

Brix J. Manoeuvring Technical Manual, Seehafen Verlag GmbH, Hamburg, 1993.

Cohen S., Beck R., Experimental and theoretical hydrodynamic forces on a mathematical model in confined waters, Journal of Ship Research, Vol. 27, No 2, 1983

Kyulevcheliev S., Georgiev S., Experimental observations of ship wavemaking at trans- and supecritical speeds, Euro-Conference HADMAR 2001, Varna, Bulgaria, 2001.

Stefan K., Svetlozar G. Hydrodynamic interaction between moving and stationary ship in a shallow canal. Bulgarian Ship Hydrodynamics Centre, Varna Department.

Vantorre M., Laforce E., Verzhbitskaya E., Model tests based formulations of ship-ship interaction forces for simulation purposes, IMSF 28th Annual General Meeting, Genova, 2001.

Varyani K., McGregor R., Wold P., Empirical formulae to predict peak of forces and moments during interactions, Hydronav'99, Gdansk-Ostrada, 1999.

Geodetic Problems in Navigational Applications

# 16. A Novel Approach to Loxodrome (Rhumb Line), Orthodrome (Great Circle) and Geodesic Line in ECDIS and Navigation in General

A. Weintrit & P. Kopacz
*Gdynia Maritime University, Gdynia, Poland*

ABSTRACT: We survey last reports and research results in the field of navigational calculations' methods applied in marine navigation that deserve to be collected together. Some of these results have often been rediscovered as lemmas to other results. We present our approach to the subject and place special emphasis on the geometrical base from a general point of view. The geometry of approximated structures implies the calculus essentially, in particular the mathematical formulae in the algorithms applied in the navigational electronic devices and systems. The question we ask affects the range and point in applying the loxodrome (rhumb line) in case the ECDIS equipped with the great circle (great ellipse) approximation algorithms of given accuracy replaces the traditional nautical charts based on Mercator projection. We also cover the subject on approximating models for navigational purposes. Moreover, the navigation based on geodesic lines and connected software of the ship's devices (electronic chart, positioning and steering systems) gives a strong argument to research and use geodesic-based methods for calculations instead of the loxodromic trajectories in general.

## 1 INTRODUCTION

A common problem is finding the shortest route across the Earth surface between two positions. Such trajectory is always a part of a geodesic (great circle, great ellipse) on the modelling globe surface. The geodesic is used by ship navigators attempting to minimize distances and the radio operators with directional antennae used to look for a bearing yielding the strongest signal. For many purposes, it is entirely adequate to model the Earth as a sphere. Actually, it is more nearly an oblate ellipsoid of revolution. The earth's flattening is quite small, about 1 part in 300, and navigation errors induced by assuming the Earth is spherical do not exceed this, and so for many purposes a spherical approximation may be entirely adequate. On a sphere, the commonly used coordinates are latitude and longitude, likewise on a spheroid, however on a spheroid one has to be more careful about what exactly one means by latitude [Williams, 1996]. The spherical model is often used in cartographic projections creating the frame of the presented chart. The trajectory of the geodesic lines and the loxodrome looks different depending on the method of the projections given by the strict formulae. Thus, many map projections are invaluable in specialized applications.

The only conformal cylindrical projection, Mercator's device was a boon to navigators from the 16th-century until the present, despite suffering from extreme distortion near the poles. We recall it has a remarkable property: any straight line between two points is a loxodrome or line of constant course on the sphere. The Mercator loxodrome bears the same angle from all meridians. Briefly, if one draws a straight line connecting a journey's starting and ending points on a Mercator map, that line's slope yields the journey direction, and keeping a constant bearing is enough to get to one's destination.

A Mercator projection is not the only one used by navigators, as the loxodrome does not usually coincides with the geodesic. This projection was possibly first used by Etzlaub (ca. 1511). However, it was for sure only widely known after Mercator's atlas of 1569. Mercator probably defined the graticule by geometric construction. E. Wright formally presented equations in 1599. Wright's work influenced, among other persons, Dutch astronomer and mathematician Willebrord Snellius, who introduced the word "loxodrome"; Adriaan Metius, the geometer and astronomer from Holland; and the English mathematician Richard Norwood, who calculated the length of a degree on a great circle of the Earth using a method proposed by Wright.

More commonly applied to large-scale maps, the transverse aspect preserves every property of Mercator's projection, but since meridians are not straight lines, it is better suited for topography than navigation. Equatorial, transverse and oblique maps offer the same distortion pattern. The transverse aspect with equations for the spherical case was presented by Lambert in his seminal paper (1772). The ellipsoidal case was developed, among others, by Carl Gauss (ca. 1822) and Louis Krüger (ca. 1912). It is frequently called the Gauss conformal or Gauss-Krüger projection.

The vessel or aircraft can reach its destination following the fixed bearing along the whole trip disregarding some obvious factors like for instance weather, fuel range, geographical obstructions. However, that easy route would not be the most economical choice in terms of distance. The two paths almost coincide only in brief routes. Although the rhumb line is much shorter on the Mercator map, an azimuthal equidistant map tells a different story, even though the geodesic does not map to a straight line since it does not intercept the projection centre. Since there is a trade-off: following the geodesic would imply constant changes of direction (those are changes from the current compass bearing and are only apparent: on the sphere, the trajectory is as straight as it can be). Following the rhumb line would waste time and fuel. So a navigator could follow a hybrid procedure [Snyder, 1987]:

- trace the geodesic on an azimuthal equidistant or gnomonic map,
- break the geodesic in segments,
- plot each segment onto a Mercator map,
- use a protractor and read the bearings for each segment,
- navigate each segment separately following its corresponding constant bearing.

## 2  GEODESIC APPROACH

For curved or more complicated modelling surfaces the metric can be used to compute the distance between two points by integration. The distance generally means the shortest distance between two points. Roughly speaking, the distance between two points is the length of the path connecting them. Most often the research and calculus in navigational literature are considered on the spherical or spheroidal models of Earth because of practical reasons. The flow of geodesics on the ellipsoid of revolution (spheroid) differs from the geodesics on the sphere. There are known different geodesics on the same surface with the same metric considered. However geodesic refers to the metric what is usually not taken into consideration in the navigational lectures. And there are different flows of geodesics on the same surface when different metrics are applied. That means we can obtain geometrically different results in navigational aspect if we change the researched modelling object with its geometrical and physical features (Kopacz, 2006).

Let us focus on two essential notions creating the base for the various fields of the mathematical research: the metric and topology. A metric space is a set with a global distance function (the metric) that, for every two points in, gives the distance between them as a nonnegative real number.

**Definition 1.** A function $g: X \times X \to [0, \infty)$ is called *a metric* (or distance) in $X$ if
(1) $g(x, y) = 0$ iff $x = y$ (positivity);
(2) $g(x, y) = g(y, x)$ for every $x, y \in X$ (symmetry);
(3) $g(x, y) \leq g(x, z) + g(z, y)$ for every $x, y, z \in X$ (triangle inequality).

Metric as a nonnegative function describes the "distance" between neighbouring points for a given set. When viewed as a tensor, the metric is called a metric tensor. We can define a metric in each non-empty set ($X \neq \varnothing$). The notion of metric has been introduced by M. Frechet in 1906. Formally the pair $(X,g)$ where $g$ is a metric in a set $X$ is called a *metric space*. Fig. 1 points out the essential role played by the metric in geodesic approach to the subject.

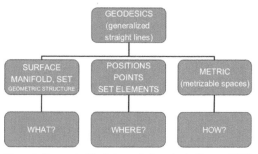

Figure 1. Geometrical basis in geodesic analysis [Kopacz, 2006]

Making one step further we can generalize the metric space to the topological space.

**Definition 2.** Let $X \neq \varnothing$ be a set and $P(X)$ the power set of $X$, i.e. $P(X) = \{U : U \subset X\}$. Let $\Omega \subset P(X)$ be a collection of its subsets such that:
(1) $(\forall \iota \in I \quad U_\iota \in \Omega) \Rightarrow \bigcup_{\iota \in I} U_\iota \in \Omega$ (the union of a collection of sets, which are elements of $\Omega$, belongs to $\Omega$);
(2) $U, V \in \Omega \Rightarrow U \cap V \in \Omega$ (the intersection of a finite collection of sets, which are elements of $\Omega$, belongs to $\Omega$);
(3) $\varnothing, X \in \Omega$ (the empty set $\varnothing$ and the whole set $X$ belong to $\Omega$).
Then
- $\Omega$ is called a topological structure or just a topology in X;
- the pair $(X, \Omega)$ is called a topological space;
- an element of X is called a point of this topological space;

– an element of $\Omega$ is called an open set of the topological space $(X, \Omega)$.

The conditions in the definition presented above are called the *axioms of topological structure*. A topology, that is a metric topology, means that one can define a suitable metric that induces it. Additionally we assume here that although the metric exists, it may be unknown. In a metric space $(X, g)$ the family of sets $\Omega$

$$\Omega = \{U \subset X : \underset{x \in U}{\forall} \ \underset{\varepsilon > 0}{\exists} \ B(x, \varepsilon) \subset U\}$$

satisfies the above mentioned axioms of topology. That means $(X, g)$ is a topological space and thus, each metric space is a topological space. There are sufficient criteria on the topology that assure the existence of such a metric even if this is not explicitly given. An example of an existence theorem of this kind is due to Urysohn who proved that a regular $T_1$-space whose topology has a countable basis is metrizable [Kelley, 1955]. Conversely, a metrizable space is always $T_1$ and regular but the condition on the basis has to be weakened since in general, it is only true that the topology has a basis which is formed by countably many locally finite families of open sets. Special metrizability criteria are known for Hausdorff spaces ($T_2$-spaces). A compact Hausdorff space is metrizable if and only if the set of all elements is a zero set [Willard, 1970]. The continuous image of a compact metric space in a Hausdorff space is metrizable. This implies in particular that a distance can be defined on every path in $T_2$-space.

ically in Fig. 2. Let us imagine that the vessels do not sail on spheroidal earth but locally torus - shaped planet. In this case the flow of geodesics and mentioned rhumb line or used charts are based on other mathematical expressions due to different geometrical object considered. The torus is topologically more simple than the sphere, yet geometrically it is a very complicated manifold indeed.

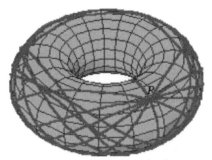

Figure 3. The geodesics on a torus $T^2 = S^1 \times S^1$ coming from an initial position $P$

The round torus metric is most easily constructed via its embedding in a Euclidean space of one higher dimension.

Taking into consideration the main theoretical aspects of the subject above mentioned as well as the practical ones influencing the base and components of the navigational algorithm to be applied we collect all of them together what has been shown in Figure 4.

Figure 4. Navigational calculations' algorithm guidelines

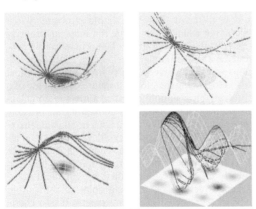

Figure 2. Flows of geodesics (distance functions) on locally modelling surfaces of differing curvatures

The mathematical formulae used in approximation of the navigational calculations are being studied and are based on spherical (spheroidal) model. However if we consider different shape of the surface the formulae change considerably. The examples of the flows of geodesics on locally modelling surfaces of differing curvature are presented graph-

The notion of geodesics makes sense not only for surfaces in space $R^3$ but also for abstract surfaces and more generally (Riemannian) manifolds. We also refer to [Funar, Gadgil, 2001] where the notion of a topological geodesic in a 3-manifold have been introduced. Geodesics in Riemannian manifolds with metrics of negative sectional curvature play an essential role in geometry. It is shown there that, in the case of 3-dimensional manifolds, many crucial properties of geodesics follow from a purely topological characterization in terms of *knotting* as well as proved basic existence and uniqueness results for topological geodesics under suitable hypotheses on the fundamental group. For further reading we send the reader to the wide literature on Riemannian and Finsler geometry and topology, in particular the geodesic research.

## 3 PLANE MODEL

The surface of revolution as the Earth's model - sphere $S^2$ or the spheroid is locally approximated by the Euclidean plane tangent in a given position. Generally, we approximate locally the curved surface by the Euclidean plane. For some applications such approximation is allowed and sufficient for practical need of research. That is satisfactory if we do not exceed the required accuracy of provided calculations. Hence the boundary conditions of applying the Euclidean plane or spherical geometry ought to be strictly defined. The mathematical components of the plane Euclidean geometry applied in navigational device are widely known and there is a common Euclidean metric used in the calculus as the distance function. We emphasize that the geodesics may look different even on the plane if different metrics are considered. For the practical reasons and the ease of use there is Euclidean plane tangent to the modelled surface used in many applications, for instance in dynamic positioning (DP) software. The plane model enables the satisfactory accuracy in a local approximation. In the local terrain geodesic research the area can be considered flat if it is inside the circle of a radius of ca. 15.5 km. This corresponds to the area of spherical circle which diameter equals ca. 17' of the great circle [Kopacz, 2010]. Practically such an approximation allows the direct geodesic measurements without considering the curvature of the modelled Earth surface and presenting the results on the plane in the appropriate scale. In the global modelling of the Earth's surface (geodesy, cartography, navigation, astronomy) the Euclidean geometry becomes not sufficient for the geometric description and the calculus coming from it. Thus, the limits of application of the approximation methods based on the flat Euclidean geometry must be clearly determined [Kopacz, 2010].

In a field of flat chart projections scale distortions on a chart can be shown by means of ratio of the scale at a given point to the true scale (a *scale factor* - *SF*). Scale distortions exist at locations where the scale factor differs from 1. For instance, a scale factor at a given point on the map is equal to 0.99960 signifies that 1000 m on the reference surface of the Earth will actually measure 999.6 m on the chart. This is a contraction of 40 cm per 1 km.

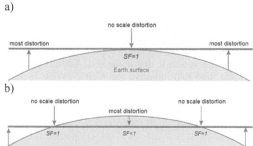

Figure 5. Scale distortions on a tangent (a) and a secant (b) map surface [Knippers, 2009]

Distortions increase as the distance from the central point (tangent plane) or closed line(s) of intersection increases. Scale distortions for tangent and secant map surfaces are illustrated in the Fig. 5. On a secant map projection - the application of a scale factor of less than 1.0000 to the central point or the central meridian has the effect of making the projection secant - the overall distortions are less than on one that uses a tangent map surface. Most countries have derived there map coordinate system from a projection with a secant map surface for this reason [Knippers, 2009].

The curved Earth is navigated using flat maps or charts, collected in an atlas. Similarly, in a calculus on manifolds a differentiable manifold can be described using mathematical maps, called coordinate charts, collected in a mathematical atlas. It is not generally possible to describe a manifold with just one chart, because the global structure of the manifold is different from the simple structure of the charts. For example, no single flat map can properly represent the entire Earth. When a manifold is constructed from multiple overlapping charts, the regions where they overlap carry information essential to understanding the global structure. In the case of a differentiable manifold, an atlas allows to do calculus on manifolds. The atlas containing all possible charts consistent with a given atlas is called the maximal atlas. Unlike an ordinary atlas, the maximal atlas of a given atlas is unique. Though it is useful for definitions, it is a very abstract object and not used directly for example in calculations. Charts in an atlas may overlap and a single point of a manifold may be represented in several charts. If two charts overlap, parts of them represent the same region of the manifold. Given two overlapping charts, a transition function can be defined which goes from an open ball in $R^n$ to the manifold and then back to another (or perhaps the same) open ball in $R^n$. The resultant map is called a change of coordinates, a coordinate transformation, a transition function or a transition map.

# 4 SPHERICAL AND SPHEROIDAL MODEL

As the Earth's global model an oblate spheroid is used providing the navigational calculations i.e. distances and angles or the sphere for the ease of use. A sphere, spheroid or a torus surface are examples of 2-dimensional manifolds. Manifolds are important objects in mathematics and physics because they allow more complicated structures to be expressed and understood in terms of the relatively well understood properties of simpler spaces. The study of manifolds combines many important areas of mathematics: it generalizes concepts such as curves and surfaces as well as ideas from linear algebra and topology. Certain special classes of manifolds also have additional algebraic structure. They may behave like groups, for instance. To measure distances and angles on manifolds, the manifold must be Riemannian. We recall that a Riemannian manifold is an analytic manifold in which each tangent space is equipped with *an inner product* in a manner which varies smoothly from point to point. This allows one to define various notions such as length, angles, areas (or volumes), curvature, gradients of functions and divergence of vector fields. More general geometric structure a Finsler manifold allows the definition of distance, but not of angle. It is an analytic manifold in which each tangent space is equipped with *a norm*, in a manner which varies smoothly from point to point. This norm can be extended to a metric, defining the length of a curve; but it cannot in general be used to define an inner product. Any Riemannian manifold is a Finsler manifold. Manifold theory has come to focus exclusively on these intrinsic properties (or invariants), while largely ignoring the extrinsic properties of the ambient space.

Triaxial ellipsoid as the 2-dimensional submanifold $M$ in $R^3$ is defined by the equation $\Phi = 0$ where

$$\Phi(x,y,z) = \frac{x^2}{a^2} + \frac{y^2}{b^2} + \frac{z^2}{c^2} - 1.$$

Let $N$ be the non-vanishing normal vector field on $M$. Then

$$N(x,y,z) = 0,5 \, grad\Phi = \frac{x}{a^2}e_1 + \frac{y}{b^2}e_2 + \frac{z}{c^2}e_3$$

where the $\{e_1, e_2, e_3\}$ is the canonical basis of the vector space $R^3$. The Gaussian curvature $K$ of the modelling triaxial ellipsoid equals $K = \dfrac{1}{a^2 b^2 c^2 \|N\|^4}$.

The Gaussian curvature is the determinant of the shape operator. For the sphere $a=b=c=r$ and then $\|N\| = \dfrac{1}{r}, K = \dfrac{1}{r^2}$ where $r$ denotes a radius of the modelling sphere. Thus, we conclude here that the curvature affects the geometry of the locally approximating surfaces essentially and in particular their geodesic trajectories.

2-dimensional sphere $S^2$ is widely considered to model globally the surface of the Earth. As a calculating tool the spherical trigonometry is used which states the base for the comparison analysis and algorithms implemented in the software of navigational aids e.g. receivers of the positioning systems, ECDIS. The surface of the Earth may be taken mathematically as a sphere instead of ellipsoid for maps at smaller scales. In practice, maps at scale 1:5000 000 or smaller can use the mathematically simpler sphere without the risk of large distortions. At larger scales, the more complicated mathematics of ellipsoids are needed to prevent these distortions in the map. A sphere can be derived from the certain ellipsoid corresponding either to the semi-major or semi-minor axis, or average of both axes or can have equal volume or equal surface than the ellipsoid [Knippers, 2009].

We recall the great circle is the equivalent of the Euclidean straight line, it has the finite distance and it is closed. The geodesics starting from a given position on three main modelling surfaces (2-dimensional modelling manifolds of positive curvature), i.e. sphere, spheroid and triaxial ellipsoid are presented in Fig. 6. Obviously the disadvantage of orthodromic sailing is bound with continuous course alteration. That is why the loxodromic line is mainly sailed only or mainly used in the approximation of the great circle sailing. Thus the combination of these two lines create the base for planned and monitored trajectories while at sea.

The general question we ask affects the range and point of usage of the rhumb line in case the ECDIS systems equipped with the great circle / great ellipse approximation algorithms of given accuracy replaces the traditional paper charts based on Mercator projection. Moreover, the navigation based on geodesic lines and connected software of the ship's device (electronic chart, positioning and steering systems) gives a strong argument to use this method for calculations instead of the loxodromic one in general. Although the basic solutions for navigational purposes have already been known and widely used there are still the new approaches and efforts made to the subject. The examples of the spherical and spheroidal approach have been found recently in the literature reviewed further in the article. The main efforts affect the optimization and approximation methods which potentially may give the practical benefits for the navigators.

a)

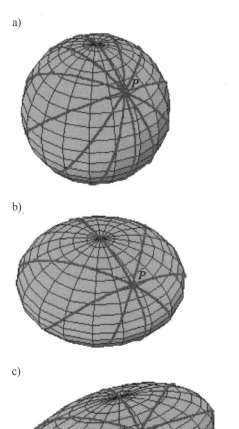

b)

c)

Figure 6. Geodesics coming from an initial position *P* on 2-dimensional modelling manifolds of positive curvature a)sphere, b) spheroid (ellipsoid of revolution), c) triaxial ellipsoid.

As we mentioned above the shortest path between two points on a smooth surface is called a geodesic curve on the surface. On a flat surface the geodesics are the straight lines, on a sphere they are the great circles. Remarkably the path taken by a particle sliding without friction on a surface will always be a geodesic. This is because a defining characteristic of a geodesic is that at each point on its path, the local centre of curvature always lies in the direction of the surface normal, i.e. in the direction of any constrained force required to keep the particle on the surface. There are thus no forces in the local tangent plane of the surface to deflect the particle from its geodesic path. There is a general procedure, using the calculus of variations, to find the equation for geodesics given the metric of the surface [Williams,

1996]. Obviously the Earth is not an exact ellipsoid and deviations from this shape are continually evaluated. The *geoid* is the name given to the shape that the Earth would assume if it were all measured at mean sea level. This is an undulating surface that varies not more than about a hundred meters above or below a well-fitting ellipsoid, a variation far less than the ellipsoid varies from the sphere. The choice of the reference ellipsoid used for various regions of the Earth has been influenced by the local geoid, but large-scale map projections are designed to fit the reference ellipsoid, not the geoid. The selection of constants defining the shape of the reference ellipsoid has been a major concern of geodesists since the early 18th century. Two geometric constants are sufficient to define the ellipsoid itself e.g. the semimajor axis and the eccentricity. In addition, recent satellite-measured reference ellipsoids are defined by the semimajor axis, geocentric gravitational constant and dynamical form factor which may be converted to flattening with formulas from physics.

Between 1799 and 1951 there were 26 determinations of dimensions of the Earth. There are over a dozen other principal ellipsoids, however, which are still used by one or more countries. The different dimensions do not only result from varying accuracy in the geodetic measurements (the measurements of locations on the Earth), but the curvature of the Earth's surface (geoid) is not uniform due to irregularities in the gravity field. Until recently, ellipsoids were only fitted to the Earth's shape over a particular country or continent. The polar axis of the reference ellipsoid for such a region, therefore, normally does not coincide with the axis of the actual Earth, although it is assumed to be parallel. The same applies to the two equatorial planes. The discrepancy between centres is usually a few hundred meters at most. Satellite-determined coordinate systems are considered geocentric. Ellipsoids for the latter systems represent the entire Earth more accurately than ellipsoids determined from ground measurements, but they do not generally give the best fit for a particular region. The reference ellipsoids used prior to those determined by satellite are related to an initial point of reference on the surface to produce a datum, the name given to a smooth mathematical surface that closely fits the mean sea-level surface throughout the area of interest. The initial point is assigned a latitude, longitude, elevation above the ellipsoid, and azimuth to some point. Satellite data have provided geodesists with new measurements to define the best Earth-fitting ellipsoid and for relating existing coordinate systems to the Earth's centre of mass. For the mapping of other planets and natural satellites, Mars is treated as an ellipsoid. Other bodies are taken as spheres, although some irregular satellites have been treated as triaxial ellipsoids and are mapped orthographically [Snyder, 1987].

## 5 ECDIS APPROACH

In the course of navigation programmes for ECDIS purposes it became apparent that the standard text books of navigation were perpetuating a flawed method of calculating rhumb lines on the Earth considered as an oblate spheroid. On further investigation it became apparent that these incorrect methods were being used in programming a number of calculator/computers and satellite navigation receivers. Although the discrepancies were not large, it was disquieting to compare the results of the same rhumb line calculations from a number of such devices and find variations of a few miles when the output was given, and therefore purported to be accurate, to a tenth of a mile in distance and/or a tenth of a minute of arc in position. The problem was highlighted in the past and the references at the end of this paper show that a number of methods have been proposed for the amelioration of this problem.

This paper presents and recommends the guidelines that should be used for the accurate solutions. Most of these may be found in standard geodetic text books, such as, but also provided are new formulae and schemes of solution which are suitable for use with computers or tables. The paper also takes into account situations when a near-indeterminate solution may arise. The data for these problems do not refer to actual terrestrial situations but have been selected for illustrative purposes. There is a need for creating the algorithm for precise navigational calculations implemented in the software of practical navigational devices. The references cited at the end of this paper presenting different detailed approaches to the subject also confirm such a need and a unification of the method.

The paper presents the review of different approaches to contact formulae for the computation of the position, the distance, and the azimuth along a great ellipse. The proposed alternative formulae are to be primarily used for accurate sailing calculations on the ellipsoid in a GIS environment as in ECDIS and other ECS. Among the ECDIS requirements is the need for a continuous system with a level of accuracy consistent with the requirements of safe navigation. At present, this requirement is best fulfilled by the Global Positioning System (GPS). The GPS system is referenced to World Geodetic System 1984 Datum (WGS 84). Using the ellipsoid model instead of the spherical model attains more accurate calculation of sailing on the Earth. Therefore, we aim to construct a computational procedure for solving the length of the arc of a geodesic, the waypoints and azimuths along a geodesic. We aspire to provide the straightforward formulae involving the great elliptic sailing based on two scenarios. The first is that the departure point and the destination point are known. The second is that the departure point and the initial azimuth are given (direct and inverse geodetic problems on reference ellipsoids or locally modelling geometric structure).

### 5.1 ECDIS Calculations

As a minimum, an ECDIS system must be able to perform the following calculations and conversions [Weintrit, 2009]:

- geographical coordinates to display coordinates, and display coordinates to geographical coordinates;
- transformation from local datum to WGS-84;
- true distance and azimuth between two geographical positions;
- geographic position from a known position given distance and azimuth (course);
- projection calculations such as great circle and rhumb line courses and distances;
- "RL-GC" difference between the rhumb line and great circle in sailing along the great circle (or great ellipse?).

### 5.2 Route planning calculations

The ECDIS allows the navigator to create waypoints and routes including setting limits of approach and other cautionary limits. Both rhumb line and great circle routes can be defined. Routes can be freely exchanged between the ECDIS and GPS or ARPA. Route checking facility allows the intended route to be automatically checked for safety against limits of depth and distance as defined by the navigator.

The mariner can calculate and display both a rhumb line and a great circle line and verify that no visible distortion exists between these lines and the chart data. Authors predict the early end of the era of the rhumb line. This line in the natural way will go out of use. Nobody after all will be putting the navigational triangle to the screen of the ECDIS. Our planned route need not be a straight line on the screen. So, why hold this line still in the use? Each ship's position plotted on the chart can be the starting point of new updated great circle, or saying more closely, great ellipse.

### 5.3 Most important questions

It is an important question whether in the ECDIS time Mercator projection is still essential for marine navigation. We really need it? And what about loxodrome? So let start navigation based on geodesics. It is high time to forget the rhumb line navigation and great circle navigation, too. But the first we need clear established methods, algorithms and formulas for sailing calculations. But it is already indicating the real revolution in navigation. We will be forced to make the revision of such fundamental notions as the course, the heading and the bearing.

And another very important question: do you really know what kind of algorithms and formulae are used in your GPS receiver and your ECS/ECDIS systems for calculations mentioned in chapter 5.1? We are almost sure your answer is negative. So we generally have a problem – a serious problem.

## 6  REVIEW OF RECENTLY PUBLISHED PAPERS

Since 1950 till 2010 the following professional magazines and journals published some papers about navigation on the great ellipse and on the spheroidal Earth: *The Journal of Navigation* [Bennett, 1996; Bourbon, 1990; Carlton-Wippern, 1992; Chen, Hsu, & Chang, 2004; Earle, 2000, 2005, 2006, 2008; Hickley, 1987; Hiraiwa, 1987; Nastro & Tancredi, 2010; Pallikaris & Latsas, 2009; Prince & Williams, 1995; Sadler, 1956; Tseng & Lee, 2007; Tyrrell, 1955; Walwyn, 1999; Williams, 1950; Williams, 1996; Williams & Phythian, 1989, 1992; Zukas, 1994], *International Hydrographic Review* [Pallikaris, Tsoulos & Paradissis, 2009a], *Coordinates* [Pallikaris, Tsoulos & Paradissis, 2010], *Navigation - The Journal of The Institute of Navigation* [Kaplan, 1995; Miller, Moskowitz & Simmen, 1991], *Bulletin Geodesique* [Bowring, 1983, 1984; Rainsford, 1953, 1955; Sodano, 1965], *Journal of Marine Science and Technology* [Tseng & Lee, 2010], *The Journal of the Washingtin Academy of Sciences* [Lambert, 1942], *The Canadian Surveyor* [Bowring, 1985], *Survey Review* [Vincenty, 1975, 1976], *Surveying and Mapping* [Meade, 1981], *The Professional Geographer* [Tobler, 1964], *College Mathematics Journal* [Nord, Muller, 1996; Schechter, 2007].

The following particular problems were discussed among the others:
- practical rhumb line calculations on the spheroid [Bennet, 1996],
- geodesic inverse problem [Bowring, 1983],
- direct and inverse solutions for the great elliptic and line on the reference ellipsoid [Bowring, 1984],
- loxodromic navigation [Carlton-Wippern, 1992],
- formulae for the solution of direct and inverse problems on reference ellipsoids using pocket calculators [Meade, 1981],
- geometry of loxodrome on the ellipsoid [Bowring, 1985],
- geometry of geodesics [Busemann, 1955],
- geodesic line on the surface of a spheroid [Bourbon, 1990],
- great circle equation [Chen, Hsu & Chang, 2004],
- novel approach to great circle sailing [Chen, Hsu & Chang, 2004],
- vector function of traveling distance for great circle navigation [Tseng & Lee, 2007],
- great circle navigation with vectorial methods [Nastro & Tancredi, 2010],
- vector solution for great circle navigation [Earle, 2005],
- vector solution for navigation on a great ellipse [Earle, 2000],
- navigation on a great ellipse [Tseng & Lee, 2010],
- great ellipse solution for distances and headings to steer between waypoints [Walwyn, 1999],
- great ellipse on the surface of the spheroid [Williams, 1996],
- vector solutions for azimuth [Earle, 2008],
- sphere to spheroid comparisons [Earle, 2006],
- great circle versus rhumb line cross-track distance at mid-longitude [Hickley, 1987],
- modification of sailing calculations [Hiraiwa, 1987],
- practical sailing formulae for rhumb line tracks on an oblate Earth [Kaplan, 1995],
- distance between two widely separated points on the surface of the Earth [Lambert, 1942],
- traveling on the curve Earth [Miller, Moskowitz & Simmen, 1991],
- new meridian arc formulas for sailing calculations in GIS [Pallikaris, Tsoulos & Paradissis, 2009a],
- new calculations algorithms for GIS navigational systmes and receivers [Pallikaris, Tsoulos & Paradissis, 2009b],
- improved algorithms for sailing calculations [Pallikaris, Tsoulos & Paradissis, 2010],
- new algorithm for great elliptic sailing (GES) [Pallikaris & Latsas, 2009],
- shortest paths [Lyusternik, 1964],
- sailing in ever-decreasing circles [Prince & Williams, 1995],
- long geodesics on the ellipsoid [Rainsford, 1953, 1955],
- spheroidal sailing and the middle latitude [Sadler, 1956],
- general non-iterative solution of the inverse and direct geodetic problems [Sodano, 1965],
- comparison of spherical and ellipsoidal measures [Tobler, 1964],
- navigating on the spheroid [Tyrrell, 1955; Williams, 2002],
- direct and inverse solutions of geodesics on the ellipsoid with application of nested equations [Vincenty, 1975, 1976],
- loxodromic distances on the terrestrial spheroid [Williams, 1950],
- Mercator's rhumb lines: a multivariable application of arc length [Nord, Muller, 1996],
- navigating along geodesic paths on the surface of a spheroid [Williams & Phythian, 1989],
- shortest distance between two nearly antipodean points on the surface of a spheroid [Williams & Phythian, 1992],

- shortest spheroidal distance [Zukas, 1994],
- navigating on a spheroid [Schechter, 2007].

# 7 CONCLUSIONS

This article is written with a variety of readers in mind, ranging from practising navigators to theoretical analysts. It also was our goal to present a current and uniform approaches to sailing calculations highlighting recent developments. Much insight may be gained by considering the examples that have recently proliferated in the literature reviewed above. We present our approach to the subject and place special emphasis on the geometrical base from a general point of view. Of particular interest are geodesic lines, in particular great ellipse calculations. The geometry of modelling structures implies the calculus essentially, in particular the mathematical formulae in the algorithms applied in the navigational electronic device and systems. Thus, is the spherical or spheroidal model the best fit in the local approximations of the Earth surface? We show that generally in navigation the essential calculating procedure refers to the distance and angle measurement what may be transferred to more general geometrical structures, for instance metric spaces, Riemannian manifolds. The authors point out that the locally modelling structure has a different "shape" and thus the different curvature and the flow of geodesics. That affects the calculus provided on it. The algorithm applied for navigational purposes, in particular ECDIS should inform the user on actually used mathematical model and its limitations. The question we also ask affects the range and point in applying the loxodrome sailing in case the ECDIS equipped with the great circle (great ellipse) approximation algorithms of given accuracy replaces the traditional nautical charts based on Mercator projection. The shortest distance (geodesics) depends on the type of metric we use on the considered surface in general navigation. The geodesics can look different even on the same plane if different metrics are taken into consideration. For instance the diameter of the parallel of latitude conical circle does not pass its centre. That differs from both the plane and spherical case. Our intuition insists on the way of thinking to look at the diameter as a part of geodesic of the researched surface crossing the centre of a circle. However the diameter depends on applied metric, thus the shape of the circles researched in the metric spaces depends on the position of the centre and the radius. It is also important to know how the distance between two points on considered structure is determined, where the centre of the circle is positioned and how the diameter passes. Changing the metric causes the differences in the obtained distances. For example $\pi$ as a number is constant and has the same value in each geometry it is used in calculations. However $\pi$ as a

ratio of the circumference to its diameter can achieve different values in general, in particular $\pi$ [Kopacz, 2010]. The navigation based on geodesic lines and connected software of the ship's devices (electronic chart, positioning and steering systems) gives a strong argument to research and use geodesic-based methods for calculations instead of the loxodromic trajectories in general. The theory is developing as well what may be found in the books on geometry and topology. This motivates us to discuss the subject and research the components of the algorithm of calculations for navigational purposes.

# REFERENCES

Admiralty Manual of Navigation, 1987, Volume 1, General navigation, Coastal Navigation and Pilotage, Ministry of Defence (Navy), London, The Stationery Office.

Bennett, G.G. 1996. *Practical rhumb line calculations on the* spheroid, The Journal of Navigation, Vol. 49, No. 1.

Bowditch, N. 2002. *American Practical Navigator*, Pub. No. 9, The Bicentennial Edition, National Imagery and Mapping Agency.

Bowring, B.R. 1983. *The Geodesic Inverse Problem*, Bulletin Geodesique, Vol. 57, p. 109 (Correction, Vol. 58, p. 543).

Bowring, B.R. 1984. *The direct and inverse solutions for the great elliptic and line on the reference ellipsoid*, Bulletin Geodesique, 58, p. 101-108.

Bowring, B.. 1985. *The geometry of Loxodrome on the Ellipsoid*, The Canadian Surveyor, Vol. 39, No. 3.

Busemann, H. 1955. *The geometry of geodesics*, Elsevier Academic Press, Washington.

Bourbon, R. 1990. *Geodesic line on the surface of a spheroid*, The Journal of Navigation, Vol. 43, No. 1.

Carlton-Wippern, K. 1992. *On Loxodromic Navigation*, The Journal of Navigation, Vol. 45, No. 2., p.292-297.

Chen, C.L., Hsu, T.P., Chang, J.R. 2004. *A Novel Approach to Great Circle sailing: The Great Circle Equation*, The Journal of Navigation, Vol. 57, No. 2, p. 311-325.

Cipra, B. 1993. *What's Happening in the Mathematical Sciences*, Vol. I. RI: American Mathematical Society, Providence.

Earle, M.A. 2000. *Vector Solution for Navigation on a Great Ellipse*, The Journal of Navigation, Vol. 53, No. 3., p. 473-481.

Earle, M.A. 2005. *Vector Solution for Great Circle Navigation*, The Journal of Navigation, Vol. 58, No. 3, p. 451-457.

Earle, M.A. 2006. *Sphere to spheroid comparisons*, The Journal of Navigation, Vol. 59, No. 3, p. 491-496.

Earle, M.A. 2008. *Vector Solutions for Azimuth*, The Journal of Navigation, Vol. 61, p. 537-545.

Funar, L., Gadgil, S. 2001. *Topological geodesics and virtual rigidity*, Algebraic & Geometric Topology, Vol. 1, pp. 369-380.

Goldberg, D. 1991. *What Every Computer Scientist Should Know About Floating-Point Arithmetic*, ACM Computing Surveys, March.

Gradshteyn, I.S., & Ryzhik, I.M. 2000. *Tables of Integrals, Series and Products*, 6th ed. CA Academic Press, San Diego.

Hickley, P. 1987. *Great Circle Versus Rhumb Line Cross-track Distance at Mid-Longitude*. The Journal of Navigation, Vol. 57, No. 2, p. 320-325 (Forum), May.

Hiraiwa, T. 1987. *Proposal on the modification of sailing calculations*. The Journal of Navigation, Vol. 40, 138 (Forum).

Hohenkerk, C., & Yallop B.D. 2004. *NavPac and Compact Data 2006 – 2010 Astro-Navigation Methods and Software for the PC*. TSO, London.

Kaplan, G.H. 1995. *Practical Sailing Formulas For Rhumb-Line Tracks on an Oblate Earth*, Navigation, Vol. 42, No. 2, pp. 313-326, Summer 1995;

Kelley, J. L. 1955. *General Topology*. New York: Van Nostrand.

Knippers, R. 2009. *Geometric aspects of mapping*, International Institute for Geo-Information Science and Earth Observation, Enschede, http://kartoweb.itc.nl/geometrics.

Kopacz, P. 2006. **On notion of foundations of navigation in maritime education**, 7th Annual General Assembly and Conference AGA-7, The International Association of Maritime Universities IAMU, Dalian, China.

Kopacz, P. 2007. **On Zermelo navigation in topological structures**, 9th International Workshop for Young Mathematicians "Topology", pp. 87–95, Institute of Mathematics, Jagiellonian University, Cracow.

Kopacz, P. 2010. **Czy π jest constans?**, „Delta" - Matematyka - Fizyka – Astronomia – Informatyka, No 12/2010, Instytut Matematyki, Uniwersytet Warszawski, Warszawa (in Polish).

Lambert, W.D. 1942. *The Distance Between Two Widely Separated Points on the Surface of the Earth*, The Journal of the Washingtin Academy of Sciences, Vol. 32, No. 5, p. 125, May.

Lyusternik, L.A. 1964. *Shortest Paths*, Pergamon Press.

Meade, B.K. 1981. *Comments on Formulas for the Solution of Direct and Inverse Problems on Reference Ellipsoids using Pocket Calculators*, Surveying and Mapping, Vol. 41, March.

Miller, A.R., Moskowitz, I.S., Simmen J. 1991. *Traveling on the Curve Earth*, The Journal of The Institute of Navigation, Vol. 38.

Nastro, V., Tancredi U. 2010. *Great Circle Navigation with Vectorial Methods*, Journal of Navigation, Vol.63, Issue 3.

Nord, J., Muller, E. 1996. *Mercator's rhumb lines: A multivariable application of arc length*, College Mathematics Journal, No. 27, p. 384–387.

Pallikaris, A., Tsoulos, L., Paradissis, D. 2009a. *New meridian arc formulas for sailing calculations in GIS*, International Hydrographic Review.

Pallikaris, A., Tsoulos, L., Paradissis, D. 2009b. *New calculations algorithms for GIS navigational systmes and receivers*, Proceeidngs of the European Navigational Conference ENC - GNSS, Naples, Italy.

Pallikaris, A., Tsoulos L., Paradissis D. 2010. *Improved algorithms for sailing calculations*. Coordinates, Vol. VI, Issue 5, May, p. 15-18.

Pallikaris, A., Latsas, G. 2009. New algorithm for great elliptic sailing (GES), The Journal of Navigation, vol. 62, p. 493-507.

Prince, R., Williams, R. 1995. *Sailing in ever-decreasing circles*, The Journal of Navigation, Vol. 48, No. 2, p. 307-313.

Rainsford, H.F. 1953. *Long Geodesics on the Ellipsoid*, Bulletin Geodesique, Vol. 30, p. 369-392.

Rainsford, H.F. 1955. *Long Geodesics on the Ellipsoid*, Bulletin Geodesique, Vol. 37.

Sadler, D.H. 1956. *Spheroidal Sailing and the Middle Latitude*, The Journal of Navigation, Vol. 9, issue 4, p. 371-377.

Schechter, M. 2007. *Which Way Is Jerusalem? Navigating on a Spheroid*. The College Mathematics Journal, Vol. 38, No. 2, March.

Silverman, R. 2002. *Modern Calculus and Analytic Geometry*. Courier Dover Publications, Dover.

Sinnott, R. 1984. *Virtues of the Haversine*. Sky and Telescope. Vol. 68, No. 2.

Snyder, J.P. 1987. *Map Projections: A Working Manual*. U.S. Geological Survey Professional Paper 1395.

Sodano, E.M. 1965. *General Non-iterative Solution of the Inverse and Direct Geodetic Problems*, Bulletin Geodesique, Vol. 75.

Tobler, W. 1964. *A comparison of spherical and ellipsoidal measures*, The Professional Geographer, Vol. XVI, No 4, p. 9-12.

Torge, W. 2001. *Geodesy*. 3rd edition. Walter de Gruyter, Berlin, New York.

Tseng, W.K., Lee, H.S. 2007. *The vector function of traveling distance for great circle navigation*, The Journal of Navigation, Vol. 60, p. 158-164.

Tseng, W.K., Lee, H.S. 2010. *Navigation on a Great Ellipse*, Journal of Marine Science and Technology, Vol. 18, No. 3, p. 369-375.

Tyrrell, A.J.R. 1955. *Navigating on the Spheroid*, The Journal of Navigation, vol. 8, 366 (Forum).

Vincenty, T. 1975. *Direct and Inverse Solutions of Geodesics on the Ellipsoid with Application of Nested Equations*. Survey Review, Vol. XXII, No. 176, p. 88-93.

Vincenty, T. 1976. *Direct and Inverse Solutions of Geodesics on the Ellipsoid with Application of Nested Equations – additional formulas*. Survey Review Vol. XXIII, No. 180, p. 294.

Walwyn, P.R. 1999. *The Great ellipse solution for distances and headings to steer between waypoints*, The Journal of Navigation, Vol. 52, p. 421-424.

Weintrit, A. 2009. *The Electronic Chart Display and Information System (ECDIS). An Operational Handbook*. A Balke-ma Book. CRC Press, Taylor & Francis Group, Boca Raton – London - New York - Leiden, pp. 1101.

Weintrit, A., Kopacz, P. 2004. *Safety Contours on Electronic Navigational Charts*. 5th International Symposium 'Information on Ships' ISIS 2004, organized by German Institute of Navigation and German Society for Maritime Technology, Hamburg.

Weisstein, E. *Great Circle*, MathWorld - A Wolfram Web Resource, http://mathworld.wolfram.com/GreatCircle.html;

Willard, S. 1970. *General Topology*, Addison-Wesley. Reprinted by Dover Publications, New York, 2004 (Dover edition).

Williams, J.E.D. 1950. *Loxodromic Distances on the Terrestrial Spheroid*, The Journal of Navigation, Vol. 3, No. 2, p. 133-140.

Williams, E. 2002. *Navigation on the spheroidal Earth*, http://williams.best.vwh.net/ellipsoid/ellipsoid.html.

Williams, R. 1996. *The Great Ellipse on the surface of the spheroid*, The Journal of Navigation, Vol. 49, No. 2, p. 229-234.

Williams, R. 1998. *Geometry of Navigation*, Horwood series in mathematics and applications, Horwood Publishing, Chichester,

Williams, R., Phythian, J.E. 1989. *Navigating Along Geodesic Paths on the Surface of a Spheroid*, The Journal of Navigation, Vol. 42, p. 129-136.

Williams, R., Phythian, J.E. 1992. *The shortest distance between two nearly antipodean points on the surface of a spheroid*, The Journal of Navigation, Vol. 45, p. 114.

Wippern, K.C.C. 1988. *Surface Navigation and Geodesy, A Parametric Approach*, ASFSPA-CECOM Technical Note, March, with addendums;

Wylie, C.R., Barrett, L.C. 1982. *Advanced Engineering Mathematics*,McGraw-Hill, p. 834.

Zukas, T. 1994. *Shortest Spheroidal Distance*, The Journal of Navigation, Vol. 47, p. 107-108.

# 17. Approximation Models of Orthodromic Navigation

S. Kos & D. Brčić

*University of Rijeka – Faculty of Maritime Studies, Croatia*

ABSTRACT: The paper deals with two different approaches to orthodromic navigation approximation, the secant method and the tangent method. Two ways of determination of orthodromic interposition coordinates will be presented with the secant method. In the second, tangent method unit change of orthodromic course ($\Delta K$) will be used.

## 1 INTRODUCTION

Navigation on the surface of the Earth is possible in two ways: by orthodrome and loxodrome. Orthodrome is a minor arc of the great circle bounded by two positions, and corresponds to their distance on a surface of the Earth, representing also the shortest distance between these positions on the Earth as a sphere. The ship, travelling in orthodromic oceanic navigation, has her bow directed towards the port of arrival all the time. The orthodorme is the curve of a variable course – it intersects meridians at different angles. When navigating by the orthodrome, course should be constantly changed, which is unacceptable from the navigational point of view. On the other hand, loxodrome (rhumb line) intersects all meridians at the same angle, and it is more suitable in maintaining the course. However, loxodromic path is longer that the orthodromic one. Sailing by loxodrome, the bow of the ship will be directed toward the final destination just before arrival. Due to the mentioned facts, it is necessary to use the advantages of both curves – the shorter path of the orthodrome and the rhumb line conformity.

Orthodrome navigation is, as mentioned, inconvenient. Therefore, only approximation of orthodrome navigation can be taken into account, reducing the number of course changes to an acceptable number – always bearing in mind that if the number of course alteration is greater, the navigation is closer to the great circle. After defining elements for course and distance determination on an orthodrome curve, navigation between the derived points is carried out in loxodromic courses.

Applying spherical trigonometry, the proposed paper elaborates models of approximation for the orthodrome navigation with the secant method and the tangent method. The secant method provides two models of navigation. In the first model, the orthodrome is divided into desired waypoints – interpositions between which the ship sails in loxodromic courses. The second model of the method implies the path between two positions divided into specific intervals of unit distances, which then define other elements of navigation (interposition coordinates and loxodromic courses). In these two models, navigation has been approximated with the secants of the orthodrome curve on which the vessel sails. The tangent method gives an approximation model by determining the unit changes of orthodromic courses, and defining the tangent on a curve, after which other navigational elements needed for navigation are performed.

## 2 IMPORTANT RELATIONS BETWEEN ORTHODROMIC AND LOXODROMIC DISTANCES FOR THE EARTH AS A SPHERE

As described above, the rhumb line, i.e. loxodrome, represents a constant course, spiral-shaped curve, asymptotically approaching the Pole. The orthodrome represents a variable course curve, the minor arc of the great circle between two positions. For the Earth as a sphere, between positions $P_1$ and $P_2$, these two distances are equal in two situations only (Figure 1):

1 if the positions are placed on the same meridian, then $\Delta\lambda=0$, $\Delta\varphi\neq0$,
2 if the positions are placed on the Terrestrial Equator, then $\Delta\varphi=0$, $\Delta\lambda\neq0$

In all other situations, the orthodrome distance is always smaller, or $D_O \neq D_L$.

Maximum difference between distances $D_O$ and $D_L$ occurs when $\varphi_1 = \varphi_2 \neq 0$, $\Delta\varphi = 0$, $\Delta\lambda = 180°$ between $P_1$ and $P_2$ is applied. In this case, the function extremum should be determined [Wippern, 1992]:

$$f_{(\varphi)} = \Delta\lambda\cos\varphi - 2\left(90° - \varphi\right)$$

$$f_{(\varphi)} = \Delta\lambda\cos\varphi - \pi + 2\varphi$$

$$f'_{(\varphi)} = -\Delta\lambda\sin\varphi + 2$$

$$f''_{(\varphi)} = -\Delta\lambda\cos\varphi < 0$$

For $\varphi$ from $0°$ to $\pm 90°$ the $\cos\varphi$ function is positive, so the second derivation $f''_{(\varphi)} < 0$. Therefore, the function has an extremum, maximum:

$$f'_{(\varphi)} = 0 - \Delta\lambda\sin\varphi + 2 = 0$$

$$\varphi = \sin^{-1}\left(\frac{2}{\Delta\lambda}\right)$$

$$f_{(\varphi)}'' = -\Delta\lambda\cos\varphi = -\Delta\lambda\sqrt{1 - \sin^2\varphi}$$

$$f'_{(\varphi)} = -\Delta\lambda\sqrt{1 - \frac{4}{\Delta\lambda^2}} > 0$$

If $\Delta\lambda = 180° = \pi$, then $\varphi = 39° 32' 24,8''$

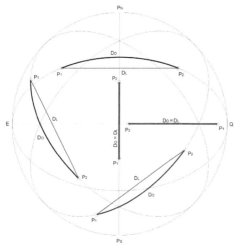

Figure 1: Orthodrome and Loxodrome relations on Earth as a sphere
Source: Made by authors

From the completely theoretical point of view, it follows that the strongest difference between orthodromic and loxodromic distance appears if positions $P_1$ and $P_2$ are placed on a geographic parallel of $\varphi = 39° 32' 24,8''$, and on anti-meridians, i.e. where $\Delta\lambda = 180°$. Then, the function value $f_{(\varphi)}$ amounts 2273, 5475...M [Benković et al]; this value represents the difference between orthodromic and loxodromic path. Maximum numerical saving of 2273,5 M in the distance, expressed in percentage counts 37,5%[5]. In

most cases, the distance saving in percentages in navigational practice reachs up to 10%.

In Equator/Meridian sailing, as well as heading close to the corresponding courses, the distance saving is minimal, given that the curves are more and more closer. In other cases, that are headings in the 090°/270° sector, particularly when sailing on the same parallel (with appropriate distance between positions), approximating the navigation could save up to one day of navigation, which nowadays represents an important element of the navigation venture.

## 3 APPROXIMATION OF ORTHODROMIC NAVIGATION BY SECANT METHODS

### 3.1 *The first secant method – Orthodrome interposition division*

In the first secant method the problem is approached in a way that the orthodrome is divided into interpositions, between which the vessel sails in loxodromic courses.

Interpositions differ in their longitude every 5° or 10° (mostly), while the division begins from the Vertex of the orthodrome, under the condition that this point is placed between the departure and arrival position. If Vertex is situated outside of the specific positions, interpositions can be defined from the point of departure, $P_1$. In the following text, Vertex interposition division is explained. In Figure 2. the required relations between the elements are shown.

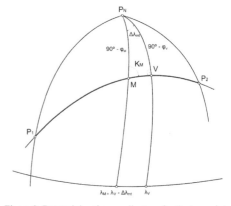

Figure 2: Determining the coordinates of orthodrome interpositions
Source: Made by authors

V – Vertex of the orthodrome, defined by the coordinates $\varphi_V$ i $\lambda_V$
$\Delta\lambda_{mt}$– the selected difference of longitude for which interpositions are required

[5] Theoretically speaking, expressed saving can reach up to 57%, but it has no practical importance for navigation, because these are very short paths between positions on the parallel at the near Pole (e.g. $\varphi=88°$).

M– orthodrome interposition, defined by the coordinates $\varphi_M$ i $\lambda_M$

The navigator selects the interposition longitude:

$$\lambda_M = \lambda_V - \lambda_{mt}$$

The latitude is obtained by applying spherical trigonometry for the right-angled triangle $\Delta P_N M V$ [Kos et al, 2010]:

$$\cos \Delta \lambda_{mt} = \operatorname{ctg}[90° - (90° - \varphi_V)]\operatorname{ctg}(90° - \varphi_M)$$

$$\cos \Delta \lambda_{mt} = \operatorname{ctg}\varphi_V \operatorname{tg}\varphi_M$$

$$\operatorname{tg}\varphi_M = \cos \Delta \lambda_{mt} \operatorname{tg}\varphi_V$$

$$\varphi_M = \operatorname{arc\,tg}(\cos \Delta \lambda_{mt} \operatorname{tg}\varphi_V)$$

In case that the Vertex lies outside positions $P_1$ and $P_2$, the division begins from the point $P_1$. Here, the inclination of the orthodrome ($i$) should be determined first. The inclination of orthodrome represents the angle at which orthodrome intersects the Equator of the Earth, resulting in a right triangle of the point of departure, $P_1$.

$$\cos i = \sin(90° - \varphi_1)\sin \alpha$$

$$\cos i = \cos \varphi_1 \sin \alpha$$

$$i = \operatorname{arc\,sin}(\cos \varphi_1 \sin \alpha)$$

The longitude of the intersection, $\lambda_S$, is defined as follows:

$$\lambda_S = \lambda_1 + \Delta \lambda_S \quad \text{where}$$

$$\operatorname{tg}\Delta \lambda_S = -\sin \varphi_1 \operatorname{tg}\alpha$$

$$\Delta \lambda_S = \operatorname{arc\,tg}(-\sin \varphi_1 \operatorname{tg}\alpha) \quad {}^{6}$$

In a right-angled triangle $\Delta SAM$, equatorial leg ($\Delta \lambda_S + \Delta \lambda_{mt}$) and the angle of inclination $i$ are known. The following relation are a result of this triangle (Figure 3) [Kos et al, 2010]:

$$\cos[90° - (\Delta \lambda_S + \Delta \lambda_{mt})] = \operatorname{ctg} i \operatorname{ctg}(90° - \varphi_M)$$

$$\sin(\Delta \lambda_S + \Delta \lambda_{mt}) = \operatorname{ctg} i \operatorname{tg}\varphi_M$$

$$\operatorname{tg}\varphi_M = \sin(\Delta \lambda_S + \Delta \lambda_{mt})\operatorname{tg} i$$

$$\varphi_M = \operatorname{arc\,tg}[\sin(\Delta \lambda_S + \Delta \lambda_{mt})\operatorname{tg} i]$$

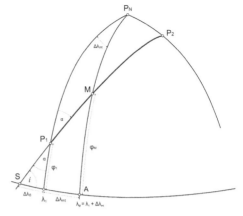

Figure 3: The inclination of the orthodrome
Source: Made by authors

### 3.1.1 Loxodromic intercourse and distances determination

The loxodromic courses between the positions are calculated from the loxodromic triangle [Benković et al, 1986]:

$$\operatorname{tg}K = \frac{\Delta \lambda}{\Delta \varphi_M}$$

$$K = \operatorname{arc\,tg}\left(\frac{\Delta \lambda}{\Delta \varphi_M}\right)$$

The first course ($K_1$), by which the orthodrome navigation begins (in position $P_1$), is calculated on the basis of $\Delta \lambda$, that is, the longitude difference between $P_1$ and the first interposition, $M_1$, and the Mercator latitudes difference between the same points. The second course ($K_2$) in $M_2$ is calculated on the basis of analogic $\Delta \lambda$ and $\Delta \varphi_M$ points $M_1$ and $M_2$, etc. The Figure 4. shows graphic determination of loxodromic courses and distances.

Besides loxodromic courses between two interpositions, to determine the distances, one needs to know the latitude difference between the positions, beginning at $P_1$ and $M_1$, then $M_1$ and $M_2$ and so on to the point of arrival $P_2$;

$$D_L = \frac{\Delta \varphi}{\cos K}$$

---

[6] Some of mentioned elements perhaps require additional explanation, mathematical derivation respectively. Bearing in mind the length limitation of the paper, as well as the extensive nature of the matter, the reader is referred to the additional literature [Kos et al, 2010].

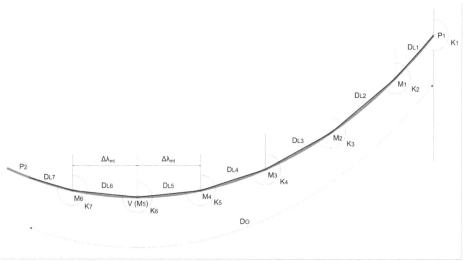

Figure 4: Orthodromic navigation approximation by the interposition division – The first secant method
Source: Made by authors

### 3.2 The second secant method – Division of the orthodrome in unit distance intervals

This method is based on the theoretical assumption that the orthodrome, which passes through two positions on the surface of the Earth as a sphere, is composed of an infinite number of infinitively small loxodromes [Kos, 1996], i.e.

$$D_O = \int_0^L \Delta d_L$$

It follows that the final greatness of the orthodrome passing through two positions that are sufficiently distant from each other, can be replaced with the infinitivelly small number of loxodromes, i.e. $dD_L = dD_O$, respectively:

$$D_O = \int_0^{D_O} dD_O$$

Given that the greatness of infinitively small loxodrome cannot be dimensionally defined, the loxodrome could be defined by the approximation of the greatness of orthodromic unit distance intervals ($dD_O$), which is then approximately equal with the loxodromic distance ($dD_L$). In this way, the inconvenient orthodrome navigation is replaced with the loxodrome sailing. The intention is that the course alternations are reduced to a navigationally acceptable amount. The smaller the greatness of orthodromic unit distance, the minor the error of orthodromic

approximation. However, it requires more frequent course alternation, which is in contradiction with practical navigation. Therefore, it is proposed as follows:

- if two positions on Earth (approximated by the shape of the sphere) are distant one from another $\leq$ 30' = 30 M, the following approximation can be introduced:

DL $\cong$ DO = 30'

Based on the above mentioned, the concept of unit distance interval is introduced, and it is 30', i.e. 30 M.

The process of orthodromic navigation performing is as follows:

$P_1$ ($\varphi_1, \lambda_1$) – departure position coordinates
$P_2$ ($\varphi_2, \lambda_2$) – arrival position coordinates

Orthodromic distance between $P_1$ and $P_2$ is calculated, using the equation which is derived from the nautical – positioning spherical triangle:

$$D_O = \cos^{-1}(\sin\varphi_1 \sin\varphi_2 + \cos\varphi_1 \cos\varphi_2 \cos\Delta\lambda)$$

$$\Delta\lambda = \lambda_2 - \lambda_1 \; ; \; 0° < \Delta\lambda < 180°$$

$\Delta\lambda$ represents the difference between the longitudes of departure and arrival positions.

Orthodromic distance ($D_O$), expressed in degrees, is then divided into orthodromic unit distances of 0,5° from the point of departure to the point of arrival.

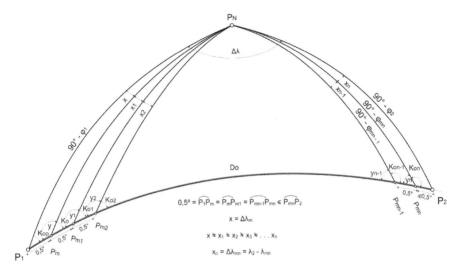

Figure 5: Division of the orthodrome in unit distance intervals of $0,5° \cong 30$ M – The second secant method
Source: Made by authors

### 3.2.1 Orthodromic interposition coordinates determination

Applying spherical trigonometry, the interposition coordinates can be determined in different ways. The following equation can be derived from the nautical – positioning spherical triangle:

$$K' = \cos^{-1}\left(\frac{\sin\varphi_2 - \sin\varphi_1 \cos D_O}{\cos\varphi_1 \sin D_O}\right)$$

The Initial Orthodromic Course ($K_{OP}$) is determined by the spherical angle $K'$ with the following relations:

$$K_{OP} = K' \quad \text{if} \quad \Delta\lambda > 0$$
$$K_{OP} = 360° - K' \quad \text{if} \quad \Delta\lambda < 0$$

The following relations can be derived from the spherical triangle $P_1 P_N P_2$ shown in Figure 5:

$$\varphi_m = \sin^{-1}\left(\cos 0,5° \sin\varphi_1 + \sin 0,5° \cos\varphi_1 \cos K_{OP}\right)$$

$$x = \cos^{-1}\left(\frac{\cos 0,5° - \sin\varphi_1 \sin\varphi_m}{\cos\varphi_1 \cos\varphi_m}\right)$$

where:

$x = \Delta\lambda_m$ – angle in terrestrial Pole enclosed by the meridians of two adjacent interpositions
$\Delta\lambda_m > 0$ – eastward navigation (E)
$\Delta\lambda_m < 0$ – westward navigation (W)

Interposition coordinates $P_m$ ($\varphi_m$, $\lambda_m = \lambda_1 + x$)

$$y = \cos^{-1}\left(\frac{\sin\varphi_1 - \cos 0,5° \sin\varphi_m}{\sin 0,5° \cos\varphi_m}\right)$$

$y$ – spherical triangle in the respective orthodromic interposition

$$\left.\begin{aligned} K_0 &= 180° \pm y \\ K_0 &= 360° - y \end{aligned}\right\}$$

$K_0$ – orthodromic course in interposition $P_m$, which depends on the hemisphere on which the ship is sailing (N or S) and sailing direction (E or W)

The coordinates of other orthodromic interpositions from $P_{m1}$ to $P_{mn}$ can be determined with the following equations:

$$\varphi_{mn} = \sin^{-1}\left(\cos 0,5° \sin\varphi_{mn-1} + \sin 0,5° \sin\varphi_{mn-1} \cos K_{on-1}\right)$$

$$x_n = \Delta\lambda_{mn} = \cos^{-1}\left(\frac{\cos 0,5° - \sin\varphi_{mn-1} \sin\varphi_{mn}}{\cos\varphi_{mn-1} - \cos\varphi_{mn}}\right)$$

$$P_{mn}\left(\varphi_{mn}, \lambda_{mn} = \lambda_{mn-1} + x_n\right)$$

$$y_n = \cos^{-1}\left(\frac{\sin\varphi_{mn-1} - \cos 0,5° \sin\varphi_{mn}}{\sin 0,5° \cos\varphi_{mn}}\right)$$

$$\left.\begin{aligned} K_{on} &= 180° \pm y_n \\ K_{on} &= 360° - y_n \end{aligned}\right\}$$

$K_{on}$ – orthodromic course in interposition $P_{mn}$

### 3.2.2 Loxodromic course determination

From one interposition to another, the ship sails in unaltered loxodromic course ($K_L$), calculated by the equation derived from the III. Loxodromic Triangle (the Course Triangle) [Kos, 1996]:

$$tgK = \frac{\Delta\lambda_m}{\Delta\varphi_{Mm}}$$

where:

$\Delta\lambda_m = \lambda_{mn} - \lambda_{mn-1}$ – longitude difference between two adjacent orthodromic interpositions, expressed in angular minutes

$\Delta\varphi_{Mm} = \varphi_{Mmn} - \varphi_{Mmn-1}$ – Mercator latitudes difference between two adjacent orthodromic interpositions, expressed in angular minutes

If the shape of the Earth is approximated by the shape of the sphere, then:

$$\varphi_{Mm} = 7915{,}7044667898 \log\left[\operatorname{tg}\left(45° + \frac{\varphi_{mn}}{2}\right)\right]...[']$$

If the shape of the Earth is approximated by the shape of the biaxial rotation ellipsoid, then [Benković et al, 1986]:

$$\varphi_{Mm} = 7915{,}7044667898 \log\left[\operatorname{tg}\left(45° + \frac{\varphi_{mn}}{2}\right)\left(\frac{1 - e\sin\varphi_{mn}}{1 + e\sin\varphi_{mn}}\right)^{\frac{e}{2}}\right]...[']$$

where:
e – the first numerical eccentricity of the ellipsoid
K – the angle in III. loxodromic triangle
$K_L$ – general loxodromic navigation course

The following quadrant navigation cases are possible, which then define loxodromic courses (Figure 6) [Wippern, 1982]:
1  I. navigation quadrant;   $\Delta\lambda_m > 0$, $\Delta\varphi_{Mm} > 0$, $K_L = 360° + K = K$
2  II. navigation quadrant;   $\Delta\lambda_m > 0$, $\Delta\varphi_{Mm} < 0$, $K_L = 180° + K$
3  III. navigation quadrant;   $\Delta\lambda_m < 0$, $\Delta\varphi_{Mm} < 0$, $K_L = 180° + K$
4  IV. navigation quadrant;   $\Delta\lambda_m < 0$, $\Delta\varphi_{Mm} > 0$, $K_L = 360° + K$

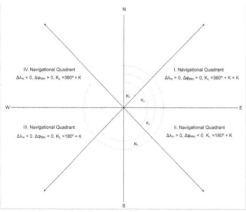

Figure 6: Four possibilities of Quadrant Navigation
Source: Made by authors

From the initial position $P_1$ to the interposition $P_m$ the ship sails in loxodromic course $K_L$. Then, between $P_m$ and $P_{m1}$ in course $K_{L1}$, between $P_{m1}$ and

$P_{m2}$ in $K_{L2}$, ... from the interposition $P_{mn-1}$ to $P_{mn}$ the ship sails in loxodromic course $K_{Ln}$, and finally, from $P_{mn}$ to the arrival position $P_2$ in the last loxodromic course. The length of the last stage of navigation between $P_{mn}$ and $P_2$ is always ≤ 30 M.

If, while navigating, the ship is not placed on the planned orthodromic path[7], new orthodrome is calculated from exact current position towards the position of arrival, and the procedure is then repeated, dividing the new orthodrome in unit distance intervals of 0,5°, and calculating navigation elements again [Kos, 1996].

## 4 APPROXIMATION OF ORTHODROMIC NAVIGATION BY THE TANGENT METHOD – ORTHODROME DIVISION IN UNIT COURSE ALTERATIONS

Instead of secants determined by the interpositions (the first secant method), or the unit distance intervals (the second secant method), the navigation is here approximated by the tangent lines of the orthodrome, i.e. unit orthodromic course alterations ($\Delta K$) are derived as follows [Zorović et al, 2010]:
– the Initial Orthodromic Course in position $P_1$ ($K_{OP}$) and the Final Orthodromic Course ($K_{OK}$) in position $P_2$ are calculated. The following values are then calculated:

$$x = \frac{(K_{OK} - K_{OP})}{\Delta K}$$

$$D_X = \frac{D_O}{x}\,[M]$$

$$\text{for } \Delta K = 1° \rightarrow D_X = \frac{D_O}{(K_{OK} - K_{OP})}$$

where:
x – total amount of orthodromic course alteration
$\Delta K$ – 1°, 2°, 3°... arbitrarily selected orthodromic unit course alteration value
$D_O$ – orthodromic distance between positions $P_1$ and $P_2$
$K_{OP}$ – the initial orthodromic course in departure position $P_1$
$K_{OK}$ – the final orthodromic course in arrival position $P_2$
$D_X$ – unit orthodromic distance

---

[7] For example, by the ship's drift due to the sea currents, the wind, waves, the collision avoidance, etc.

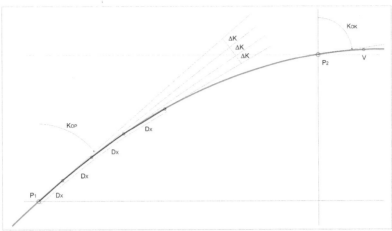

Figure 7: Approximation of orthodromic navigation by the tangent method
Source: Made by authors

## 5 CONCLUSION

From a theoretical point of view, it is not possible to navigate on great circle. The orthodrome navigation is the shortest, while the loxodrome is acceptable nautically, given that the navigation here is obtained in constant, general loxodromic course (with the longer distance travelled) [Bowditch, 1984]. Using the combination of this navigation curves, the problem is solved in a way that the features of the orthodrome as a shortest distance between two points on Earth are maximally utilized. The proposed goal of navigation is thus fulfilled from the practical point of view, given that the great circle is divided in unit values of the specific elements, depending of the method used, on which the navigation is then carried out in loxodromic courses, and loxodromic distances respectively.

Three approximation models of orthodromic navigation have been elaborated in the paper. The first model determines the orthodromic interpositions, which can be calculated from the orthodrome vertex or the initial position, depending on the position of the Vertex, whether it is placed inside the position of departure and arrival or not. With this method, the ship sails in unequal distance intervals. The second model implies the division of the orthodrome in unit distance intervals with the amount of $0.5° \cong 30$ M. Hereby, the orthodromic interposition coordinates are determined, and the ship sails in constant loxodromic courses between them. This method is the most accurate of all of the three elaborated. In the third method, the orthodromic navigation is approximated by the determination of the orthodromic unit course alterations $\Delta K$. Here, it is first necessary to calculate the unit course alterations, after which unit orthodromic distances are defined, expressed in nautical miles, representing the navigation of the vessel

in the specific course, in a way that the required alteration $\Delta K$ would appear.

The extent to which the navigation will be orthodrome – like, depends on several parametres – considering a specific navigation case, and taking navigation courses and distances between two positions into account. In the Equator and Meridian sailing, the orthodrome and the loxodrome overlap – their distances are equal. This also applies to smaller distances between positions, where there are no discrepancies between these curves. However, in certain cases, the difference between these two distances reaches noticeable values, and then, by approximating the orthodrome, the time spent in navigation can be significantly reduced.

## ACKNOWLEDGEMENTS

The authors acknowledge the support of research project "Research into the correlations of maritime-transport elements in marine traffic" (112-1121722-3066) funded by the Ministry of Science, Education and Sports of the Republic of Croatia.

## REFERENCES

[1] Benković, F. And others: *Terestrička i elektronska navigacija*, Hidrografski institut JRM, Split, 1986.
[2] Bowditch, N.: *The American Practical Navigator*, Vol I, US Defence Mapping Agency, Bethesda, 1984.
[3] Kos, S.: *Aproksimacija plovidbe po ortodromi*, Zbornik radova Pomorskog fakulteta, Rijeka, 1996.
[4] Kos, S., Zorović, D., Vranić D.: *Terestrička i elektronička navigacija*, Pomorski fakultet u Rijeci, Rijeka, 2010.
[5] Wippern, K.C.C.: *On Loxodromic Navigation*, The Journal of Navigation, Royal Institute of Navigation, 45, Cambridge, 1992.
[5] Zorović, D. And others: *Vademecum Maritimus*, Pomorski fakultet u Rijeci, Rijeka, 2002.

# 18. Solutions of Direct Geodetic Problem in Navigational Applications

A.S. Lenart

*Gdynia Maritime University, Gdynia, Poland*

ABSTRACT: Solutions of such navigational problems as positions from ranges, bearings and courses without any simplifications for a plane or a sphere, by an application of solutions of direct geodetic problem are presented. The rigorous, rapid, non-iterative solution of the direct geodetic problem according to Sodano, for any length of geodesics, is attached.

## 1 INTRODUCTION

The navigational problem is as follows (Fig. 1):
- we have known geographic coordinates of position $P'$– $\varphi_{P'}$, $\lambda_{P'}$,
- we have ranges, bearings or courses from position $P'$ to position P,
- we search for geographic coordinates of position $P$ – $\varphi_P$, $\lambda_P$,

The most common solution of such a navigational problem is a rather strange combination of flat and ellipsoidal calculations:
- conversion ranges, bearings and courses, by solving flat triangles, to $\delta x$, $\delta y$ increments in a flat rectangular coordinate frame (with the y-axis pointing north),
- conversion rectangular $\delta x$, $\delta y$ flat increments to geographic coordinates increments $\delta\varphi$, $\delta\lambda$, on the reference ellipsoid, by the equations

$$\delta\varphi = \frac{\delta y}{R_M(\varphi_{P'})} \tag{1}$$

$$\delta\lambda = \frac{\delta x}{R_N(\varphi_{P'})\cos(\varphi_{P'})} \tag{2}$$

where $R_M(\varphi_{P'})$ = the radius of curvature in meridian for $P'$; and $R_N(\varphi_{P'})$ = the radius of curvature in the prime vertical for $P'$
given by the equations

$$R_M = \frac{a_0(1-e^2)}{\sqrt{(1-e^2\sin^2\varphi_{P'})^3}} \tag{3}$$

$$R_N = \frac{a_0}{\sqrt{1-e^2\sin^2\varphi_{P'}}} \tag{4}$$

where $a_0$ = the semi-major axis of the reference ellipsoid; and e = eccentricity
and finally

$$\varphi_P = \varphi_{P'} + \delta\varphi \tag{5}$$

$$\lambda_P = \lambda_{P'} + \delta\lambda \tag{6}$$

if east longitude and north latitude are considered positive and west longitude and south latitude are considered negative.

Apart from the obvious errors of assuming the part of the ellipsoid a plane there are also errors hidden in Equations 1 and 2 – although Equations 3 and 4 are accurate for the ellipsoid – the errors of assuming the main radii of curvature constant at points $P'$ and P.

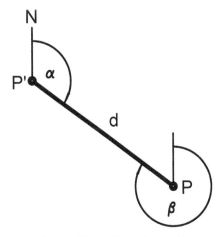

Figure 1.Definition of the problem and the position from range and bearings

## 2 ERROR OF ASSUMING THE PART OF ELLIPSOID TO BE A PLANE

The spherical excess in an equilateral spherical triangle with sides d and spherical radius R is

$$\varepsilon = \frac{d^2 \sqrt{3}}{4R^2} \tag{7}$$

For d = 22 km ($\approx$ 12 n.m.) and R = 6370 km, we get

$$\varepsilon \approx 5 \times 10^{-6} \text{ rad}$$

This gives linear changes in the range of 11 cm.

## 3 ERRORS OF ASSUMING THE MAIN RADII CONSTANT AT POINTS P′ AND P

A better approximation of Equation 1 should be (Lenart 1985)

$$\delta\varphi^* = \frac{\delta y}{R_M\left(\dfrac{\varphi_{P'}+\varphi_P}{2}\right)} = \frac{\delta y}{R_M(\varphi_{P'})+\dfrac{1}{2}\delta\varphi\dfrac{dR_M}{d\varphi}} \tag{8}$$

Defining the resulting error as

$$\Delta\delta\varphi = \delta\varphi - \delta\varphi^* \tag{9}$$

we get

$$\Delta\delta\varphi = \delta y \frac{\dfrac{1}{2}\delta\varphi\dfrac{dR_M}{d\varphi}}{R_M(\varphi_{P'})\left(R_M(\varphi_{P'})+\dfrac{1}{2}\delta\varphi\dfrac{dR_M}{d\varphi}\right)} \tag{10}$$

which with

$$\delta y = \delta\varphi\left(R_M(\varphi_{P'})+\frac{1}{2}\delta\varphi\frac{dR_M}{d\varphi}\right) \tag{11}$$

yields

$$\Delta\delta\varphi = \frac{\dfrac{1}{2}(\delta\varphi)^2\dfrac{dR_M}{d\varphi}}{R_M(\varphi_{P'})} \tag{12}$$

After substitution

$$\frac{dR_M}{d\varphi} \approx \frac{3}{2}e^2\sin 2\varphi R_M \tag{13}$$

we finally get

$$\Delta\delta\varphi \approx \frac{3}{4}e^2(\delta\varphi)^2\sin 2\varphi \tag{14}$$

where, if $\delta\varphi$ is in radians the result is also in radians.
For example, if

$$\delta\varphi = 12'' \approx 35 \times 10^{-4} \text{ rad}$$

$$\varphi = 45°$$

we get

$$\Delta\delta\varphi \approx \frac{3}{4} \times \frac{1}{150} \times 1225 \times 10^{-8} \text{ rad} \approx 6.125 \times 10^{-8} \text{ rad}$$

$$\approx .0126'' \approx 39 \text{ cm}$$

In the case of error in longitude we have, in accordance with Equation 2,

$$\delta\lambda^* = \frac{\delta x}{\left(R_N(\varphi_{P'})+\dfrac{1}{2}\delta\varphi\dfrac{dR_N}{d\varphi}\right)\cos\left(\varphi_{P'}+\dfrac{\delta\varphi}{2}\right)} \tag{15}$$

Then

$$\Delta\delta\lambda = \delta\lambda - \delta\lambda^* \tag{16}$$

and, after substitution

$$\cos\left(\varphi_{P'}+\frac{\delta\varphi}{2}\right) \approx \cos\varphi_{P'} - \frac{\delta\varphi}{2}\sin\varphi_{P'} \tag{17}$$

$$\delta x = \delta\lambda\left(R_N(\varphi_{P'})+\frac{1}{2}\delta\varphi\frac{dR_N}{d\varphi}\right)\cos\left(\varphi_{P'}+\frac{\delta\varphi}{2}\right) \tag{18}$$

After simplification

$$\Delta\delta\lambda \approx \frac{\dfrac{1}{2}\delta\varphi\delta\lambda\left(\dfrac{dR_N}{d\varphi}\cos\varphi_{P'} - R_N\sin\varphi_{P'}\right)}{R_N\cos\varphi_{P'}} \tag{19}$$

Since

$$\frac{dR_N}{d\varphi} \approx \frac{1}{2}e^2\sin 2\varphi R_N \tag{20}$$

then finally

$$\Delta\delta\lambda \approx -\frac{1}{2}\delta\varphi\delta\lambda\tan\varphi \tag{21}$$

If $\delta\varphi$ and $\delta\lambda$ are in radians, the result is also in radians.
For example, if

$$\varphi = 80°$$

$$\delta\varphi = 12' \approx 35 \times 10^{-4} \text{ rad}$$

$$\delta\lambda = 69' \approx 201 \times 10^{-4} \text{ rad (which relates to } 12'$$
 on the equator)

then

$$\Delta\delta\lambda \approx -\frac{1}{2} \times 35 \times 201 \times 10^{-8} \times 5.7 \text{ rad} \approx -41.36''$$

At that latitude, this corresponds to –222 m!

## 4  THE DIRECT GEODETIC PROBLEM

It can be seen from the above, that the errors of simplifications are neglectable or significant, depending on the required accuracy and the values of $\varphi$, $\delta\varphi$ and $\delta\lambda$, but all of them are systematic and are integrated in dead reckoning.

These simplifications have been necessary to reduce the number of calculations on the ellipsoid and justified in times of manual mechanical or electronic calculators, but are completely unnecessary and unjustified in times of computer calculations. Therefore we will directly apply the solution of the problem known in geodesy as direct geodetic problem.

In the solution of the direct geodetic problem (Fig. 2) from the given coordinates $\varphi_1$, $\lambda_1$ and azimuth $\alpha_{1-2}$ at the start of geodesics $P_1$ and their length $S$ are calculated coordinates $\varphi_2$, $\lambda_2$ of the endpoint $P_2$ and the reversed azimuth $\alpha_{2-1}$, on any reference ellipsoid.

E. M. Sodano (Sodano 1958, 1965, 1967) from Helmert's classical iterative formulae derived a rigorous non-iterative procedure, for any length of geodesics and for any required accuracy, which is attached in Appendix A. This procedure will be used in this paper in the formal form

$$\varphi_2, \lambda_2 = \mathrm{SDGP}\,(\varphi_1, \lambda_1, \alpha_{1-2}, S) \qquad (22)$$

$$\alpha_{2-1} = \mathrm{SDGP}\,(\varphi_1, \lambda_1, \alpha_{1-2}, S) \qquad (23)$$

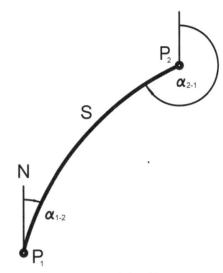

Figure 2. Direct and inverse geodetic problem

## 5  APPLICATION OF THE DIRECT GEODETIC PROBLEM

### 5.1  Position from range and bearings

We search for the position $P(\varphi_P, \lambda_P)$ for which we have the range d and the bearing $\alpha$ from, or the bearing $\beta$ to, known position $P'(\varphi_{P'}, \lambda_{P'})$ (Fig. 1).
The solution is

$$\varphi_P, \lambda_P = \mathrm{SDGP}\,(\varphi_{P'}, \lambda_{P'}, \alpha, d) \qquad (24)$$

or

$$\varphi_P, \lambda_P = \mathrm{SDGP}\,(\varphi_{P'}, \lambda_{P'}, \beta - 180°, d) \qquad (25)$$

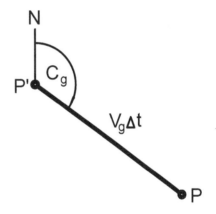

Figure 3. Dead reckoned position

### 5.3  Position from two ranges

We search for the position $P(\varphi_P, \lambda_P)$ for which we have two ranges $d_1$ and $d_2$ to known positions $P'_1(\varphi_{P'1}, \lambda_{P'1})$ and $P'_2(\varphi_{P'2}, \lambda_{P'2})$ (Fig. 4).
The solution is iterative

$$\varphi_{P1}, \lambda_{P1} = \mathrm{SDGP}\,(\varphi_{P'1}, \lambda_{P'1}, \alpha_1 = \mathrm{var}, d_1) \qquad (27)$$

$$\varphi_{P2}, \lambda_{P2} = \mathrm{SDGP}\,(\varphi_{P'2}, \lambda_{P'2}, \alpha_2 = \mathrm{var}, d_2) \qquad (28)$$

where $\alpha_1$ and $\alpha_2$ are adjusted by any small increments until e.g.

$$(\varphi_{P2} - \varphi_{P1})^2 + (\lambda_{P2} - \lambda_{P1})^2 \cos\varphi_{P2} \cos\varphi_{P1} = \min \qquad (29)$$

### 5.2  Dead reckoned position

We search for the position $P(\varphi_P, \lambda_P)$ dead reckoned from known position $P'(\varphi_{P'}, \lambda_{P'})$ with the speed over ground $V_g$ and the course over ground $C_g$ during the time interval $\Delta t$ (Fig. 3).
The solution is

$$\varphi_P, \lambda_P = \mathrm{SDGP}\,(\varphi_{P'}, \lambda_{P'}, C_g, V_g\,\Delta t) \qquad (26)$$

This iterative process, although looks as very complicated, is very fast and simple with using e.g. the Solver in Microsoft Excel.

### 5.4 Position from two bearings

We search for the position $P(\varphi_P, \lambda_P)$ for which we have the bearing $\alpha_1$ from, or the bearing $\beta_1$ to, known position $P'_1(\varphi_{P'1}, \lambda_{P'1})$ and the bearing $\alpha_2$ from, or the bearing $\beta_2$ to, known positions $P'_2(\varphi_{P'2}, \lambda_{P'2})$ (Fig. 4).

The solution is iterative

$$\varphi_{P1}, \lambda_{P1} = SDGP\ (\varphi_{P'1}, \lambda_{P'1}, \alpha_1, d_1 = var) \qquad (30)$$

$$\varphi_{P2}, \lambda_{P2} = SDGP\ (\varphi_{P'2}, \lambda_{P'2}, \alpha_2, d_2 = var) \qquad (31)$$

or

$$\varphi_{P1}, \lambda_{P1} = SDGP\ (\varphi_{P'1}, \lambda_{P'1}, \beta - 180°, d_1 = var) \qquad (32)$$

$$\varphi_{P2}, \lambda_{P2} = SDGP\ (\varphi_{P'2}, \lambda_{P'2}, \beta - 180°, d_2 = var) \qquad (33)$$

or any combination of the above, where $d_1$ and $d_2$ are adjusted by any small increments until e.g. Equation 29 is fulfilled.

### 5.5 Position from range and bearing to different positions

We search for position $P(\varphi_P, \lambda_P)$ for which we have the range $d_1$ to known position $P'_1(\varphi_{P'1}, \lambda_{P'1})$ and the bearing $\alpha_2$ from, or the bearing $\beta_2$ to, known positions $P'_2(\varphi_{P'2}, \lambda_{P'2})$ (Fig. 4).

The solution is iterative

$$\varphi_{P1}, \lambda_{P1} = SDGP\ (\varphi_{P'1}, \lambda_{P'1}, \alpha_1 = var, d_1) \qquad (34)$$

$$\varphi_{P2}, \lambda_{P2} = SDGP\ (\varphi_{P'2}, \lambda_{P'2}, \alpha_2, d_2 = var) \qquad (35)$$

or

$$\varphi_{P2}, \lambda_{P2} = SDGP\ (\varphi_{P'2}, \lambda_{P'2}, \beta - 180°, d_2 = var) \qquad (36)$$

where $\alpha_1$ and $d_2$ are adjusted by any small increments until e.g. Equation 29 is fulfilled.

### 5.6 Position from any combination of ranges and bearings

The above can be easily extended to any number of combination of ranges and bearings - we search for position $P(\varphi_P, \lambda_P)$ for which we have n ranges d or bearings $\alpha$ from, or bearings $\beta$ to, n known positions $P'(\varphi_{P'}, \lambda_{P'})$ (Fig. 5).

The solution is iterative

$$\varphi_{P1}, \lambda_{P1} = SDGP\ (\varphi_{P'1}, \lambda_{P'1}, \alpha_1 = var, d_1) \qquad (37)$$

$$\varphi_{P2}, \lambda_{P2} = SDGP\ (\varphi_{P'2}, \lambda_{P'2}, \alpha_2, d_2 = var) \qquad (38)$$

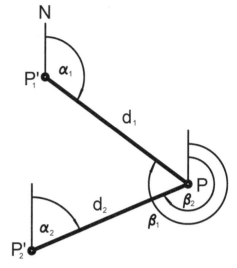

Figure 4. Positions from two ranges or bearings

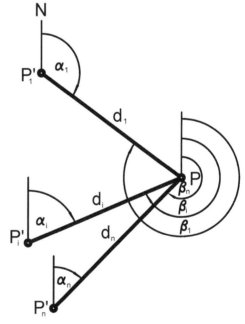

Figure 5. Position from n ranges and bearings

$$\varphi_{Pi}, \lambda_{Pi} = SDGP\ (\varphi_{P'i}, \lambda_{P'i}, \beta_i - 180°, d_i = var) \qquad (39)$$

$$\varphi_{Pn}, \lambda_{Pn} = SDGP\ (\varphi_{P'n}, \lambda_{P'n}, \beta_n - 180°, d_n = var) \qquad (40)$$

where d or $\alpha$ or $\beta$ are adjusted by any small increments until e.g.

$$\sum_{i=1}^{n} (\varphi_{Pi} - \overline{\varphi})^2 + (\lambda_{Pi} - \overline{\lambda})^2 \cos^2 \varphi_{Pi} = \min \qquad (41)$$

where

$$\overline{\varphi} = \frac{1}{n} \sum_{i=1}^{n} \varphi_{Pi} \qquad (42)$$

$$\overline{\lambda} = \frac{1}{n} \sum_{i=1}^{n} \lambda_{Pi} \qquad (43)$$

It is worth mentioning, that we will achieve the least square errors position in the case of excessive number of position lines.

### 5.7 Position lines of different accuracies

In the case of position lines of different accuracies we can extend Equation 29 or 41 with weights – e.g. reciprocal of mean square errors $m_i$

$$\sum_{i=1}^{n} \frac{(\varphi_{Pi} - \overline{\varphi})^2 + (\lambda_{Pi} - \overline{\lambda})^2 \cos^2 \varphi_{Pi}}{m_i^2} = \min \qquad (44)$$

### 5.8 Bearings for long ranges

For long ranges

$$\alpha \neq \beta - 180° \qquad (45)$$

If this difference is significant, for Section 5.1, we at first iteratively search for $\alpha$ from the equation

$$\alpha_{2\text{-}1} = \text{SDGP} (\varphi_{P'}, \lambda_{P'}, \alpha = \text{var}, d) \qquad (46)$$

until

$$\alpha_{2\text{-}1} = \beta \qquad (47)$$

and then

$$\varphi_P, \lambda_P = \text{SDGP} (\varphi_{P'}, \lambda_{P'}, \alpha, d) \qquad (48)$$

For Section 5.4 and 5.5 Equations 32, 33, 36, 39 and 40 becomes respectively to

$$\varphi_{Pi}, \lambda_{Pi} = \text{SDGP} (\varphi_{P'i}, \lambda_{P'i}, \alpha_i = \text{var}, d_i = \text{var}) \qquad (49)$$

and additionally

$$\alpha_{2\text{-}1i} = \text{SDGP} (\varphi_{P'i}, \lambda_{P'i}, \alpha_i = \text{var}, d_i = \text{var}) \qquad (50)$$

as well as Equations 29 and 41 should be supplemented by the component e.g.

$$(\beta_i - \alpha_{2\text{-}1i})^2 \left[ \left( \frac{d_i \sin \beta_i}{R_N(\varphi_{Pi}) \cos \varphi_{Pi}} \right)^2 + \left( \frac{d_i \cos \beta_i}{R_M(\varphi_{Pi})} \right)^2 \right] \qquad (51)$$

for each $\beta_i$.

## 6 ACCURACY OF THE SOLUTION OF THE DIRECT GEODETIC PROBLEM

"The accuracy of geodetic distances computed through the $e^2$, $e^4$, $e^6$ order for very long geodesics is within a few meters, centimeters and tenth of millimeters respectively. Azimuths are good to tenth, thousandths and hundreds thousandths of a second. Further improvement of results occurs for shorter lines" (Sodano 1958).

## 7 DIRECT COMPUTATION FORM SIMPLIFIED

For shorter distances (the abovementioned "very long geodesics" means even 20 000 km) or lower required accuracies we can use equations from Appendix A reduced to $e^2$ and f order. Therefore Equation A 9 becomes to

$$\Phi_0 = \Phi_S - \frac{1}{2} a_1 e'^2 \sin \Phi_S \\ + \frac{1}{4} m_1 e'^2 (-\Phi_S + \sin \Phi_S \cos \Phi_S) \qquad (52)$$

and Equation A 12 becomes to

$$L = -f \Phi_S \cos \beta_0 + \gamma \qquad (53)$$

## 8 CIRCULAR FUNCTIONS

The angles $\alpha_{2\text{-}1}$ and $\gamma$ from Equations A 10 and A 11 have to be calculated with the circular function $\tan^{-1}()$, but this function gives solutions in the range $(-90°, 90°)$. For full range $(0°, 360°)$ retrieving tables of quadrants are used in Sodano 1965.

For computer calculations a special procedure should be used to retrieve the full range $(0°, 360°)$ from the signs of the numerator N and the denominator D and to detect and support a division by zero case e.g.:

For

$$\text{angle} = \tan^{-1} \frac{N}{D}$$

IF D ≠ 0 THEN ANGLE = ATN(N/D)

    IF D < 0 THEN ANGLE = ANGLE + 180°: END IF

ELSE

    ANGLE = (2 - SIGN(N))*ABS(SIGN(N))*90°

END IF

IF ANGLE<0 THEN ANGLE=ANGLE+360°: END IF

## 9 CONCLUSIONS

Presented procedures are quite general and universal. They can be used for any number of ranges and bearings, any combinations of ranges and bearings, any ranges – from meters up to 20 000 km, with almost any required accuracy, on any reference ellipsoid and can calculate the optimal position according to any objective function.

## REFERENCES

Lenart A.S. 1985. Errors of algorithms for position determination from hyperbolic navigation systems. *Marine Geodesy* 9(1): 93-111.

Sodano E.M. 1958. A rigorous non-iterative procedure for rapid inverse solution of very long geodesics. *Bulletin Géodésique* 47/48: 13-25.

Sodano E.M. 1965. General non-iterative solution of the inverse and direct geodetic problems. *Bulletin Géodésique* 75: 69-89.

Sodano E.M. 1967. Supplement to inverse solution of long geodesics. *Bulletin Géodésique* 85: 233-236.

## APPENDIX A

*Direct computation form (Sodano 1965)*

Given: $\varphi_1$, $\lambda_1$, $\alpha_{1\text{-}2}$, S
Required: $\varphi_2$, $\lambda_2$, $\alpha_{1\text{-}2}$

Reference ellipsoid: $a_0$, $b_0$ = semi-major and semi-minor axes
Flattening

$$f = 1 - \frac{b_0}{a_0} \tag{A 1}$$

Second eccentricity squared

$$e'^2 = \frac{a_0^{\,2} - b_0^{\,2}}{b_0^{\,2}} \tag{A 2}$$

$$\tan\beta_1 = (1 - f)\tan\varphi_1 \tag{A 3}$$

$$\cos\beta_0 = \cos\beta_1 \sin\alpha_{1-2} \tag{A 4}$$

$$g = \cos\beta_1 \cos\alpha_{1-2} \tag{A 5}$$

$$\Phi_S = \frac{S}{b_0} \tag{A 6}$$

$$m_1 = (1 + \frac{e'^2}{2}\sin^2\beta_1)(1 - \cos^2\beta_0) \tag{A 7}$$

$$a_1 = (1 + \frac{e'^2}{2}\sin^2\beta_1)(\sin^2\beta_1 \cos\Phi_S + g\sin\beta_1 \sin\Phi_S) \tag{A 8}$$

$$
\begin{aligned}
\Phi_0 &= \Phi_S - \frac{1}{2}a_1 e'^2 \sin\Phi_S \\
&+ \frac{1}{4}m_1 e'^2(-\Phi_S + \sin\Phi_S \cos\Phi_S) \\
&+ \frac{5}{8}a_1^{\,2}e'^4 \sin\Phi_S \cos\Phi_S \\
&+ m_1^{\,2}e'^4(\frac{11}{64}\Phi_S - \frac{13}{64}\sin\Phi_S \cos\Phi_S \\
&- \frac{1}{8}\Phi_S \cos^2\Phi_S + \frac{5}{32}\sin\Phi_S \cos^3\Phi_S) \\
&+ a_1 m_1 e'^4(\frac{3}{8}\sin\Phi_S + \frac{1}{4}\Phi_S \cos\Phi_S \\
&- \frac{5}{8}\sin\Phi_S \cos^2\Phi_S)
\end{aligned} \tag{A 9}
$$

$$\tan\alpha_{2-1} = \frac{\cos\beta_0}{g\cos\Phi_0 - \sin\beta_1 \sin\Phi_0} \tag{A 10}$$

$$\tan\gamma_1 = \frac{\sin\Phi_0 \sin\alpha_{1-2}}{\cos\beta_1 \cos\Phi_0 - \sin\beta_1 \sin\Phi_0 \cos\alpha_{1-2}} \tag{A 11}$$

$$
\begin{aligned}
L = [&-f\Phi_S + \frac{3}{2}f^2 a_1 \sin\Phi_S \\
&+ \frac{3}{4}f^2 m_1(\Phi_S - \sin\Phi_S \cos\Phi_S)]\cos\beta_0 + \gamma_1
\end{aligned} \tag{A 12}
$$

$$\lambda_2 = \lambda_1 + L \tag{A 13}$$

$$\sin\beta_2 = \sin\beta_1 \cos\Phi_0 + g\sin\Phi_0 \tag{A 14}$$

$$\tan\varphi_2 = \frac{\tan\beta_2}{(1 - f)} \tag{A 15}$$

Route Planning in Marine Navigation

# 19. Advanced Navigation Route Optimization for an Oceangoing Vessel

E. Kobayashi, T. Asajima & N. Sueyoshi
*Kobe University, Kobe, Japan*

ABSTRACT: A new weather routing method is proposed that accounts for ship maneuvering motions, ocean currents, wind, and waves through a time domain computer simulation. The maneuvering motions are solved by differential equations of motion for every moment throughout the voyage. Moreover, the navigation route, expressed in terms of a Bézier curve, is optimized for minimum fuel consumption by the Powell method. Although the optimized route is longer than the great circle route, simulation results confirm a significant reduction in fuel consumption. This method is widely applicable to finding optimal navigation routes in other areas.

## 1 INTRODUCTION

An oceangoing ship is affected by external forces such as wind, ocean currents, and waves. Weather routing techniques are usually applied in cases where the magnitude of those external forces are very large and may pose a danger to ships. In addition, optimal navigation is required from an economical point of view. Although there have been many efforts to develop route optimization techniques, most cases do not take ship maneuvering dynamics into account. Recent improvements in computing performance have made it possible to carry out a large amount of computation, so it is even possible to take ship motion dynamics into consideration. In this study, a new weather routing method was developed that takes ship maneuvering motions, current, wind, and waves into account by time domain computer simulation.

An MMG-type mathematical model of ship maneuvering motions is introduced for the dynamic calculation. The model includes many variables such as sway, yaw, propeller thrust and torque, rudder force, and fuel consumption. The maneuvering motions are solved by differential equations of motion for every moment throughout the voyage. Moreover, the optimal navigation route is determined by minimizing the fuel consumption through Powell's method. In this paper, the mathematical models of the ship maneuvering navigation model are first shown. Next, methods for calculating the current, wind, and waves from the database are introduced and demonstrated. Then, several kinds of computer simulations for weather routing are shown. Finally, the applicability and future works are discussed.

## 2 TARGET SHIP AND ROUTE

A container ship was chosen as the subject ship of this study because it is one of the principal means of marine transportation. Moreover, container ships consume more fuel than other marine vehicles because they are run faster to accommodate tight customer schedules. The specifications of the subject ship are shown in Table 1.

Table 1. Specifications of subject container ship.

| | | |
|---|---|---|
| Length overall | Loa | 299.85 m |
| Length between perpendiculars | Lpp | 299.85 m |
| Breadth molded | Bmld | 40.00 m |
| Depth molded | Dmld | 24.30 m |
| Draft designed | d | 14.02 m |
| Propeller diameter | Dp | 9.52 m |
| Propeller pitch | Pp | 7.25 m |
| Lateral projected area | $A_{AL}$ | 8284.25 m$^2$ |
| Transverse projected area | $A_{AT}$ | 1052.18 m$^2$ |
| Gross tonnage | GT | 75,201 t |
| Service speed | Vs | 25.0 kt |

The intercontinental route between Yokohama, Japan, and San Francisco, USA, as shown in Figure 1, was used as the subject route in this study because it is a very important trade route for Japan and weather conditions along the route are sometimes rough.

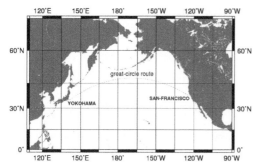

Figure 1. Subject transportation route between Yokohama and San Francisco.

## 3 WIND, WAVES, AND CURRENT ESTIMATION

Ocean surface current data from the National Oceanic and Atmospheric Administration (NOAA) of the United States Department of Commerce were used in this study. The data were five-day averages for a $1.0° × 1.0°$ mesh in the area between 0.5°E and 0.5°W and 59.5°N and 59.5°S. Sample data from 26 December 2008 are shown in Figure 2. The arrows show the eastbound and westbound current vectors, respectively.

The predicted data for wind and waves were derived from global prediction data by NCEP (National Center for Environmental Prediction); they include several parameters for wind and wave information, such as wind direction, wind velocity, and significant wave height every 3 h; the data were updated every 6 h. The range of the data were between longitudes of 0°E and 1.25°W and latitudes of 78°N and 78°S; the mesh size was 1.25° and 1° for longitude and latitude, respectively. The target data times of the voyage simulation starting in this study were midnight on 6, 9, 22, and 25 December 2008. The current, wind, and wave data were updated every 3 h.

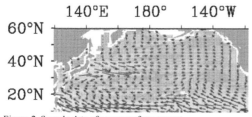

Figure 2. Sample data of ocean surface current.

## 4 MATHEMATICAL MODEL OF SHIP MANEUVERING

### 4.1 Coordinate system and basic equations

A body-fixed coordinate system whose origin is located at the ship center of gravity was adopted to express ship maneuvering motions (Figure 3).

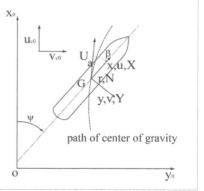

Figure 3. System coordinates in ship maneuvering motion.

Basic equations for ship maneuvering motions in the longitudinal, lateral, and yaw directions are expressed in the following equations:

$$\left.\begin{array}{l}(m + m_x)\dot{u} - (m + m_y + X_{vr})vr \\ - (u_{c0} \sin\psi - v_{c0}\cos\psi)(m_y - m_x + X_{vr}) \\ = X_H + X_P + X_R + X_A + R_{AW} \\ (m+m_y)\dot{v}+(m+m_x)ur \\ - (u_{c0}\cos\psi + v_{c0}\sin\psi)(-m_y + m_x)r \\ = Y_H + Y_P + Y_R + Y_A \\ (I_{zz}+J_{zz})\dot{r} = N_H + N_P + N_R + N_A\end{array}\right\} \quad (1)$$

where $m$ = mass of a ship; $m_x, m_x$ = added masses in the x and y directions, respectively; $I_{zz}, J_{zz}$ = mass moment of inertia and added mass moment of inertia around the z axis; $u, v$ = ship speed components in x, y coordinates; $u_{C0}, v_{C0}$ = current velocity components in x, y coordinates; $r$ = rate of turn; $\psi$ = yaw angle; $\dot{u}, \dot{v}$ = time differentiation for $u, v$; $X_H, Y_H, N_H$, $X_P, Y_P, N_P$, $X_R, Y_R, N_R$, $X_A, Y_A, N_A$ = longitudinal force in the x direction, lateral force in the y direction, and yaw moment in the z axis acting on the hull, propeller, rudder and wind, respectively; $R_{AW}$ = added resistance by waves; and $X_{vr}$ = hydrodynamic derivative.

### 4.2 Hull force

The hull forces and moment acting on the ship were expressed by the polynomial of motion variables u, v, and r in the abovementioned basic equations for maneuvering motion based on the model test results as follows:

$$X_H = -R + \frac{1}{2}\rho L d U^2 X'_{H1} \left.\vphantom{\begin{array}{c}1\\1\\1\end{array}}\right\}$$
$$Y_H = \frac{1}{2}\rho L d U^2 Y'_H \qquad\qquad (2)$$
$$N_H = \frac{1}{2}\rho L^2 d U^2 N'_H$$

$$X'_{H1} = X'_{vv}v'^2 + X'_{rr}r'^2 + X'_{vr}v'r' + X'_{vvvv}v'^4 \left.\vphantom{\begin{array}{c}1\\1\\1\end{array}}\right\}$$
$$Y'_H = Y'_v v' + Y'_r r' + Y'_{vvv}v'^3 + Y'_{vvr}v'^2 r' + Y'_{vrr}v'r'^2 + Y'_{rrr}r'^3 \qquad (3)$$
$$N'_H = N'_v v' + N'_r r' + N'_{vvv}v'^3 + N'_{vvr}v'^2 r' + N'_{vrr}v'r'^2 + N'_{rrr}r'^3$$

where $R$ is ship resistance; $X'_{H1}, Y'_H, N'_H$ are non-dimensional forces and moments acting on a wetted hull due to swaying and yawing motions in the the $x$, $y$, and yawing directions, respectively; and $X'_{vv}, X'_{rr}, \ldots, N'_{rrr}$ are the hydrodynamic derivatives.

## 4.3  Propeller force and fuel consumption

In this study, the lateral force and moment due to the propeller were neglected because they were negligible under straight-going conditions in the ocean. The longitudinal propeller force is expressed as follows:

$$X_P = (1-t)T, Y_P = 0, N_P = 0 \qquad\qquad (4)$$

where $t$ is the propeller thrust reduction factor and $T$ is propeller thrust. Moreover, the propeller thrust is expressed as follows:

$$T = \rho n^2 D_P^4 K_t \left.\vphantom{\begin{array}{c}1\\1\\1\end{array}}\right\}$$
$$K_T = c_0 + c_1 J + c_2 J^2 \qquad\qquad (5)$$
$$J = \frac{u_P}{n D_P}$$

where $n$ is the number of propeller revolutions; $K_T$ is the propeller thrust coefficient; $J$ is the advance coefficient; $c_0, c_1, c_2$ are the propeller characteristics coefficients; and $u_P$ is the propeller inflow velocity.

The propeller torque $Q$ is expressed as follows:

$$Q = \rho n^2 D_P^5 K_Q \left.\vphantom{\begin{array}{c}1\\1\end{array}}\right\}$$
$$K_Q = d_0 + d_1 J + d_2 J^2 \qquad\qquad (6)$$

where $K_Q$ is the propeller torque coefficient; and $d_0, d_1, d_2$ are the propeller open characteristics coefficients.

Then, the main engine power is calculated as follows:

$$BHP = DHP / \eta_t \left.\vphantom{\begin{array}{c}1\\1\end{array}}\right\}$$
$$DHP = 2\pi n Q \qquad\qquad (7)$$

where $BHP$ represents the brake horsepower of the main engine, $DHP$ is the delivered horsepower, and $\eta_t$ is the transmission efficiency.

Finally, the fuel oil consumption per unit time $\Delta FOC$ is calculated by the equation below:

$$\Delta FOC = BHP \times FOCR \qquad\qquad (8)$$

where $FOCR$ is the fuel oil consumption rate.

Then, the total fuel oil consumption during the voyage $FOC$ is expressed as follows:

$$FOC = \int \Delta FOC dt \qquad\qquad (9)$$

where $FOCR$ is the fuel oil consumption rate.

## 4.4  Rudder force

The rudder forces and moment are expressed as follows:

$$X_R = -(1-t_R)F_N \sin\delta \left.\vphantom{\begin{array}{c}1\\1\\1\end{array}}\right\}$$
$$Y_R = -(1+a_H)F_N \cos\delta \qquad\qquad (10)$$
$$N_R = -(x_R + a_H x_H)F_N \cos\delta$$

where $X_R, Y_R, N_R$ are nondimensional forces and moments on the rudder in the $x$, $y$, and yaw directions, respectively; $F_N$ = rudder normal force; $\delta$ = rudder angle; $t_R, a_H, x_H$ = interaction coefficients between the hull and rudder; and $x_R$ = coordinate of rudder position. The abovementioned rudder normal force is expressed as follows:

$$F_N = \frac{1}{2}\rho A_R U_R^2 f_\alpha \sin(\delta_e) \qquad\qquad (11)$$

where $A_R$ is the rudder area; $U_R$ is the rudder inflow velocity; $f_\alpha$ is the rudder normal force coefficient; and $\delta_e$ is the effective rudder angle. For the value of $f_\alpha$, the following empirical formula is used:

$$f_\alpha = \frac{6.13\lambda}{2.25 + \lambda} \qquad\qquad (12)$$

where $\lambda$ is the aspect ratio of the rudder as expressed by $f_\alpha = b/h$.

On the other hand, the rudder inflow velocity $U_R$ is expressed by

$$U_R = \sqrt{u_R^2 + v_R^2} \qquad\qquad (13)$$

where $u_R$ and $v_R$ are the velocity components in the $x$ and $y$ directions, respectively. Here, $u_R$ is expressed by the following equation as a function related to propeller thrust:

$$u_R = \varepsilon u_P \sqrt{1 + 8\frac{\kappa K_T}{\pi J^2}} \qquad\qquad (14)$$

where $\varepsilon$ and $\kappa$ are empirical or experimental coefficients for the propeller flow acceleration; $u_P$ is the propeller inflow velocity, and is expressed as $u_P = (1-w)u$ by the use of the wake fraction coefficient $w$; and $K_T$ and $J$ are the thrust coefficient and advance constant explained above, respectively. Moreover, $v_R$ is expressed as follows:

$$v_R = -\gamma(v + l'_R L_{PP} r) \qquad (15)$$

where $\gamma$ and $l'_R$ are empirical factors, $v$ is the lateral velocity component, and $r$ is the rate of turn. The effective rudder angle $\delta_e$ is expressed as follows:

$$\delta_e = \delta - v_R / u_R \qquad (16)$$

where $v_R$ and $u_R$ is the longitudinal and lateral rudder inflow velocity components.

### 4.5 Rudder control

An automatic rudder control algorithm was introduced to perform ship maneuvering simulations during the voyage for the ship passing through designated points and courses as follows:

$$\delta = -\tilde{c}_0 \Delta y - \tilde{c}_1 \Delta \psi - \tilde{c}_2 r \qquad (17)$$

where $\tilde{c}_0, \tilde{c}_1, \tilde{c}_2$ are control coefficients; $\Delta y$ is the deviation from the designated route, $\Delta \psi$ is the deviation from the designated heading angle, and $r$ is the rate of turn.

### 4.6 Wind force and moment

The forces and moment by wind are expressed as follows:

$$\left. \begin{aligned} X_A &= \frac{1}{2} \rho_A V_A^2 A_T C_{XA}(\theta_A) \\ Y_A &= \frac{1}{2} \rho_A V_A^2 A_L C_{YA}(\theta_A) \\ N_A &= \frac{1}{2} \rho_A V_A^2 A_T C_{NA}(\theta_A) \end{aligned} \right\} \qquad (18)$$

where $X_A, Y_A, N_A$ are nondimensional forces and represent the moment due to wind in the $x$, $y$, and yaw directions, respectively; $\rho_H$ is the density of air; $V_A$ is the relative wind velocity; $A_T$ is the transverse projected area; $A_L$ is the lateral projected area; $\theta_A$ is the relative wind direction; and $C_{XA}, C_{YA}, C_{NA}$ are the wind force and moment coefficients in the $x$, $y$, and yaw directions, respectively.

Wind force characteristics such as $C_{XA}, C_{YA}, C_{NA}$ were calculated by using Fujiwara's (Fujiwara 2001) method, as shown in the following figure.

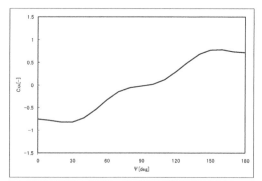
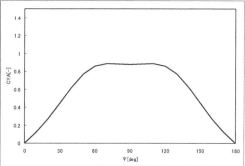
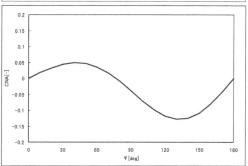

Figure 4. Wind force and moment characteristics obtained by Fujiwara's method.

### 4.7 Added resistance due to waves

The added resistance due to waves was estimated by following simple reliable equations (Sasaki 2008) because the significant wave height in most of the navigated sea area was expected to be less than 3 m:

$$\left. \begin{aligned} R_{AW} &= C_1 \frac{1}{2} \rho g \left(1 + C_2 F_{nB}^{0.8}\right) H_W^2 BB_{fcp}^2 \\ F_{nB} &= \frac{V_S}{(gB)^{0.5}} \\ B_{fcp}^2 &= \frac{1}{\left(1 + 2(1 - C_{pf}) C_{pf}\right)^2} \end{aligned} \right\} \qquad (19)$$

where $R_{AW}$ = added resistance due to waves; $\rho$ = density of seawater; $g$ = gravity acceleration; $F_{nB}$ =

Froude number by breadth of the ship; $H_W$ = significant wave height; $B$ = breadth of the ship; $B_{fcp}$ = bluntness coefficient; $V_S$ = ship speed; $C_{pf}$ = fore part of the prismatic coefficient. Moreover, the abovementioned coefficients $C_1, C_2$ are expressed as follows:

$$C_1 = 0.46 \tag{20}$$

$$C_2 = 2.0, \text{ when } B_{fcp} > 0.3 \tag{21}$$

$$C_2 = 2.0 + 60(0.3 + B_{fcp}), \text{ when } B_{fcp} \leq 0.3 \tag{22}$$

## 5 OPTIMAL METHOD

Although there are several methods such as the gradient method, which uses first-order differentiation of a function, or the Newton method, which uses second-order differentiation, they require procedures for differentiation. However, the gradient could not be obtained analytically in some complex subjects such as route optimization in this study. Therefore, Powell's method, which does not use gradients, was used in this study. Moreover, the following cost function was used in this study:

$$J = FOC \tag{23}$$

The value $FOC$ is the integrated fuel consumption during a voyage route as defined by a Bézier curve, which is an 'N - 1'th order curve defined by 'N' control points.

In the optimal procedure using Powell's method, some variables are changed through iterative calculations until convergence. Therefore, the route should be expressed by several variables.

Although there are several methods to express a curve with several variables such as a trigonometric function and multi-degree polynomials, the Bézier curve was chosen because it can be used to create a smooth curve suitable for navigation route expression. A flowchart of this optimal calculation is shown in Figure 6. Calculations were repeated automatically until the results were confirmed to converge.

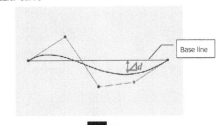

**Bezier Curve**

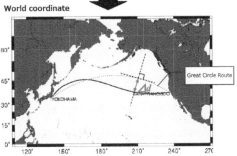

**World coordinate**

Figure 5. Bézier curve treatment for the expression of an oceangoing route. (temporary figure)

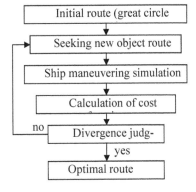

Figure 6. Flowchart of optimal calculation

## 6 RESULTS OF SIMULATION

The optimized westbound and eastbound routes for a voyage on 9 December 2008 using this method and the initial condition of the great circle route, i.e., minimum distance, for optimal calculation are shown in Figure 7.

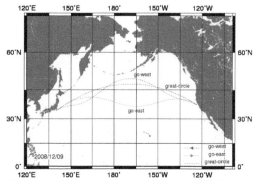

Figure 7. Optimized transportation routes between Yokohama San Francisco.

Moreover, the fuel consumption, voyage distance, and voyage time of the initial optimized routes are shown in Figures 8–10.

Although the voyage distance of the optimized route was sometimes over 200 hundred miles larger than the original great circle route, as shown in Figure 9, the fuel consumption for the optimized route was 10–50 tons less than that of the great circle route. On the other hand, the travelling times of the original and optimized routes were almost the same, as shown in Figure 10. Thus, the optimal route was concluded to be more economical. Moreover, by comparing the westbound and eastbound legs of the great circle route, the fuel consumptions of the two were found to be slightly different. This suggests that fuel consumption is affected by weather conditions such as wind and waves.

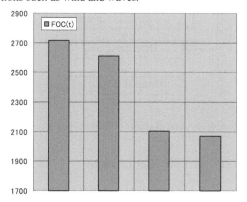

Eastbound    Eastbound  Westbound  Westbound
Great Circle  Optimized  Great Circle  Optimized

Figure 8. Comparison of fuel consumption between great circle and optimized routes.

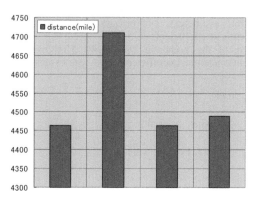

Eastbound    Eastbound  Westbound  Westbound
Great Circle  Optimized  Great Circle  Optimized

Figure 9. Comparison of voyage distance between great circle and optimized routes.

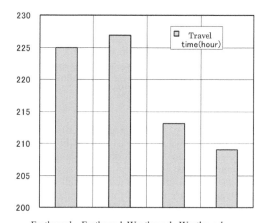

Eastbound    Eastbound  Westbound  Westbound
Great Circle  Optimized  Great Circle  Optimized

Figure 10. Comparison of voyage time between great circle and optimized routes. (to be changed, temporary figure)

## 7 CONCLUSIONS

In this study, advanced weather routing using a optimized method to minimize fuel consumption was demonstrated. The target ship of this study was a large container ship that traverses the Pacific Ocean between Yokohama and San Francisco. The route was expressed in term of a Bézier curve, and Powell's method (Fletcher and Powell 1963) was introduced to optimize the route. The results of this study can be summarized as follows:

1. A new weather routing method is proposed that uses an optimized Powell's method. This optimal method can be applied to seeking an optimal navigating route.

2 Moreover, the route was expressed by a Bézier curve. The method was shown to be widely applicable to expressing a voyage route.
3 Through computer simulations, a significant reduction in fuel consumption was obtained by tracking the optimized route, even though the distance was longer than that for the great circle route.
4 This method is widely applicable to seeking optimal navigation routes in other areas.

REFERENCES

Fletcher, R. and M.J.D. Powell 1963. "A rapidly convergent descent method for minimization," Computer Journal, Vol. 6, pp 163-168.
Fujiwara, T., Ueno, M. and Nimura, T. 2001. An estimation method of wind forces and moments acting on ships, Mini Symposium on Prediction of Ship Manoeuvring Performance, 18 October, 2001. pp. 83-92.
Sasaki, N., Motsubara, T. and Yoshida, T. 2008. Analysis of speed drop of large container ships operating in sea way, Conference Proceedings, The Japan Society of Naval Architects and Ocean Engineering, May 2008 Volume6, pp9-12.

# 20. On the Method of Ship's Transoceanic Route Planning

O. D. Pipchenko
*Odessa National Maritime Academy, Odessa, Ukraine*

ABSTRACT: In this article control of ship on a transoceanic route is represented as multicriteria optimization problem. Optimal route can be found by minimizing the objective function expressed as ship integral work for a voyage, taking into account ship's schedule, weather conditions, engine loads and risks connected with ship dynamics in waves. The risk level is represented as non-linear function with heterogeneous input variables which estimated by means of multi-input fuzzy inference system on the basis of pre-calculated or measured ship motion parameters. As the result of this research the optimal transoceanic route planning algorithm is obtained.

## 1 INTRODUCTION

Ship routing is one of traditional navigational tasks directly related to her safe and efficient exploitation. Rising fuel prices and overcapacity spurring owners to implement on more of their ships the slow steaming policy that obviously makes ocean passage stage of the voyage longer and consequentially increases risks connected with ship operation in heavy weather conditions. Hence, the problem of optimal compromise between safety and economy become even more crucial. At the present time series of routing methods exist, such as isochrones (James 1959, Bijlsma 2004, Szlapczynska, Smierzchalski 2007), graph (Vagushchenko 2004, Padhy et al. 2008), expert (Oses, Castells 2008) and intelligence methods (Nechayev et al. 2009). All of them allow to perform route optimization by number of preset criteria. However the main problem, connected to optimization process, that's still remaining, is to obtain the objective function based on formalized relationship between ship motion parameters, power inputs, needed for environmental disturbances compensation, and route economical efficiency. One of the possible solutions of this problem is given below.

## 2 OPTIMIZATION TASK & OBJECTIVE FUNCTION

The solution of the above mentioned problem is based on the next hypothesis (Pipchenko 2009): *optimal route in prescribed weather conditions is such combination of route legs and corresponding engine loads, on which expended ship power inputs are closest to minimum and predicted voyage time does not exceed scheduled one, with regard to the safety limits.*

To assess the economical efficiency of the route, one can divide overall ship costs in two categories: minimal-unavoidable costs, needed for the voyage in ideal conditions that can be expressed by minimal-unavoidable work $A_{min}$, and additional work $\Delta A$. Therefore, total work, performed during the voyage can be given as:

$$A = A_{\min} + \Delta A, \qquad (1)$$

or as voyage time integral of variable engine power:

$$A = \int_T P(t)\,dt, \qquad (2)$$

where $P$ = main engine power.

Minimal-unavoidable work can be defined from the condition of minimum work performed during specified time with constant engine power on the shortest distance between ports:

$$A_{\min} = \int_{T_v} P_{const}\,dt, \text{ with } S_v = \min\left(S(\Re)\right), \qquad (3)$$

where $S_v$ = the shortest route length; $T_v$ = scheduled voyage time; $S(\Re)$ = length depending on route $\Re$ configuration.

Thus, the main voyage optimality criterion, without risks consideration, is the minimum of additionally performed work, appeared due to weather, time and distance limitations. This work can be given as voyage length integral of additional resistance $R_W$ arisen due to environmental disturbances:

$$\Delta A = \int_S R_W \, ds \qquad (4)$$

From equations (2), (3) the additional work can be obtained as:

$$\Delta A = \int_T P(t) \, dt - A_{min} \qquad (5)$$

Therefore, the objective function representing the specified route optimality can be expressed as:

$$Z = \int_{T \le T_v} P(t) \, dt, \ Z \rightarrow A_{min} \qquad (6)$$

For the full-valued solution of the problem, it is also necessary to take into account corresponding limitations. For this purpose the risk assessment concept was used and next was formulated: *the optimal route is found if the total work for the voyage is closest to minimal, voyage time does not exceed the scheduled one, and the risk level on each route leg is less then specified limit.* Thus, the objective function will be given as:

$$Z = \int_{T \le T_p} \min \left\{ P_{max}, \begin{bmatrix} P\left(U_{safe}(R), t\right) \\ +\Delta P\left(R_W\left(U_{safe}(R)\right), t\right) \end{bmatrix} \right\} dt, \quad (7)$$

where $U_{safe}$ = maximum safe speed, at which the specified hazardous occurrence risk R is below the critical limit; $P_{max}$ = maximum engine power; $P$ = engine power needed to keep defined calm water speed; $\Delta P(R_W)$ = additional power needed to compensate the resistance due to environmental disturbances $R_W$; $R \in (0,1)$ = risk level on the specified route leg.

## 3 RISK EVALUATION

### 3.1 *Problem definition*

According to the route optimality definition, given above, the risk level conducted with ship activity in prescribed weather conditions shall be determined for each route leg. Therefore, we define the leg as the part of the route on which ship control regime (speed and heading) and weather conditions remain constant. As opposed to classical definition two or more different route legs may be situated on one line between the waypoints, depending on weather grid density.

Mathematically the risk level can be defined as product of likelihood of hazardous occurrence and its consequence. In our case we define likelihood as probability of reaching defined dynamical motion

parameters that may lead to the series of negative consequences, conducted with ship's operation in storm.

Assessing the risks of ship operation in heavy weather conditions one can define the situations connected with damages to hull structure, ship's systems and machinery and the situations arising due to violations of cargo handling technology.

For instance, the achievement of defined high amplitudes of roll may lead to the series of situations with different levels of consequences, such as shifting or loss of cargo, flooding of ship's compartments, capsizing. Therefore, next risk levels can be highlighted: *insignificant, low, practically allowable and not allowable*. The risk management should cover such measures which allow to vary the probability of definite event or to reduce the degree of its consequence. When solving the problem of safe ship control regime selection in heavy seas we assume the degree of consequence as constant. From the other hand by altering ship control settings operator can affect the probability of reaching such ship motion parameters that lay beyond the limits of practically allowable risk. In this case the risk level can be given as

$$R = f\left(p_1, p_2, ..., p_n\right), \qquad (8)$$

where $p_1$, $p_2$,...,$p_n$ = probabilities of reaching the ship motion parameters, that may lead to definite hazardous occurrence.

### 3.2 *Seaworthiness criteria*

To perform the risk assessment and to find a safe control regime in given weather conditions it's necessary to define appropriate criteria, thereupon following factors should be taken into account:

- frequency and force of slamming;
- frequency of green water;
- motion amplitudes;
- hull stresses;
- propeller racing;
- accelerations in various ship points;
- forced and controlled speed redaction.

The comparative table of general operability limiting criteria for wide variety of ships in waves combined from data of Lipis (1982) & Stevens (2002) is given in table 1. However criteria of NORDFORSK and NATO STANAG appear to be too strict, and in series cases, when ship proceeds through a heavy storm, the motion parameters may exceed these criteria.

According to inquiry of management level navigators (captains and chief mates) passing the Ship Handling course in Training & Certifying Centre of Seafarers of Odessa National Maritime Academy (TCCS ONMA) empirical values of ship operability criteria were obtained (table 2).

Table 1. General operability limiting criteria for ships.

| Criterion | Cruikshank & Landsberg (USA) | Tasaki et al. (Japan) | NORDFORSK, 87 (Europe) | NATO STANAG 4154 (USA) |
|---|---|---|---|---|
| RMS of vertical accelerations on forward perpendicular | 0.25 g | 0.8 g / p = $10^{-3}$ | 0.275g ($L_{pp}$ < 100 m) 0.05g ($L_{pp}$ > 300 m) | - |
| RMS of vertical accelerations on the bridge | 0.2 g | - | 0.15g | 0.2g |
| RMS of transverse accelerations on the bridge | - | 0.6 g / p = $10^{-3}$ | 0.12g | 0.1g |
| RMS of roll motions | 15° | 25°/ p = $10^{-3}$ | 6° | 4° |
| RMS of pitch motions | - | - | - | 1.5° |
| Probability of slamming | 0.06 | 0.01 | 0.03 ($L_{pp}$ < 100 m) 0.01 ($L_{pp}$ > 300 m) | - |
| Probability of deck wetness | 0.07 | 0.01 | 0.05 | - |
| Probability of propeller racing | 0.25 | 0.1 | - | - |

*The significant motion amplitudes ($X_{1/3}$) can be obtained by doubling the corresponding RMS (root mean square value).

Table 2. Management level navigators inquiry results.

| | Roll motion amplitude, ° | Slamming, intensity per hour | Deck wetness, intensity per hour | Speed reduction, % | Deviation from course, ° |
|---|---|---|---|---|---|
| Small | < 7 | < 5 | < 5 | < 13 | < 20 |
| Not dangerous | < 14 | < 11 | < 10 | < 24 | < 38 |
| Substantial | < 23 | < 19 | < 20 | < 46 | > 40 |
| Dangerous | > 26 | > 23 | > 23 | > 58 | - |

*The average values of inquiry data are given.
** Example: slamming probability with period of pitching 5 sec and intensity 20 times/hour: **0.028.**

Usage of last gives possibility to perform more detailed, supported by personal seagoing experience of navigators, assessment of ship state in waves.

It should be noted that risk assessment by only threshold values, defined for the series of criteria is ineffective. Therefore, we suggest to apply not two-valued state assessment function, but numerical or linguistic function, defined in range between two extreme values: «0» - «1», «best» - «not allowable» (minimal – maximal risk level).

## 4 FUZZY LOGIC ASSESSMENT

### 4.1 *Assessment algorithm*

To implement above mentioned suggestion seaworthiness assessment system consisting of two fuzzy inference subsystems (FIS) was built (fig. 1) on the basis of more complex model given in (Pipchenko, Zhukov 2010).

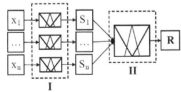

Figure 1. Multicriteria seaworthiness assessment system
$x_1...x_n$ = motion parameters, $S_1...S_n$ = corresponding rates, R = risk level.

Following algorithm was adopted in the system to define the generalized risk level from several motion parameters. Ship motion parameters, taken as the system input, pass the FIS structure of the 1st level. As the result series of rates on each criterion in form of numerical or linguistic variables (for instance, slamming impact: "small", "substantial" or "dangerous") received on its output.

In course of definition system's membership functions (MF) it is suggested to form boundary conditions on the basis of existing international operability criteria, and MF's intermediate values – by approximation of preliminary transformed expert inquiry data.

After that obtained rates pass the FIS of the 2nd level, on the output of which the general assessment on the set of conditions is obtained in the form of risk level. For defuzzification Mamdani algorithm was used in both subsystems.

### 4.2 *Membership functions evaluation*

Let's describe the FIS membership functions (MF) definition process on example of roll amplitude.

Maximum allowable roll amplitude can be determined from condition:

$$\varphi_{1/3}^{limit} = \min\left\{\varphi_{shift}, \varphi_{flood}, \varphi_{capsize}, \varphi_{operator}\right\}, \quad (9)$$

where $\varphi_{shift}$ = cargo critical angle; $\varphi_{flood}$ = flooding angle; $\varphi_{capsize}$ = capsize angle; $\varphi_{operator}$ = operator defined maximum roll amplitude. For general case the maximum angle of 30° was chosen.

For each linguistic term a numerical interval, on which a membership function is defined, can be found from condition:

$$\varphi \in \left(0, \max\left\{\varphi_T^*\right\}\right), \varphi = 0,1,2...,\max\left\{\varphi_T^*\right\} \in N, \quad (10)$$

where $\varphi_T^*$ = values declared by respondents as limits for specified terms. For roll amplitude these terms are: "Non Significant" – NS, "Not Dangerous" – ND, "Significant" – S, "Dangerous" – D.

The principal variable on which the computation of experimental membership function made in the work is *relative term repetition frequency* $\tilde{v}_T = v_T / v_T^{max}$, $v_T$ = quantity of respondents, declared specific value (i.e. roll amplitude is "non significant", if $\varphi < 5°$), $v_T^{max}$ = maximum number of value repetitions for specified term.

Basing on relative term repetition frequency experimental data for membership functions $\mu_T^*$ obtained in the way given below.

For "Non Significant" amplitude term $\mu_{NS}^*$:

$$\left.\begin{array}{l}\mu_{NS}^*(\varphi) = 1 - \tilde{v}_{NS}(\varphi), \text{for } \varphi < \varphi\left(\max\left(\tilde{v}_{NS}\right)\right)\\ \mu_{NS}^*(\varphi) = \tilde{v}_{NS}(\varphi)/2, \text{for } \varphi \geq \varphi\left(\max\left(\tilde{v}_{NS}\right)\right)\end{array}\right\} \quad (11)$$

For "Not Dangerous" amplitude term $\mu_{ND}^*$:

$$\left.\begin{array}{l}\mu_{ND}^*(\varphi) = \tilde{v}_{NS}(\varphi)/2, \text{for } \varphi < \varphi\left(\max\left(\tilde{v}_{NS}\right)\right)\\ \mu_{ND}^*(\varphi) = 1 - \tilde{v}_{ND}(\varphi), \text{for}\\ \varphi\left(\max\left(\tilde{v}_{NS}\right)\right) \leq \varphi < \varphi\left(\max\left(\tilde{v}_{ND}\right)\right)\\ \mu_{ND}^*(\varphi) = \tilde{v}_{ND}(\varphi)/2, \text{for } \varphi \geq \varphi\left(\max\left(\tilde{v}_{ND}\right)\right)\end{array}\right\} \quad (12)$$

For "Significant" amplitude term $\mu_S^*$:

$$\left.\begin{array}{l}\mu_S^*(\varphi) = \tilde{v}_{ND}(\varphi)/2, \text{for } \varphi < \varphi\left(\max\left(\tilde{v}_{ND}\right)\right)\\ \mu_S^*(\varphi) = 1 - \tilde{v}_S(\varphi), \text{for}\\ \varphi\left(\max\left(\tilde{v}_{ND}\right)\right) \leq \varphi < \varphi\left(\max\left(\tilde{v}_S\right)\right)\\ \mu_S^*(\varphi) = \tilde{v}_S(\varphi)/2, \text{for } \varphi \geq \varphi\left(\max\left(\tilde{v}_S\right)\right)\end{array}\right\} \quad (13)$$

From table 2 it can be seen that limit values for terms NS, ND & S roll amplitudes were defined from condition $\varphi < \varphi_{max}^*$. At the same time term "Dangerous" amplitude was defined from condition $\varphi > \varphi_{max}^*$, therefore:

$$\mu_D^*(\varphi) = \tilde{v}_D(\varphi) \quad (14)$$

On the basis of experimental membership functions values, following function can be approximated for application in fuzzy inference algorithm:

$$\left.\begin{array}{l}\mu(\varphi) = e^{\frac{-(\varphi/\varphi_{max} - c)^2}{2\sigma^2}}, \varphi < \varphi_{max}\\ \mu(\varphi) = 1, \varphi \geq \varphi_{max}\end{array}\right\} \quad (15)$$

where $\sigma$, $c$ = function parameters.

As result of approximation four MF's were obtained (fig. 2.).

### 4.3 Rules set definition

To make an inference or to get a determined ship state assessment applying fuzzy logic it is necessary to construct corresponding set of rules.

As input parameters roll amplitude and "maximum probability" coefficient were applied in suggested system. "Maximum probability" coefficient $K_{SGR} \in (0,1)$ can be determined as:

$$K_{SGR} = \min\left(1, \max\left(\frac{p_S}{p_S^{max}}, \frac{p_{GW}}{p_{GW}^{max}}, \frac{p_R}{p_R^{max}}\right)\right) \quad (16)$$

$$K_{SGR} \in (0,1),$$

where $p_S$, $p_{GW}$, $p_R$ = slamming, green water and propeller racing probabilities, superscript *max* means maximum allowable criterial value.

The output risk level R is divided in four linguistic terms: «non significant», «low», «allowable» and «not allowable».

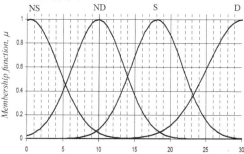

Figure 2. Roll amplitude assessment membership functions

The corresponding set of rules is given in table 3.

Table. 3. Risk evaluation rules set.

| № | | Roll amplitude, $\varphi$ | | Probability coefficient, $K_{SGR}$ | Conclusion Risk level, R |
|---|---|---|---|---|---|
| 1 | IF | Non significant | AND | Low | Non significant |
| 2 | IF | Non significant | AND | Moderate | Low |
| 3 | IF | Not dangerous | AND | Low | Non significant |
| 4 | IF | Not dangerous | AND | Moderate | Low |
| 5 | IF | Significant | AND | Low | Allowable |
| 6 | IF | Significant | AND | Moderate | Allowable |
| 7 | IF | Dangerous | OR | High | Not allowable |

Thus, the risk level for each route leg can be assessed on the basis of weather prognosis data and measured or predicted ship motion parameters. Such prediction can be made by ship dynamic model either linear or non-linear which satisfies accuracy and computational costs criteria. To meet these requirements the combination of linear and non-linear ship motion models were used for calculations in (Pipchenko 2009).

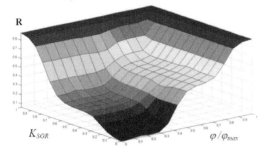

Figure 3. Function surface $\mathbf{R}(\varphi_{1/3}, K_{SGR})$.

## 5 ENGINE LOADS ESTIMATION

To estimate engine power required to keep preset safe speed the functional relationship between speed, power and additional resistance in waves shall be determined.

Ship speed with regard to environmental disturbances, basing on equality condition of propeller thrust to water resistance in calm water can be found as follows:

$$U_W = f\left(T_e - R_W\right); \tag{17}$$

where $T_e$ = propeller thrust in calm water; $R_W$ = average additional resistance due to wind and waves, calculated in this work using methods of Boese (1970) and Isherwood (1973).

Engine load, required to keep specified speed undergoing the wind and waves influence can be determined as:

$$T_w = f\left(U\right) + R_W\left(U\right)$$
$$= c_1 \cdot U^2 + c_2 \cdot U + c_3 + R_W\left(U\right), \tag{18}$$

$$P_w = \frac{T_w \cdot U}{\eta}, \tag{19}$$

where $P_w$ = engine power; $c$ = approximation coefficients, determined from experimental data.

Additional resistance in constant weather conditions can be represented as function of ship speed. Therefore if required speed cannot be reached due to lack of engine power and wave impacts, maximum possible speed can be found applying next recursive procedure:

$E(0) = U'(0)$, $U'(0) = U_{max}$;

**WHILE** $E > \varepsilon$, $\varepsilon \to 0$

$$T_w' = c_1 \cdot U'^2 + c_2 \cdot U' + c_3 - R_W\left(U'\right);$$

$$U'' = \max\left\{0, \min\left\{U_{max}, c_1' \cdot e^{c_2' \cdot T_w'} + c_3' \cdot e^{c_4' \cdot T_w'}\right\}\right\};$$

$$E = \left|U'' - U'\right|; \quad U'' = U'.$$

**END OF CYCLE**

Where $U'$ = calm water speed; $U''$ = predicted maximum speed in waves, defined as inverse function of $T_w$; $c'$ = approximation coefficients, determined from experimental data.

## 6 ROUTE OPTIMIZATION ALGORITHM

The route optimization is performed by following algorithm.
- Ship motion parameters in specified load condition are calculated for defined range of speeds and courses in wave frequency domain. The result of such calculation is a group of four-dimensional arrays $X = f(U, \mu, \omega)$, where $X$ – specified motion parameter.
- Initial transoceanic route is given as great circle line, on which the optimal engine load and corresponding minimal work $A_{min}$ needed to perform the voyage in calm water are estimated.
- Weather prognosis for the voyage is given as multidimensional array with discrecity 1-2° φ x λ.
- After indexing of cells containing weather data, correspondence between route legs and chart grid shall be defined.
- On each route leg (1:N): ship motion parameters for specified wind and wave conditions are recalculated using spectral analysis techniques; risk level and corresponding safe speed are determined.
- If the safe speed on any route leg is less then specified minimum threshold, algorithm switches to route variation stage, if no – engine power inputs and additional work are calculated.
- Optimization task is reached if the minimum additional work in given weather conditions is found, and the maximum risk level on the route is less then specified threshold.

In isochrones method proposed by James (1959) the engine power is considered as constant, where speed is only changed due to wind and waves effect. Thus it's not applicable with the objective function (7). From the other hand directed graph method

(Vagushchenko 2004) allows to control the ship by both speed & course. But to get the accurate solution the dense waypoint matrix shall be built that leads to high computational costs. Therefore we suggest to make generation of alternative routes by setting additional waypoints – "poles". In this case, pole it is intermediate point inserted for avoidance of adverse weather conditions. Positions of poles may be changed either manually or by optimization algorithm.

Poles shall be set as:

$$\begin{bmatrix} Pole_1 \\ Pole_2 \\ ... \\ Pole_m \end{bmatrix} = \begin{bmatrix} \varphi_1 & \lambda_1 \\ \varphi_2 & \lambda_2 \\ ... & ... \\ \varphi_m & \lambda_m \end{bmatrix}, m = 1, 2, ..., M \quad (20)$$

Position of each pole shall satisfy following conditions (fig. 4.):

1 Length of perpendicular, dropped to the orthodromy line between start and destination points must not exceed specified threshold:

$$d(m) \le d_{margin} \quad (21)$$

$$d(m) = \arctan \frac{\cos\left(\arctan \frac{\cos l_{AP}}{\cot(\Psi - \Psi_A)}\right)}{\cot l_{AP}} \quad (22)$$

2 Absolute difference between courses put from pole to start and destination points must exceed 90°. It provides that pole stays in the space between start and destination points:

$$\Psi_P(m) \ge 90^o \quad (23)$$

3 Distance from start point to each next pole shall increase:

$$l_{AP}(m) > l_{AP}(m-1) \quad (24)$$

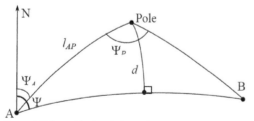

Figure 4. Pole position in relation to the route.

Route legs are rebuilt depending on poles positions. Quantity of waypoints is determined proportionally to the distances between poles. Route legs before or after poles normally built as great circles.

If M≤4, optimization is carried out by Nelder-Mead method. If M>4, optimization is carried out by Genetic Algorithm method, because of Nelder-Meads coefficient quantity limitations.

## 7 EXAMPLE

As an implementation example of the given route optimization method planning of imitated transatlantic route for handymax container vessel (L = 200 m, B = 30 m, GM = 1.0 m) is illustrated on figure 5. The comparison of initial great circle and alternative routes is given in table 4.

Table. 4. Initial and alternative routes comparison.

| Parameter | Great circle | | Alternative route | |
|---|---|---|---|---|
| | Min | Max | Min | Max |
| Distance, nm | 2125 | | 2212 | |
| Poles positions | 42.0 N 45.0 W // 42.0 N 40.0 W // 43.0 N 30.0 W | | | |
| Specified voyage time, hours | 106 | | | |
| Voyage speed, knots | 20 | | 20.8 | |
| ΔA, % from $A_{min}$ | 14.9 | | 11.2 | |
| Risk level, % | 7 | 88 | 6 | 56 |
| Engine load, % | 61 | 95 | 67 | 67 |
| Rolling amplitude, deg | 0 | 18 | 1 | 22 |
| Slamming intensity, times/hour | 0 | 3 | 0 | 0 |
| Green water intensity, times/hour | 0 | 103 | 0 | 0 |

As can be seen from this data, optimization leads to quite good results both for safety and efficiency of the route. Thereupon the time of the voyage remains the same, with even less total engine loads (and obviously less fuel consumption). From the other hand on alternative route slamming and green water probabilities reduced to a minimum. However the rolling amplitude remains high on separate parts of alternative route, but it should be taken into account that there are not many good choices to make as the imitated weather conditions are almost everywhere adverse.

## 8 CONCLUSION

To optimize the transoceanic route objective function which represents ship work expended for the voyage was suggested. The work in that case represented as time integral of main engine power-inputs needed to keep specified speed and to compensate the additional resistance arisen due to environmental disturbances with regard to ship's safety.

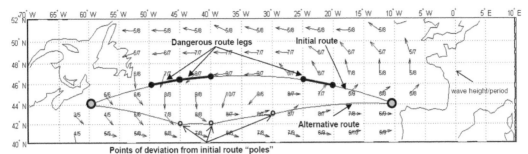

Figure 5. Example of imitated transatlantic route optimization.

To perform the safety assessment a fuzzy logic system which represents relation between ship motion parameters and corresponding risks was developed. That allowed to perform the general risk level value evaluation on the basis of multi input data.

Both these results give the opportunity to perform effective transoceanic route planning on the basis of formalized safety and efficiency assessment with regard to specified weather conditions.

## REFERENCES

Bijlsma, S. J. 2004. A computational method in ship routing using the concept of limited maneuverability. *The Journal of Navigation*. Vol.57, No3: 357-370.

Isherwood, R.M. 1973. Wind resistance of merchant ships. *Transactions of Royal Institution of Naval. Architects, Vol. 115*: 327-338.

James, R.W. 1959 Application of wave forecasts to Marine Navigation. U.S. Navy Hydrographic office.

Boese, P. 1970. Eine Einfache Methode zur Berechnung der Wiederstandserhöhung eines Schioes in Seegang. *Technical Report 258, Institüt für Schiobau der Universität Hamburg, BRD.*

Lipis, V. & Remez, Yu. 1982. Safe ship control regimes in storm. M.: Transport.

Nechaev, Yu., Pipchenko, O. & Sizov, V. 2009. Selection of optimal course in adverse weather conditions. *Navigation: ONMA scientific journal, Issue 16*. Odessa: Izdatinform: 105-118.

Oses, F. & Castells, M. 2008. Heavy Weather in European Short Sea Shipping: It's Influence on Selected Routes. *The Journal of Navigation. Vol.61, No1, Jan 2008*: 165-176.

Padhy, C. P., Sen, D. & Bhaskaran, P. K. 2008. Application of wave model for weather routing of ships in the North Indian Ocean. *Nat Hazards 44*: 373–385.

Pipchenko, O. D. & Zhukov, D. S. 2010. Ship control optimization in heavy weather conditions. *Proceedings of the 11th AGA, IAMU, Busan*: 91-95.

Pipchenko, O. D. 2009. Method for optimal ship routing in adverse weather conditions. *Proceedings of scientific and technical conference «Modern problems of navigation safety enhancement».* Odessa, Ukraine: ONMA: 85-87.

Stevens, S. & Parsons, M. 2002. Effects of Motion at Sea on Crew Performance: A Survey. *Marine Technology, Vol. 39, N1, Jan 2002*: 29-47.

Szlapczynska, J. & Smierzchalski, R. 2007. Adopted Isochrones Method Improving Ship Safety in Weather Routing with Evolutionary Approach. *International Journal of Reliability and Safety Engineering. Vol 14; # 6, 2007*: 635 – 646.

Vagushchenko, O. L. 2004. Method for transoceanic routing. *Navigation: ONMA scientific journal, Issue 8*. Odessa: Latstar: 12-21.

# 21. Weather Hazard Avoidance in Modeling Safety of Motor-driven Ship for Multicriteria Weather Routing

P. Krata & J. Szlapczynska
*Gdynia Maritime University, Gdynia, Poland*

ABSTRACT: Weather routing methods find the most suitable ocean's route for a vessel, taking into account changeable weather conditions and navigational constraints. In the multicriteria approach based on the evolutionary SPEA algorithm one is able to consider a few constrained criteria simultaneously. The approach applied for a ship with hybrid propulsions has already been presented by one of the authors on previous Trans-Nav'2009. This time a motor-driven version of the solution is presented. The paper is focused especially on a proposal of ship safety measure, based on restricting the impact of weather hazards on the ship. Besides the weather conditions and navigational restraint the safety of a vessel is one of more important factors to be considered. The new approach towards a safety factor modeling is described and implemented.

## 1 INTRODUCTION

Weather routing methods and tools deal with a problem of finding the most suitable vessel route. During the route optimization process they take into account changeable weather conditions and navigational constraints. Such a problem is mostly considered for ocean going ships where adverse weather conditions may impact both, often contradictory, economic and security aspects of voyage. Most of recent scientific researches in weather routing focus on shortening the passage time or minimization of fuel consumption alone.

One of the first weather routing approaches was a minimum time route planning based on weather forecasted data. Proposed by R.W. James (James 1957) an isochrone method, where recursively defined time-fronts are geometrically determined, was in wide use through decades. In late seventies based on the original isochrone method the first computer aided weather routing tools were developed. Numerous improvements to the method were proposed since early eighties, with (Hagiwara 1989, Spaans 1986, Wisniewski 1991) among others. Nonetheless, even the improved method has recently been displaced by genetic algorithms.

Evolutionary approach as a natural successor of genetic approach has become popular in the last two decades and has been successfully applied to anti-collision maneuver modeling. Modern weather routing tools often utilize evolutionary algorithms (Wisniewski et al. 2005) instead of the deprecated isochrone time-fronts. Due to multiobjective nature

of weather routing the multicriteria versions of evolutionary algorithms have been also recently applied to the ship routing problem (Marie et al. 2009, Szlapczynska et al. 2009)

One of the authors has already proposed a multicriteria weather routing algorithm - MEWRA (Szlapczynska et al. 2009) designed especially for a ship with hybrid propulsion. In this paper an adjusted for a motor-driven ship and revised version of the algorithm is presented. One of the key amendments is related with modeling the safety measure. Here a new measure, based on reducing the impact of weather hazards on ship is proposed. The new approach towards modeling of ship safety is based on dynamical phenomena taking place while sailing in rough sea. As the ship behavior is strongly nonlinear and difficult for exact prediction (Landrini 2006) a sort of generalization is used. The proposed method is based on the IMO Circ. 1228, concisely comprising significant hazards resulting from complex interactions of ship's hull and waves, especially following and quartering seas.

The paper is organized as follows: section 2 recalls the optimization model and key technical background of original MEWRA algorithm for a hybrid propulsion ship. Section 3 describes the amendments required to suit MEWRA to motor-driven ship model. Section 4 addresses ship stability and seakeeping performance as optimization factors. Then again in section 5 the new measure of safety for MEWRA is introduced. Finally, section 6 summarizes the material presented.

## 2 MULTICRITERIA WEATHER ROUTING FOR A SHIP WITH HYBRYD PROPULSION

The following subsections recall the optimization model and the general framework of the Multicriteria Evolutionary Weather Routing Algorithm - MEWRA (Szlapczynska et al. 2009) designed for a ship with hybrid propulsion.

### 2.1 Optimization model

A proposed set of goal functions in the weather routing optimization process is presented below:

$$f_{passage\_time}(t_r) = t_r \rightarrow \min \qquad (1)$$

$$f_{fuel\_consumption}(v_{fc}) = v_{fc} \rightarrow \min \qquad (2)$$

$$f_{voyage\_risks}(i_{safety}) = (1 - i_{safety}) \rightarrow \min \qquad (3)$$

where:

$t_r$ – [h] passage time for given route and ship model,
$v_{fc}$ – [t] total fuel consumption for given route and ship model,
$i_{safety}$ – [-] safety coefficient for given route and ship model. It is defined as a value ranging [0;1], describing a level to which the route is safe to be passed. "0" depicts totally impassable route and "1" absolutely safe route.

Exact formulas for goal functions (1) – (3) strongly depend on the assumed ship model. Thus, the explicit formulas for a ship model with hybrid-propulsion can be found in (Szlapczynska et al. 2009).

Set of constraints in the considered optimization problem includes the following elements:
– landmasses (land, islands) on given route,
– predefined minimum acceptable level of safety coefficient for given route,
– shallow waters on given route (defined as waters too shallow for given draught of ship model),
– floating ice bergs expected on given route during assumed ship's passage,
– tropical cyclones expected on given route during assumed ship's passage.

### 2.2 Mutlicriteria Evolutionary Weather Routing Algorithm (MEWRA)

The Multicriteria Weather Routing Algorithm (MEWRA), presented in Figure 1, searches for an optimal route (according to goal functions (1) – (3)) for the assumed ship model. The input data for the algorithm are:
– geographical coordinates of route's origin & destination,
– weather forecasts (wind, wave and ice) for considered ocean area and time period of the voyage being planned.

Figure 1. Multicriteria Evolutionary Weather Routing Algorithm (MEWRA)

The algorithm starts with a generation of initial population i.e. a diversified set of routes including the outermost elements of the searching space (Figure 2). The modified isochrone method (Hagiwara 1989) with extensions described in (Szlapczynska et al. 2007) is a source of single-criterion time-optimal and fuel-optimal routes. The routes are then a base for random generation of initial population. Also the original routes are included in the population.

In the next step SPEA algorithm iteratively proceeds the evolution on the initial population towards achieving Pareto-optimal set of routes. Once the evolution cannot improve on the Pareto set anymore the first optimization procedure is stopped. Then, from the set of Pareto-optimal routes (Figure 3) a single route must be selected, becoming a route recommendation.

Yet another problem might be encountered: how to decide which route should be recommended? To solve this problem decision-maker's (e.g. captain's) preferences to the given criteria set should be defined. Hence a tool for sorting the Pareto-optimal set is provided – Fuzzy TOPSIS method. First the decision-maker has to set their preferences for given criteria set. In MEWRA these preferences are expressed by means of linguistic values (Table 1) with fuzzy sets assigned accordingly. The decision-maker selects a linguistic value of the predefined set to each of optimization criteria. Then the corresponding fuzzy sets build a weight vector for the ranking method. The last step of MEWRA – Fuzzy TOPSIS – is responsible to apply given weight vector to the decision matrix built of the goal function values of the Pareto-optimal routes. The route having the highest value of ranking automatically becomes then a route recommendation (Figure 4). Exemplary MEWRA results (Figures 2 – 4) have been obtained for hybrid propulsion ship model, Miami-Lisbon voyage on 2008-02-15 (departure time 12:00 pm) and decision-maker preferences given in Table 2.

Table 1. Linguistic values and corresponding triangular fuzzy values, utilized to express decision-maker's preferences to the criteria set

| Linguistic value | Triangular fuzzy set |
|---|---|
| very important | (0.7; 1.0; 1.0) |
| important | (0.5; 0.7; 1.0) |
| quite important | (0.2; 0.5; 0.8) |
| less important | (0.0; 0.3; 0.5) |
| unimportant | (0.0; 0.0; 0.0) |

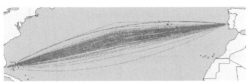

Figure 2. Initial population generated by MEWRA for Miami-Lisboa voyage on 2008-02-15 (departure time 12:00 pm)

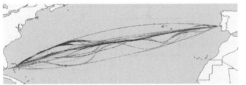

Figure 3. Pareto-optimal set of routes generated by MEWRA for Miami-Lisbon voyage on 2008-02-15 (departure time 12:00 pm)

Table 2. Linguistic values assigned by a decision-maker to the criteria set

| Criterion name | Linguistic value | Triangular fuzzy set |
|---|---|---|
| Passage time | Important | (0.5; 0.7; 1.0) |
| Fuel consumption | quite important | (0.2; 0.5; 0.8) |
| Voyage safety | very important | (0.7; 1.0; 1.0) |

Figure 4. Recommended route selected by MEWRA according to the preferences given in Table 2 for Miami-Lisbon voyage on 2008-02-15 (departure time 12:00 pm)

## 3 MULTICRITERIA WEATHER ROUTING FOR MOTOR-DRIVEN SHIPS

MEWRA application for the motor only propulsion has been constructed based on its hybrid propulsion predecessor. The key differences between application versions for hybrid and motor driven ship model are as follows.

1 The safety coefficient $i_{safety}$ (utilized by goal function (3) and appropriate constraints) in case of the hybrid propulsion model is based on wind speed and heading only (assuming strict correlation be-

tween wind and wave conditions). In the motor-driven model the coefficient has been redefined as a percentage part of a route that is free from disturbances caused by weather hazards. However, regions with severe wave threat are still bypassed by means of fulfilling a new constraint for restricted course sectors. More details on both these elements are given in the following sections.

2 Weather input forecasts for the motor-driven model has been enlarged by wave period and wave angle. Also MEWRA's graphical user interface (GUI) has been changed accordingly to allow displaying the new data on the screen.

3 In the hybrid propulsion case there is a possibility to use one of the three propulsion modes, namely "motor only", "sails only" or "motor & sails". The "sails only" mode for a route segment requires the engine to be temporarily switched off, which in return may significantly decrease the fuel consumption. In the motor-driven case there is only one propulsion mode i.e. "motor only", which drastically limits possible fuel savings.

## 4 SHIP STABILITY AND SEAKEEPING PERFORMANCE AS OPTIMIZATION FACTORS

The most straightforward attitude towards ocean voyage routing and route optimization is about to find the shortest way from a point of origin to a destination. However, getting rid of any other important aspects of navigation seems to be too simple and incomplete. Thus, a set of objectives has been implemented in the MEWRA's optimization model with voyage safety as the most important element, with the highest degree of significance as given in Table 2, of the goal function set Thus, it is required to elaborate on safety of the vessel modeling.

The desirable course of ship exploitation requires not only fast steaming of a vessel but also a lack of ship and cargo damage. The analysis of historical data regarding LOSA casualties reveals that their causes may be attributed to interacting elements presented at the Venn diagram in Figure 5 (Kobyliński 2007a).

Figure 5. Four-fold Venn diagram for ship stability system (Kobylinski 2007a)

The interaction of all four groups of conditions from Venn diagram can lead to stability accident of a vessel and perturb a voyage. To avoid such a situation a set of stability standards are worked out. The ship's loading condition of insufficient stability may induce a list, a strong heel and even a capsizing.

Vessels' stability calculation and evaluation, performed on-board nowadays, is based on the stability criteria published by the ship's classification societies. These criteria are mainly based on the A749(18) Resolution of International Maritime Organization. The resolution and their later amendments are known as the Intact Stability Code.

The ship stability criteria are to ensure the relevant level of safety against capsizing and strong heel. The capsizing is not often occurring phenomenon at the sea, although it cause considerable number of fatalities. The most significant feature influencing the capsizing rate is the size of a vessel (Krata 2007). Generally the smaller is the vessel, the bigger is the risk of capsizing. This is due to the scalability in vessel stability. It depends on the square-cubed rule; i.e. the heeling forces, which depends on water and wind impact areas, go up with the square of the dimensions, but the righting moment which depends on the displacement, goes up with the cube of the dimensions (Womack 2002).

Taking into account the square-cubed rule of ship stability and simultaneously the same stability standards for all size of vessels, the Intact Stability Code cannot be applied as a basis for safety factor estimation for the purpose of route optimization. Moreover, ISC stability standards are related to one state of weather and one state of a sea, described by the wind pressure and ships rolling amplitude, while the optimization procedure has to follow variable weather and sea conditions to be passed by a vessel on her way.

The state-of-the-art solution to the stability assessment problem is a risk based approach. The core idea of this approach is presented in Figure 6.

The risk based approach to ship stability assessment reveals a list of advantages but the simplicity comparing to proscriptive stability standards based on IS Code. It comprises a seakeeping performance of a ship and a wide range of possible weather and sea conditions, however cannot be computed relatively easy. The researches are still underway and no complete practical tool has been established yet. Furthermore, application of such an approach would be extremely time-consuming which is unacceptable for the purpose of multicriteria evolutionary weather routing algorithm described in the paper.

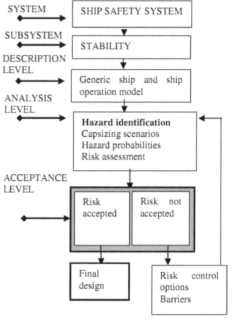

Figure 6. Block diagram of risk based approach related to ship stability (Kobylinski 2007b)

## 5 NEW MEASURE OF SHIP SAFETY BASED ON WEATHER HAZARD AVOIDANCE

Consideration of characteristic features of contemporary methods of ship stability assessment (described in section 4) and the main aim of the study i.e. weather routing optimization, leads to the conclusion that there is a need for new measure of ship safety. The index or coefficient of safety should comprise safety aspects related to ship stability and seakeeping performance and simultaneously it has to be reasonably applicable. Both, time of computation and reliability of computation results need to be considered and balanced.

The limitation of ISC-based approach is its disregard of size of a vessel and consideration of only one hydrometeorological conditions. Correspondingly the limitation of risk based approach towards ship stability is a lack of fully practical computational tools and a long time of processing. A sort of a trade-off between these two methods may be the new approach based on weather hazard avoidance.

The proposal of a new measure of ship safety or just a method for safety index calculation is derived form the *Revised guidance to the master for avoiding dangerous situation in adverse weather and sea conditions*, published by the International Maritime Organization as MSC Circular 1228. The relatively up to date publication (in comparison to dr. Rahola's findings from 1939 being the foundation of IS

Code) comprises a set of remarks and advices regarding avoidance of following dangerous dynamical phenomena at sea (IMO 2007):
- surf-riding and broaching-to;
- reduction of intact ship stability when riding a wave crest amidships;
- synchronous rolling motion;
- parametric roll motions.

According to the IMO Circ. 1228 some combinations of wave length and wave height under certain operation conditions may lead to dangerous situations for ships complying with the IS Code. As the sensitivity of a ship to dangerous phenomena depends on the actual stability parameters, hull geometry, ship size and ship speed, the vulnerability to dangerous responses, including capsizing, and its probability of occurrence in a particular sea state may differ for each ship (IMO 2007). In addition the wave encounter period and the wave length and height depend on actual hydrometeorological conditions. These features are strictly advantageous for the purpose of weather routing applications and as a consequence they are desirable for safety index calculation in the course of route optimization.

The dangerous surf-riding zone cay be obtained from a graph presented in IMO Circ. 1228 which is shown in Figure 7. The symbol $V$ denotes velocity of a vessel in knots and $L$ – ship's length between perpendiculars.

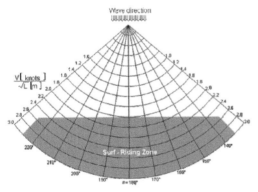

Figure 7. Dangerous zone of surf-riding in following or quartering seas (IMO 2007)

It is found out that ship surf-riding and broaching-to may occur when:
- the angle of encounter $\alpha$ is in the range $135° < \alpha < 225°$;
- and the ship speed is higher than $(1.8 \cdot L) \cdot \cos(180 - \alpha)$ (knots).

To avoid surf riding, and possible broaching the ship velocity should be taken outside the dangerous region reported in Figure 7 (IMO 2007). The alternation of the speed, the course or both is advised.

The next group of dangerous dynamic phenomena taken into account in IMO guidance 1228 is related to resonance gain of rolling motion. This may occur due to nonlinearity of ship response in resonance conditions, i.e. when the encounter wave frequency is similar to first or second harmonic frequency of natural roll motion of a ship (Landrini 2006).

The period of encounter $T_E$ can be calculated by the formula:

$$T_E = \frac{3 \cdot T_W^2}{3 \cdot T_W + V \cdot \cos(\alpha)} \ [s] \qquad (4)$$

where:
$T_W$ – [s] wave period,
$V$ – [knots] ship's speed,
$\alpha$ – [deg] angle of encounter (angle between keel direction and wave direction).

The prevention against a synchronous rolling motion consists in avoiding such combinations of ship speed and course which result in the encounter wave period $T_E$ nearly equal to the natural rolling period of ship $T_R$.

For avoiding parametric rolling in following, quartering, head, bow or beam seas the course and speed of the ship should be selected in a way to avoid conditions for which the encounter period is close to the ship roll period ($T_E \approx T_R$) or the encounter period is close to one half of the ship roll period ($T_E \approx 0{,}5 \ T_R$) (IMO 2007).

Another dangerous zone presented in IMO 1228 guidance is established for high wave attack and a list of phenomena related to sailing in following and quartering rough sea. This dangerous zone is presented in Figure 8.

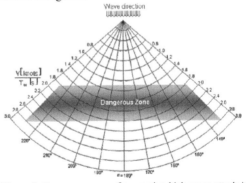

Figure 8. Dangerous zone of successive high wave attack in following and quartering seas (IMO 2007)

A ship master is advised to avoid entering dangerous zone indicated in Figure 8 when:
- the average wave length is larger than 0.8 ship's length between perpendiculars;
- and the significant wave height is larger than 0.04 ship's length;

and at the same time some indices of dangerous behavior of the ship can be seen (IMO 2007). In case of actual velocity vector violating the dangerous zone, the ship speed should be reduced or the ship course should be changed to prevent successive attack of high waves, which could induce the danger due to the reduction of intact ship stability, synchronous rolling motions, parametric rolling motions or combination of various phenomena (IMO 2007).

The core idea of new safety index calculation with regard to weather hazard avoidance is to include all the restrictions described in IMO *Revised guidance to the master for avoiding dangerous situation in adverse weather and sea conditions*. For the purpose of practical application of new concept, the appropriate software was developed. All the restrictions resulting from the guidance are computed and presented in the form of polar plots.

The polar plots presenting dangerous configurations of speed and course of a vessel need to be computed for every single ship and even for every loading conditions of a ship. The length of a ship is a variable and furthermore, the hull beam, draft and the metacentric height influences her natural rolling period. Thus, for a given ship in actual loading conditions a set of polar plots should be prepared and used.

The use of such polar plot presenting zones to be avoided is quite simple. The only condition is to keep the velocity vector away from the restricted zone. For the sake of presentation an exemplary case was computed and plotted. The characteristics of a ship taken into consideration are following:
– length between perpendiculars 180 m;
– beam 30 m;
– draft 11 m;
– metacentric height 1.9 m.
The fully developed ocean wave system was considered.

The dangerous zone where an exemplary ship is exposed to surf-riding is plotted in Figure 9.

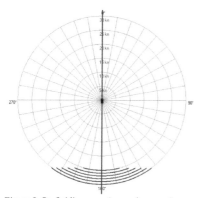

Figure 9. Surf-riding zone (exemplary case)

Then, the zones describing ship course / speed configuration where she is likely to experience synchronous roll motion or parametric rolling are shown in Figure 10.

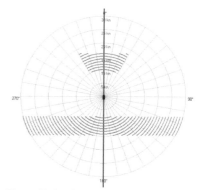

Figure 10. Synchronous rolling and parametric rolling zones (exemplary case)

The last dangerous zone is related to conditions comprising high wave action and phenomena associated with such waves in following and quartering seas. The polar plot of this zone is computed for an exemplary case and shown in Figure 11.

All the results of computation of dangerous zones can be plotted together as one graph. Some of the zones may overlap for a number of conditions. The cumulative graph computed and plotted for exemplary ship characteristics is shown in Figure 12.

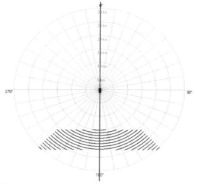

Figure 11. High wave attack zone (exemplary case)

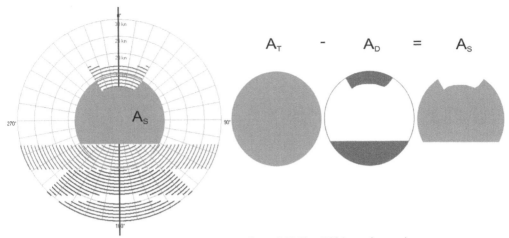

Figure 12. Cumulative graph presenting dangerous zones according to IMO Circ. 1228 (exemplary case)

The core idea of the proposed new ship safety index calculation is based on the cumulative graph comprising all dangerous zones described in IMO Circ. 1228. The safety index $SI$ is defined by the following formulas:

$$SI = \frac{A_S}{A_T} \ [\text{-}] \tag{5}$$

$$A_S = A_T - A_D \tag{6}$$

where:
$A_T$ – total area of ship maximum speed circle,
$A_S$ – area of non dangerous zone inside ship maximum speed circle,
$A_D$ – area of all dangerous zones included in ship maximum speed circle.

The total area of ship maximum speed circle depicts just all possible speeds and courses of a ship. It depends on ship's hull and propulsion characteristics and does not take into account any potential danger. Thus, this total area is determined by maximum speed of a vessel.

The area of all dangerous zones (exemplary shown in Figure 12) depicts restricted configuration of ship speed and course with regard to adverse weather. The greater dangerous area is the more constraints faces a vessel.

The area of non dangerous zone represents allowed configurations of ship velocity vector which are found as safe in terms of IMO Circ. 1228. This area reflects the potential possibility of course and speed choice and alternation or in other words a kind of freedom in safe maneuvering.

The safety index $SI$ defined by the formula (5) varies from 0 to 1. The value $SI=0$ expresses a lack of safe choice of vessel speed and course. The value $SI=1$ stands for a complete absence of any re-

strictions, which is typical for good weather conditions. The safety index based on weather hazard avoidance can be apply as an element of optimization process according to the formula (3) i.e. one of the goal functions in multicriteria weather routing process.

According to the procedure described in the paper it is clearly apparent that the safety index is not a fixed value for the entire voyage of a vessel, even for one type of the vessel and fixed lading conditions (however unchanging loading conditions are rather impossible to maintain in long term due to fuel consumption and ballasting). The shape of dangerous zones depends on wave characteristics therefore all the defined zones needs to be updated consequently after every reception of up to date weather forecast. It needs to be emphasized that weather routing and described weather routing optimization tool are designed for planning purposes. Thus, the processing has to be performed prior encountering adverse weather conditions.

## 6 SUMMARY

The multicriteria evolutionary weather routing algorithm (MEWRA) presented in the paper offers a constrained, three element goal function route optimization. Especially for a motor-driven ship the rough sea impact on ship must be taken into account when safety of people and cargo on board are in stake. Thus the authors have proposed a brand new safety measure that becomes an element of the goal function set in the optimization process.

The new safety index refers to dynamical phenomena taking place in rough sea conditions which advantages this approach among contemporary ones. The consideration of dynamical behavior of a vessel instead an application only Intact Stability Code based requirement is a significant step forward. The authors believe that the MEWRA solution together with the safety index can be successfully applied to a motor-driven ship route planning and optimization.

# REFERENCES

James, R.W. 1957. *Application of wave forecast to marine navigation*, Washington: US Navy Hydrographic Office.

Hagiwara, H. 1989. *Weather routing of (sail-assisted) motor vessels*, PhD Thesis, Delft: Technical University of Delft.

IMO. 2007. Revised guidance to the master for avoiding dangerous situation in adverse weather and sea conditions, International Maritime Organization, MSC.1/Circ.1228.

Kobylinski L. 2007a. System and risk approach to ship safety, with special emphasis of stability, *Archives of Civil and Mechanical Engineering. Vol. VII. No. 4.* Wrocław, Poland.

Kobylinski L. 2007b. Goal-based Stability Standards, *10th International Ship Stability Workshop*, Hamburg, Germany.

Krata P. 2007. Total loses of fishing vessels. *Advances in marine navigation and safety of sea transportation*, Gdynia, Poland.

Landrini M. 2006. Strongly Nonlinear Phenomena in Ship Hydrodynamics, *Journal of Ship Research, Vol. 50, No. 2.*

Marie S. & Courteille E. 2009. Multi-Objective Optimization of Motor Vessel Route. *Marine Navigation and Safety of Sea Transport. Balkema Book, Taylor & Francis Group.* London.

Spaans, J.A., 1986, *Windship routeing*, Technical University of Delft.

Szlapczynska J., Smierzchalski R. 2007. Adopted isochrone method improving ship safety in weather routing with evolutionary approach. *International Journal of Reliability, Quality and Safety Engineering, Vol. 14, No. 6. 635-645.* World Scientific Publishing Company.

Szlapczynska J. & Smierzchalski R. 2009. Multicriteria Optimisation in Weather Routing. *Marine Navigation and Safety of Sea Transport. Balkema Book, Taylor & Francis Group.* London.

Wisniewski, B. 1991. *Methods of route selection for a sea going vessel* (in Polish), Gdansk: Wydawnictwo Morskie.

Wisniewski B. & Chomski J. 2005. Evolutionary algorithms and methods of digraphs in the determination of ship time-optimal route, *2nd International Congress of Seas and Oceans*, Szczecin-Świnoujście.

Womack J. 2002. Small commercial fishing vessel stability analysis, Where are we now? Where are we going?, *Proceedings of the 6th International Ship Stability Workshop*, Washington.

# 22. Evolutionary Sets of Safe Ship Trajectories: Evaluation of Individuals

R. Szlapczynski
*Gdansk University of Technology, Gdansk, Poland*

J. Szlapczynska
*Gdynia Maritime University, Gdynia, Poland*

ABSTRACT: The paper presents a description of the evaluation phase of the Evolutionary Sets of Safe Ship Trajectories method. In general, the Evolutionary Sets of Safe Ship Trajectories method combines some of the assumptions of game theory with evolutionary programming and finds an optimal set of cooperating trajectories of all ships involved in an encounter situation. While developing a new version of this method, the authors decided to use real maps instead of a simplified polygon modelling and also to focus on better handling of COLREGS. The upgrade to the method enforced re-designing the evaluation phase of the evolutionary process. The new evaluation is thoroughly described and it is shown how evaluation affects final solutions returned by the method.

## 1 INTRODUCTION

A desired solution to a multi-ship encounter situation would include a set of planned, optimal trajectories for all the ships involved in an encounter, such that no collision or domain violations occur when these ships follow the trajectories. When solving this situation the key difficulty is that even a single course change performed by one ship involved in the encounter may force one or even more the other ships to manoeuvre. Thus the optimisation method utilized to find a solution to the problem should be flexible enough to efficiently look through the vast search space and handle even minor changes in the ship's behaviour e.g. in its motion parameters.

There is a number of approaches to solving a multi-ship encounter situation. Two basic trends are either utilization of differential games (Lisowski 2005) or searching for a single trajectory (for the own ship) by evolutionary algorithms (Smierzchalski et al. 2000). The former method assumes that the process of steering a ship in multi-ship encounter situations can be modeled as a differential game played by all ships involved, each having their strategies. Unfortunately, high computational complexity is its serious drawback. The latter approach is the evolutionary method focused on finding only a single trajectory of the own ship. In short, the evolutionary method uses genetic algorithms, which, for a given set of pre-determined input trajectories find a solution that is optimal according to a given fitness function. However, the method's limitation is that it assumes targets motion parameters not to change

and if they do change, the own trajectory has to be recomputed. This limitation becomes a serious one on restricted waters. If a target's current course collides with a landmass or another target of a higher priority, there is no reason to assume that the target would keep such a disastrous course until the crash occurs. Consequently, planning the own trajectory for the unchanged course of a target will be futile in the majority of such cases. Also, the evolutionary method does not offer a full support to VTS operators, who might face the task of synchronizing trajectories of multiple ships with many of these ships manoeuvring.

Therefore, the authors have proposed a new approach, which combines some of the advantages of both methods: the low computational time, supporting all domain models and handling stationary obstacles (all typical for evolutionary method), with taking into account the changes of motion parameters (changing strategies of the players involved in a game). Instead of finding the optimal own trajectory (from the own ship's perspective) for the unchanged courses and speeds of targets, an optimal set of safe trajectories of all ships involved is searched for (from the coast, e.g. VTS, perspective). The method is called evolutionary sets of safe trajectories and its early version has been presented by one of the authors in (Szlapczynski 2009).

The newly developed version of the method uses real maps instead of simplified polygon modelling and focuses on COLREGS compliance. The upgrade to the method enforced changes in all phases of the evolutionary process including evaluation. The paper presents a description and a discussion of the new evaluation phase.

The rest of the paper is organized as follows. In the next section a brief description of the problem is given, including basic constraints of the optimization problem as well as the additional constraints - the COLREGS rules, which are taken into account. Section 3 covers the issue of detecting various constraints violations. This is followed by a Section 4, where it is shown, how, on the basis of previous sections, the fitness function is formulated. In section 5 different evaluation approaches and the consequences of applying them are compared by means of simulation experiments. Finally the summary and conclusions are given in Section 6.

## 2 SOLVING MULTI-SHIP ENCOUNTER SITUATIONS AS AN OPTIMIZATION PROBLEM

It is assumed that we are given the following data:
– stationary constraints (such as landmasses and other obstacles),
– positions, courses and speeds of all ships involved,
– ship domains,
– times necessary for accepting and executing the proposed manoeuvres.

Ship positions and ship motion parameters are provided by ARPA (Automatic Radar Plotting Aid), or, if there is no reliable identification assured, AIS (Automatic Identification System) systems. A ship domain can be determined based on the ship's length, its motion parameters and the type of water region. Since the shape of a domain is dependent on the type of water region, the authors have assumed and used a ship domain model by Davis (Davis et al. 1982), which updated Goodwin model (Goodwin 1975), for open waters and to use a ship domain model by Coldwell (Coldwell 1982), which updated Fuji model (Fuji et al. 1971), for restricted waters.

As for the last parameter – the necessary time, it is computed on the basis of navigational decision time and the ship's manoeuvring abilities. By default an assumed 6-minute value is used here.

Knowing all the abovementioned parameters, the goal is to find a set of trajectories, which minimizes the average way loss spent on manoeuvring, while fulfilling the following conditions:
– none of the stationary constraints are violated,
– none of the ship domains are violated,
– the minimal acceptable course alteration is not lesser than 15 degrees (assumed to eliminate slow and insignificant turns),
– the maximal acceptable course alteration is not to be larger than assumed 60 degrees,

– speed alteration are not to be applied unless necessary (collision cannot be avoided by course alteration up to 60 degrees),
– a ship manoeuvres, if and only if she is obliged to,
– it is assumed that manoeuvres to starboard are favoured over manoeuvres to port board.

The first two conditions are obvious: all obstacles have to be avoided and the ship domain is an area that should not be violated by definition. All the other conditions are either imposed by COLREGS (IMO 1977) and good marine practice or by the economics. In particular, the course alterations lesser than 15 degrees might be misleading for the ARPA systems (and therefore may lead to collisions) and the course alterations larger than 60 degrees are not recommended due to efficiency reasons. Also, ships should only manoeuvre when necessary, since each manoeuvre of a ship makes it harder to track its motion parameters for the other ships ARPA systems (Wawruch 2002). Apart from these main constraints, additional constraints – selected COLREGS rules have to be directly handled.

The COLREGS rules, which are of interest here are:
– Rule 13 – overtaking: an overtaking vessel must keep well clear of the vessel being overtaken.
– Rule 14 - head-on situations: when two power-driven vessels are meeting head-on both must alter course to starboard so that they pass on the port side of the other.
– Rule 15 - crossing situations: when two power-driven vessels are crossing, the vessel, which has the other on the starboard side must give way.
– Rule 16 - the give-way vessel: the give-way vessel must take early and substantial action to keep well clear.
– Rule 17 - the stand-on vessel: the stand-on vessel may take action to avoid collision if it becomes clear that the give-way vessel is not taking appropriate action.

There are also some additional COLREGS-related assumptions, namely:
– there are always good visibility conditions,
– all considered ships are equally privileged,
– all considered ships have motor engine (no sailing ships taken into account),
– no narrow passages are taken into account
– no port board manoeuvres are assumed when overtaking,
– no manoeuvres to bypass navigational signs are taken into account.

In the following sections it will be analysed how these constraints violations can be detected, in what order should they be taken into account and how severely should they be penalized during the evaluation phase by the fitness function of the evolutionary method.

# 3 DETECTING CONSTRAINTS VIOLATIONS

Below it is described how the constraints violations can be detected and, in case of various possible approaches, which one has been chosen by the authors and why.

## 3.1 Detecting static constraints violations (collisions with landmasses and safety isobate)

In the first version of the method (Szlapczynski 2009) simplified polygon modelling of the static constraints have been applied, instead of using real maps. Therefore it was natural to find collisions by detecting all crossings of the ships' trajectories with polygons' edges. This is shown in Figure 1. A number of operations that the algorithm has to perform to find collisions in such situation is proportional to the number of the edges of all polygons in a given area.

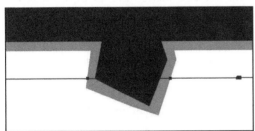

Figure 1. A ship's trajectory crossing a landmass modeled as a polygon. The geometrical crossings of the trajectory and polygon edges are marked in black

However, the current version of the method uses a vector map of a given area. While vector maps also uses polygons defined by coordinates of their vertices, the number of vertices and thus the edges rises drastically, when compared to the simplification used before. Even after limiting the map to a certain area, the numbers of the edges that have to be checked for possible crossings are still much larger. This is shown in Figure 2.

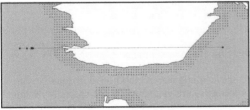

Figure 2. A ship's trajectory crossing a landmass on a bitmap

Therefore the authors have decided not to process vector map directly for crossing detection, but to use it for generating bitmap of an area. Although it is a time-taking operation, fortunately, it is enough to generate such bitmaps offline and only once for each area. Then, when the method is running in real time, instead of checking the edges for geometrical crossings, each bitmap cell, which the trajectory of a ship traverses, is read and checked if it belongs to landmass, water or safety isobate. For a bitmap, whose detail level reflects this of a given vector map, the computational time would be much shorter: proportional to the number of traversed cells, instead of a number of all vertices. This approach is also more flexible in terms of future implementation of bathymetry: if every cell contained information on the water depth, it would be easy to check, whether a cell is passable or not for a particular ship.

## 3.2 Detecting collisions with other ships

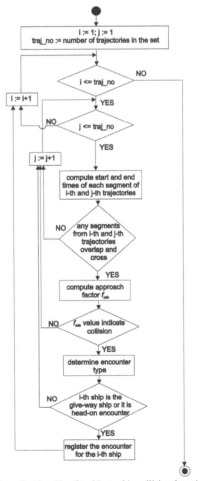

Figure 3. Algorithm for ship-to-ship collision detection

The algorithm for detecting ship-to-ship collisions (Figure 3) is as follows. Each ship's trajectory is checked against all other ships. For each pair of ships, the start time and end time of each trajectory's segments are computed. If two segments of the two trajectories overlap in time, they are checked for geometrical crossing. In case of a crossing, the approach factor value is computed. Then, if the approach factor value indicates collision, the type of an encounter (head-on, crossing or overtaking) is determined on the basis of the ships' courses and it is decided, which ship is to give way (both ships in case of head-on). The collision is only registered for the give way ship and the information on the collision are stored in the trajectory data structure.

### 3.3 Detecting COLREGS violations

Detecting COLREGS violations is much more difficult than violations described in the previous two sub-sections. In general, there may be three types of COLREGS violations:
- a ship does not give way, when it should,
- a ship gives way, when it should not, because it is a stand-on ship,
- a ship manoeuvres to port-board when it should manoeuvre to starboard.

Each of these three situations may happen on either open or restricted waters, which gives us a total of six cases to handle. The difficulty with deciding, whether a ship has acted lawfully, or not, lies in the nature of evolutionary algorithms as well as in the nature of the problem itself: COLREGS specify only the procedures for ship-to-ship encounters. Looking at a set of ship trajectories for a multi-target encounter it is sometimes impossible to tell, what was the reason for a particular manoeuvre: which ship was given way intentionally, and which one benefited from it only as a side effect. A partial solution to this problem is storing in the trajectory data the information on the reasons of the manoeuvres. The possible reasons might be:
- landmass avoidance or other static constraint violation avoidance,
- giving way to a privileged ship,
- any other, e.g. due to the ship's passage plan.

However, the course alterations that each trajectory contains may be made intentionally – as a result of applying a collision avoidance operator or unintentionally – as a result of crossing or mutation. A manoeuvre which resulted accidentally from crossing or mutation may be just as good as the one being the effect of a specialised operator's more 'conscious' work. Therefore the 'any other' manoeuvre's reason cannot always be registered as COLREGS violation. All this considered, the authors have decided to limit the used types on the manoeuvre's rea-

sons to: obstacle avoidance and any other. The final COLREGS violations detection rules are:
1 On open waters:
  a) if a ship is not obliged to give way to any other ship, any manoeuvre (other than the manoeuvres given by the passage plan) it performs is registered as COLREGS violation,
  b) if a ship is obliged to give way, and does not perform a manoeuvre it is registered as COLREGS violation,
  c) all manoeuvres to port board are registered as COLREGS violations.

The c) point may raise some doubts, but it must be emphasized that COLREGS violations registration is done for the sake of future penalizing of violations, when the final fitness function values is being computed. Therefore, the only effect of penalizing the manoeuvres to port board will be additional favouring of manoeuvres to starboard, which are already favoured by domain models. In no way does penalizing make it impossible to choose a manoeuvre to port board. It is only less profitable for most cases.

2 On restricted waters: here, as explained before every trajectory node, which is a part of a manoeuvre, contains special information on the reason why this particular node has been inserted or shifted: land or other stationary obstacle avoidance, target avoidance or accidental manoeuvre generated by evolutionary mechanisms. Based on this, COLREGS violations are registered as follows:
  a) if a ship does not initially have to give way to any target and its first manoeuvre has reason other than static constraint violation avoidance, it is registered as COLREGS violation,
  b) any manoeuvre to port board of reason other than static constraint violation avoidance is registered as COLREGS violation.

Point b) means that occasionally the correct manoeuvres introduced by crossing or mutation and avoiding static constraint violation will be penalized unjustly. However, it is not a problem, as long as penalties for static constraint violations will be larger and trajectories avoiding them will still be selected for next generations. After all, we are interested in the final sets of trajectories themselves much more than in their slightly imprecise fitness function values.

## 4 FORMULATING FITNESS FUNCTION

In the evolutionary method all individuals (sets of trajectories) are evaluated by the specially designed fitness function, which should reflect optimisation criteria and constraints (Michalewicz et al. 2004). In this section it is shown, how, on the basis of previous sections, this fitness function is formulated.

### 4.1 Basic criterion – minimizing way loss

The basic criterion is the economic one – minimizing way losses of trajectories in a set. For each of the trajectories, a *trajectory_economy_factor* is computed according to the formula (1).

$$trajectory\_economy\_factor_i = \left( \frac{trajectory\_length_i - way\_loss_i}{trajectory\_length_i} \right) \quad (1)$$

where:
$i$ – the index of the current ship [/],
*trajectory_length*$_i$ – the total length of the $i$-th ship's trajectory [nautical miles],
*way_loss*$_i$ – the total way loss of the $i$-th ship's trajectory [nautical miles] computed as a difference between the trajectory length and length of a segment joining trajectory's start point and endpoint.

As can be seen, the *trajectory_economy_factor* is always a number from a (0,1] range.

### 4.2 Penalizing static constraint violation

After the trajectory economy factor has been computed the static constraints are handled by introducing penalties for violating them. For each trajectory its static constraint factor *scf*$_i$ is computed. The static constraints are always valid and their violations must be avoided at all cost, therefore penalties applied here are the most severe – hence the square in the formula (2).

$$scf_i = \left( \frac{trajectory\_length_i - trajectory\_cross\_length_i}{trajectory\_length_i} \right)^2, \quad (2)$$

where:
*trajectory_cross_length*$_i$ – the total length of the parts of the $i$-th ship's trajectory, which violate stationary constraints [nautical miles].

The static constraint factor is a number from a [0,1] range, where "1" value means no static constraint violation (no landmasses or other obstacles are crossed) and "0" value is for trajectories crossing landmasses on their whole length.

### 4.3 Penalizing collisions with other ships

Analogically to the static constraint factor, collision avoidance factor *caf*$_i$ is computed to reflect the ship's collisions with all other privileged ships as shown by (3).

$$caf_i = \prod_{j=1, j \neq i}^{n} \left( \min(fmin_{i,j}, 1) \right) \quad (3)$$

where:
$n$ – the number of ships [/],
$j$ – the index of a target ship [/],

*fmin*$_{i,j}$ – the approach factor value for an encounter of ships $i$ and $j$, if $i$-th ship is the privileged one, the potential collision is ignored and the approach factor value is equal to 1 by definition. [/].

The collision avoidance factor is a number from a [0,1] range, where "1" value means no ship domain violation and "0" means a crash with at least one of the targets.

### 4.4 Penalizing COLREGS violations

The COLREGS violations are secondary to static constraint violations and to collisions with other ships and therefore the authors have decided to penalize it moderately, to make sure that constraints from the previous two points are met first. COLREGS compliance factor *ccf*$_i$ is computed according to the following formula (4).

$$ccf_i = 1 - \sum_{k=1}^{m} \left[ COLREGS\_violation\_penalty_k \right], \quad (4)$$

where:
$m$ – the number of COLREGS violation registered for the current ship as has described in section 3.3 [/],
$k$ – the index of a registered violation [/],
*COLREGS_violation_penalty*$_k$ – the penalty for the $k$-th of the registered COLREGS violation [/].

The penalty values for all registered COLREGS violations described in section 3.3 by points 1. a) - c) and 2. a) - b) are configurable in the method and are set to 0.05 by default.

### 4.5 Fitness function value

Once all aforementioned factors have been computed, the fitness function value is calculated. The authors wanted the fitness function to be normalized, which is convenient for further evolutionary operations, mostly for selection purposes. When fitness function values are normalized, we do not need any additional operations on them and they can directly be used for random proportional and modified random proportional selection in the reproduction and succession phases of the evolutionary algorithm. We can also easily measure and see progress we make with each generation. However, normalized fitness function is harder to obtain, because we have to make sure that we keep the high resolution of evaluating the individuals, namely that we differ between various levels of penalties: stationary constraints, being more important than collision avoidance and collision avoidance being more important than COLREGS compliance.

Here, we succeeded in formulating a normalized fitness function, while keeping relatively high resolution of evaluation: minor stationary constraints violations are penalized similarly as major collisions with other ships and minor collisions with other ships are penalized similarly as multiple COLREGS violations. The final fitness function is as follows:

$$fitness = \sum_{i=1}^{n} \frac{trajectory\_fitness_i}{n},$$ (5)

where:

$$trajectory\_fitness_i =$$
$$= trajectory\_economy\_factor_i * scf_i * caf_i * ccf_i,$$ (6)

The final fitness function value assigned to an individual is an arithmetical average of fitness function values computed for all trajectories. It is discussable, whether all trajectories should have the same impact on final fitness function value (as it is done here), or should the trajectory fitness function values be taken with weights proportional to the trajectory lengths. When combined with the formula for trajectory economy factor, the current approach means that we are trying to minimize average relative way loss computed over all trajectories, instead of total absolute way loss (with weights being used). However, experiments have shown, that minimizing total absolute way loss leads to discrimination of ships, whose basic trajectories are shorter and to their large relative way losses (section 5.2).

## 5 COMPARING DIFFERENT EVALUATION APPROACHES

In the following subsections different evaluation approaches and the consequences of applying them are compared.

### 5.1 Penalizing COLREGS violations: how it affects solutions returned by the method

Even when a domain model, which favours COLREGS is applied, it is possible to find an encounter situation, where additional COLREGS violations penalties must be used, as has been described in section 4.4 or otherwise the method will return incorrect solution. A simple example is a head-on encounter of two ships, whose parameters are shown in Figure 4. In this scenario, following the Rule 14 of COLREGS for head-on situations, it is required that:
*"(...) both (vessels) must alter course to starboard so that they pass on the port side of the other".*

| Ship Parameters | | | | |
|---|---|---|---|---|
| | Initial pos | Goal pos | Velocity [kn] | Turn [deg./sec.] |
| Ship1 | (7.03 ; -10.56) | (-1.37 ; 12.91) | 12.47 | 1.00 |
| Ship2 | (-2.67 ; 13.58) | (8.22 ; -11.29) | 13.57 | 1.00 |

Figure 4. Parameters of two ships in a head-on encounter

Because the method tends to propose manoeuvres no lesser than 15 degrees, a manoeuvre from one ship only would be enough to avoid collision. From the way loss minimization point of view, the extra manoeuvre from the second ship is redundant. Consequently, individuals containing trajectories with manoeuvres from both ships would be ranked lower than those with only one ship manoeuvring and the final solution will have only one ship manoeuvring, which is shown in Figure 5.

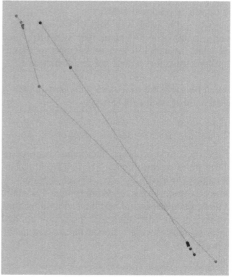

Figure 5. An incorrect solution to a head-on encounter situation returned by the method without COLREGS violations penalties

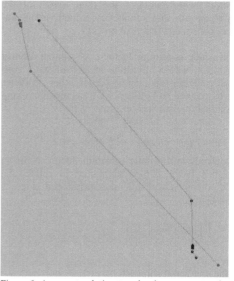

Figure 6. A correct solution to a head-on encounter situation returned by the method with COLREGS violations penalties applied

Thus we need to additionally penalize the individuals for COLREGS violations to favour the individuals with both ships manoeuvring and larger way loss. The default penalties of 0.05 are sufficient for the correct solution to be chosen. This is shown in Figure 6.

### 5.2 Optimization criterion: total absolute way loss or average relative way loss

Another question already raised before (section 4.5) is whether we should minimize total absolute way loss or average relative way loss. An example scenario of an encounter of 6 ships on restricted waters is presented below. Ship parameters are gathered in Figure 7. The results of minimizing total absolute way loss are shown in Figure 8, the results for the minimization of average relative way loss – in Figure 9.

| Zero longitude 20.5 | | | Zero latitude 58.5 | |
|---|---|---|---|---|
| **Ship Parameters** | | | | |
| | Initial pos | Goal pos | Velocity [kn] | Turn [deg./sec.] |
| Ship1 | (58.28 ; 163.55) | (67.26 ; 138.75) | 13.19 | 1.00 |
| Ship2 | (54.19 ; 139.40) | (71.34 ; 162.89) | 14.54 | 1.00 |
| Ship3 | (79.25 ; 157.53) | (46.29 ; 144.76) | 17.67 | 1.00 |
| Ship4 | (73.59 ; 145.04) | (51.93 ; 157.24) | 12.43 | 1.00 |
| Ship5 | (70.05 ; 138.46) | (55.65 ; 163.90) | 14.61 | 1.00 |
| Ship6 | (48.01 ; 144.18) | (71.99 ; 155.82) | 13.32 | 1.00 |

Figure 7. Parameters of six ships in an encounter on restricted waters

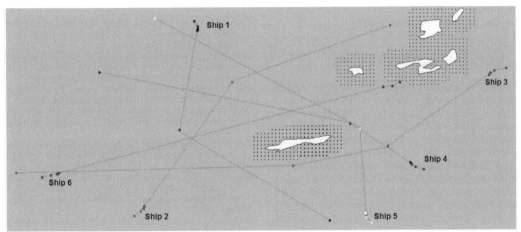

Figure 8. A solution to a multi-ship encounter situation returned by the method with minimization of total absolute way loss

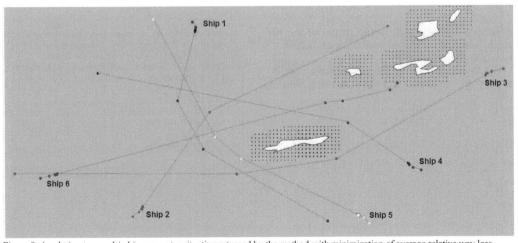

Figure 9. A solution to a multi-ship encounter situation returned by the method with minimization of average relative way loss

As can be seen above, minimizing average relative way loss (Figure 9) results in smoother trajectories for ship 1 and ship 5. Ship 5 also has considerably lesser way loss because it passes the island on its left side (Figure 9), instead of right side (Figure 8). Other trajectories (the longer ones) have no major visual differences between them in Figures 8 and 9, though fitness function values of some ships are slightly larger for Figure 8, because of their (insignificantly) lesser way losses. Unfortunately, it is impossible to formally compare the solutions returned by the two variants of the method, which use different formulas for global fitness function and thus aim at different goals. However, after a series of simulation experiments, the authors are of the opinion that in general the minimization of average relative way loss brings more balanced and intuitive results for most cases and therefore have chosen it to be the default option of the current version of the Evolutionary Sets of Safe Trajectories method.

## 6 SUMMARY AND CONCLUSIONS

The paper documents the research on the evaluation phase of the Evolutionary Sets of Safe Ship Trajectories method. For some of the optimisation constraints, gathering the data on their violations for evaluation purposes is time consuming (collisions with other ships and static obstacles), while for others it is discussable in some cases, whether a constraint has been met or not (COLREGS rules), which seriously limits detection possibilities. Even such a seemingly simple issue as main optimisation criterion (way loss minimisation) becomes a problem, when a particular fitness function value is to be formulated. The authors have explored various possibilities of gathering the data on constrain violation, as well as using them in the fitness functions and have presented in the paper their conclusions: the techniques and formulas that, in the course of the research, occurred to be most useful for evaluation of the sets of ship trajectories.

The chosen elements of the method have been illustrated by simulation examples showing how a change in the evaluation phase affects the final solutions returned by the method. The authors' search for the optimal evaluation is being continued, as the whole method's functional scope is expanding. The current works are focused on handling Traffic Separation Schemes directly in the Evolutionary Sets of Safe Trajectories method, which brings new evaluation issues.

ACKNOWLEDGEMENTS

The authors thank the Polish Ministry of Science and Higher Education for funding this research under grant no. N N516 186737.

REFERENCES

Coldwell T.G. 1982. Marine Traffic Behaviour in restricted Waters, The Journal of Navigation, 36, 431-444. Cambridge: Cambridge Journals.
Davis P.V. & Dove M.J. & Stockel C.T. 1982. A Computer Simulation of multi-Ship Encounters. The Journal of Navigation, 35, 347-352. Cambridge: Cambridge Journals.
Fuji J., Tanaka K. (1971). Traffic Capacity. The Journal of Navigation, 24, 543-552. Cambridge: Cambridge Journals.
Goodwin E.M. 1975. A Statistical Study of Ship Domains. The Journal of Navigation, 28, 329-341. Cambridge: Cambridge Journals.
IMO. 1977. Convention on the International Regulations for Preventing Collisions at Sea. The International Maritime Organization.
Lisowski J. 2005. Dynamic games methods in navigator decision support system for safety navigation, Proceedings of the European Safety and Reliability Conference, vol. 2, 1285 – 1292.
Michalewicz Z. & Fogel D.B. 2004. How To Solve It: Modern Heuristics, Springer-Verlag.
Smierzchalski, R., Michalewicz, Z. 2000. Modeling of a Ship Trajectory in Collision Situations at Sea by Evolutionary Algorithm, IEEE Transactions on Evolutionary Computation No. 3 Vol. 4, pp. 227-241.
Szlapczynski R. 2009. Solving multi-ship encounter situations by evolutionary sets of cooperating trajectories, Marine Navigation and Safety of Sea Transportation. 437-442. CRC Press / Taylor & Francis Group / Balkema.
Wawruch R. 2002. ARPA zasada działania i wykorzystania, Gdynia: WSM Gdynia.

# 23. Development of a 3D Dynamic Programming Method for Weather Routing

S. Wei & P. Zhou
*Department of Naval Architecture and Marine Engineering, University of Strathclyde, Glasgow, UK*

ABSTRACT: This paper presents a novel forward dynamic programming method for weather routing to minimize ship fuel consumption during a voyage. Compared with the traditional two dimensional dynamic programming (2DDP) methods which only optimize the ship's heading, while the engine power or propeller rotation speed are set as a constant throughout the voyage, this new method considers both the ship power setting and heading control. A float state technique is used to reduce the iteration on the process of optimization for computing time saving. This new method could lead to a real global-optimal routing in a comparison with a tradition weather routing method which results in a sub-optimal routing.

## 1 INTRODUCTION

Ship weather routing is defined as an optimum track of ship route with an optimum engine speed and power for an ocean voyage based on en-route weather forecasts and ship's characteristics. Within specified limits of weather and sea conditions, the term optimum means a maximum of safety and crew comfort, a minimum of fuel consumption and time underway, or any desired combination of these factors. It can be clearly seen that, the accuracy of determining the optimum route depends on three aspects.
- The accuracy of the prediction of the ship's hydrodynamic behavior under different weather conditions.
- The accuracy of the weather forecast.
- The capability and practicability of the optimization algorithm.

The focus of this study is on research of the optimization algorithm. Many optimization algorithms have been developed for solving ship routing problems in which ship fuel consumption and/or passage time are minimized. Most popular methods include calculus of variations (Bijlsma S.J 1975), modified isochrone method (Hagiwara H 1989, Hagiwara H & Spaans JA 1987), two dimensional dynamic programming (2DDP) method (De Wit C 1990, Calvert S et al. 1991,) and isopone method (Klompstra MB et al. 1992, Spaans JA 1995) .

The method of calculus of variation treats the ship routing as a continuous optimization problem. Inaccuracy in the solution may arise in the functions where second order differentials are required. The errors could be expanded to an unacceptable level at the end of the calculation.

The modified isochrone method is a recursive algorithm. The route with the minimum of passage time is obtained by repeatedly computing isochrones (or time fronts) which are defined as the outer boundaries of the attainable region from the departure point after a certain time. This method offers a route with minimum fuel consumption by keeping the propeller revolution speed as a constant during the voyage first, then applying the modified isochrone method to determine the minimum time of passage. By varying propeller rotation speed this method is able to find the propeller rotation speed at which the minimum time of passage satisfies with the desired arrival time. This minimum time route will be treated as the minimum fuel route. Thus, the fuel consumption of this route itself is not minimized.

The 2DDP method based on Bellman's principle of optimality is similar to the modified isochrone method. It uses a recursive equation to solve ship routing problems formulated as a discrete optimization problem. The accuracy of the solution depends on the fineness of the grid system used. Compared with the modified isochrone method, the advantage of the 2DDP method is that it allows the operators to take into account of navigation boundaries by means of an appropriate selection of the grid points. Both the modified isochrone and 2DDP methods assume that ship sails at a constant propeller rotation speed or a constant engine power for the entire voyage.

The isopone method is an extension of the modified isochrone method. An isopone is the plane of equal fuel consumption that defines the outer boundary of the attainable region in a three-dimensional space, i.e. position and time. This method enables the operators to consider the variations of ship engine power to optimize the route. SPOS, a weather routing software, adopted the isopone method at the beginning of the software development. Although the isopone method is mathematically more elegant and theoretically offers better results than that of the modified isochrone method, finally SPOS had nevertheless abandoned the isopone method and applied the modified isochrone method. The main reason for this change was due to the fact that the isopone method appeared to be more difficult to understand by the operators onboard ships, whereas the modified isochrone method is straightforward and easy to understand.

Besides these methods, there are many other methods that have been used for weather routing in recently years, like iterative dynamic programming algorithm (Kyriakos Avgouleas 2008), augmented Lagrange multiplier (Masaru Tsujimoto, Katsuji Tanizawa 2006), Dijkstra algorithm (Chinmaya Prasad Padhy et al. 2008), genetic algorithm (Harries S, Hinnenthal J 2004) and so on.

Weather routing was first developed for determining shipping courses during a voyage with a minimum of passage time. However, nowadays shipping companies began to show more interesting in reducing fuel consumption driven by the fuel oil prices, environmental considerations and maintaining a certain time schedule which is specified in the chartering contract of a merchant vessel. In this paper, a new forward three dimensional dynamic programming (3DDP) method is presented for minimizing fuel consumption during a voyage. It is an extension of the traditional 2DDP method, allowing change heading and speed with both time and position, thus, it is able to realize a real global optimum result. Compared with the isopone method, the 3DDP method is straightforward and easy for programming.

## 2 PROBLEM STATEMENT

Ship engine power and shipping course directly decide the shipping route in the ocean. Ship speed over the ground depends on the engine power. There is a one-to-one relationship between them. Thus, both of them can be equally treated as the control variables in a weather routing process. In this paper, ship speed over the ground and shipping course measured from the true north are chosen as the control variables. The control variables are denoted as a control vector $\vec{U}$, $\vec{U} = U(u, \psi)$, where $u$ represents ship speed over the ground and $\psi$ is shipping course measured from the true north. Ship position $\vec{X}$ is al-

so a vector, specified by longitude $\varphi$ and latitude $\theta$, $\vec{X} = X(\varphi, \theta)$.

Ship position $\vec{X}$ and voyage time $t$ determine the ship trajectory. Using $\vec{E}$ to denote weather conditions (speed and direction of wind, significant height, direction and peak frequency of wave and swell), $\vec{E}$ is a function of position $\vec{X}$ and time $t$,

$$\vec{E} = E(\vec{X}, t) \tag{1}$$

During a voyage, constraints $\vec{C}$ must be met. The constraints include geographic constraints, control constraints and safety constraints.

Thus, ship position $\vec{X}$ at time $t$ can be described by the function below:

$$\vec{X} = f_0 (\vec{X}', \vec{U}', \vec{E}', \vec{C}') \tag{2}$$

Where $\vec{X}'$, $\vec{U}'$, $\vec{E}'$, $\vec{C}'$ correspond to time $t'$, $t - t' = \Delta t$, $\Delta t$ is a time step used in calculation.

Because $\vec{E}'$ is a function of $\vec{X}'$ and $t'$, so $\vec{X}$ can also be described by:

$$\vec{X} = f (\vec{X}', \vec{U}', t', \vec{C}') \tag{3}$$

This function can be explained that while comply with the constraints, the ship will arrive at the present position $\vec{X}$ at the present time $t$ from $\vec{X}'$ under control of $\vec{U}'$ during $\Delta t$ time step.

Instantaneous fuel consumption rate $q$ can be obtained by:

$$q = q (\vec{X}, \vec{U}, t, \vec{C}) \tag{4}$$

The total fuel consumption for a voyage can be obtained by

$$C = \int_{t_{end}}^{t_s} q(\vec{X}, \vec{U}, t, \vec{C}) \, dt \tag{5}$$

Where:
Initial conditions: $\vec{X}_s = (\varphi_s, \theta_s)$, $t = t_s$
Final conditions: $\vec{X}_{end} = (\varphi_{end}, \theta_{end})$, $t = t_{end}$
The constraints $\vec{C}$: Geographic constraints, control constraints, safety constraints.

## 3 DYNAMIC PROGRAMMING

### 3.1 Advantages of the 3DDP method

Dynamic programming is a method which can solve complex problems by breaking them down into many simpler sub-problems. A stage is defined as the division of sequence of the sub-problems in the optimization procedure. The procedure of this method is to solve the sub-problems stage by stage. The variables used to define a stage must be parameters which are monotonically increasing with the progress of problem solving going-on. There are two choices of variable selection to specify a stage for ship routing problems, i.e. time and a measure of the progress of the vessel from departure (voyage progress). Each stage consists of many states which can be defined as a specific measurable condition of the ship operation, such as time and location. If time is chosen as the stage variable, the state can be defined by possible locations where the ship could pass. If voyage progress is chosen as the stage variable, states should be defined by time and possible positions away from the great circle.

The 2DDP method chose voyage progress as the stage variable. Because this method assumes that ships sail at a constant propeller rotation speed or a constant engine power for the entire voyage, there is a one-to-one relation between ship position and time. Thus, time variable is not needed to specify states in this method. Several authors have already attempted to solve the weather routing problem by using 3DDP treating both engine power and shipping course as the control variables during a voyage. Aligne, F. et al. (1998) chose time as the stage variable and used the forward algorithm; Henry Chen (1978) and Simon Calvert (1990) employed the voyage progress as the stage variable and used the backward algorithm. The method presented in this paper employs the voyage progress as the stage variable together with the use of the forward algorithm.

The advantages of using the forward algorithm can be stated as the following: When optimizing a route, the initial departure time is fixed, the arrival time can be treated as a flexible parameter, allowing a set of route with a minimum fuel consumption to be obtained corresponding to different specified arrival times in one calculation.

To compare using voyage time as stage variables, the advantages of using voyage progress as stage variables are:, ship headings are pre-defined by voyage progress on grid points, so that ship speed over the ground becomes the only explicitly defined control variable to be optimized during the routing optimization process. This method doesn't need a finer grid system. It can save much more computing time than the methods which choose voyage time as a stage variable.

## 3.2 Grid design

Since the great circle is an optimum route under calm water conditions from the departure to the destination, it is chosen as a reference for the construction of the grid system used in the route optimization.

As describe above, states are of three dimensions, i.e. time and geographic location with a unit spacing $\Delta Y$ located on a stage, perpendicularly away from the great circles. The farthest states on a stage from the great circle are the possible locations the ship may pass to avoid a bad weather or certain sea conditions. Unlike the tradition dynamic programming, the variable of voyage time t of states are determined as the optimization procedure is progressing. Grids

should be deleted when a shipping route crosses islands/rocks.

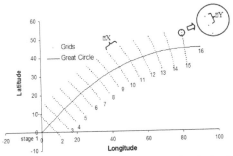

Figure 1. Projections of space grid system on a longitude × latitude plane.

Fig. 1 shows an example of stage projections on a longitude × latitude plane where 16 stages have been allocated (1, 2, 3…16) from the departure to the destination of a shipping route. The distance between two stages can be equally spaced $\Delta X$. The total number of stages is determined according to the total distance of the route and the availability of computing capacity.

## 3.3 Estimate of fuel consumption between two stages

Fuel consumption is a function of ship hydrodynamics. The hydrodynamics of ship are modeled and simulated in the process of fuel consumption estimation. Thus, the accuracy of the modeling of the ship hydrodynamics is critical in the accuracy of fuel consumption estimation. As a consequence of added resistance due to wind and waves as well as the increased hull roughness over time, ship speed is often smaller than the designed speed so called involuntary speed reduction. Besides that, a voluntary speed reduction is needed to insure ship safety to minimize or avoid slamming, deck wetness, propeller racing, parametric rolling, motion sickness, engine overloading and so on. All these factors need to be considered in routing optimization as the constraints. Since the focus of this paper is to discuss the optimization algorithm of fuel consumption during a ship voyage, how to accurately predict ship hydrodynamic at the sea is not discussed in depth or further. However, the procedure of prediction of fuel consumption between two stages is presented here. This procedure can be treated as a sub-problem of a dynamic programming problem. The optimized fuel consumption during an entire voyage is obtained by adding up of all individual fuel consumption between two stages along a route with the newly developed 3DDP method.

As a ship voyage follows a predefined grid system, her heading is fixed between two stages. The ship speed over the ground is the only control variable which directly determines the fuel consumption between any two stages during a course of shipping.

Fig. 2 shows the procedure determining fuel consumption between two stages. In detail, it is as the following:

Step 1: Calculation of ship resistance. Ship resistance is calculated based on the ship speed over the ground, draft, trim and the weather condition. Ship resistance can be divided into three main components: a). the calm water resistance; b). the added resistance due to wave; c). the added resistance due to wind. Ship sea trial data, model test data and numerical simulation results are used to estimate these resistances.

Step 2: Estimation of engine power. The engine power is calculated to overcome the above calculated resistances based on the propeller characteristics.

Step 3: If the engine power is more than MCR (maximum continuous rate), the ship speed will be reduced by $\Delta u$, and then go back to step 1.

Step 4: Calculation of probability of slamming, deck wetness, and propeller racing. To ensure the ship safety, if these constrain values exceed certain pre-set limits, the ship speed will be reduced $\Delta u$, and then go back to step 1.

Step 5: Calculation of fuel consumption and ship position for next time interval $\Delta t$.

Step 6: Execute step 1 to step 5 repetitively in a fixed time interval $\Delta t$ between two stages until the ship (simulation step) arrives at the next stage or final destination.

The time interval $\Delta t$ for calculation is normally chosen at the frequency of the reception of weather forecasting onboard which is usually every 6 hours.

### 3.4 Algorithm description

The backward recursive algorithm has been used in most dynamic programming of weather routing. However, the forward dynamic programming offers more convenience in programming. The forward dynamic programming can be interpreted as that a path is optimal if and only if, for any intermediate stages, the choice of the foregoing path is optimum for this stage. By using this principle, the weather routing procedure can be broken down into a sequence of simpler problem solving. Notations defined in the programming are as follows.

- $K$: total number of stage.
- $N(k)$: total number of state projection on the latitude × longitude plane on stage $k$, where: $k = 1, 2, 3... K$. $N(1) = 1, N(K) = 1$.
- $P(i, k)$: state project position on stage $k$, where: $i = 1, 2, 3..., N(k)$. $P(1, 1)$ is the departure position; $P(1, K)$ is the destination position.
- $J$: total number of time interval between states on a geographical position.

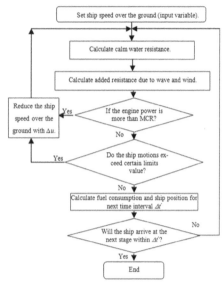

Figure 2. Estimate of fuel consumption between two stages.

- $\Delta \bar{t}$ : time interval between states on a stage.
- $X(i, j, k)$: a state on stage $k$, where: $i = 1, 2, 3...,$ $N(k), j = 0, 1, 2 ..., J - 1, k = 1, 2, 3..., K$. The geographic position of the state $X(i, j, k)$ is $P(i, k)$ on stage $k$, the time variable of the state is $t_{i,j,k}$, $\bar{t}_j \leqq t_{i,j,k} \leqq \bar{t}_{j+1}$, and $\bar{t}_j = \Delta \bar{t} \times j, \bar{t}_{j+1} = \Delta \bar{t} \times (j + 1)$. The state $X(i, j, k)$ at stage $k$ is a floating point between position $(P(i, k), \bar{t}_j)$ and $(P(i, k), \bar{t}_{j+1})$.
- $X(1, 0, 1)$: the initial state. Time variable $t_{1,0,1}$ of the initial state is 0.
- $X(1, j, K)$: the states on the final stage $K, j = 0, 1, 2..., J - 1$, the position of these states is $P(1, K)$.
- $F_{opt}(X(i, j, k))$: the minimum fuel consumption from the initial state to the state $X(i, j, k)$.
- $u(m)$: ship speed over the ground varying between 5 to 30 knot, $u(m) = 5 + 0.1 \times (m - 1)$, where: $m = 1, 2, 3..., M$.

The recursion procedure of the forward dynamic programming can be described as follows:

Step 1: Set stage variable $k = 1$.

Step 2: Iterate step 3 to 6 below for each attainable state $X^* = X(i, j, k)$ on stage $k$, where: $i = 1, 2, 3..., N(k), j = 0, 1, 2..., J$. if a state $X^* = X(i, j, k)$ is unattainable due to constraints, its fuel consumption $F_{opt}(X(i, j, k))$ is set to infinitive.

Step 3: Calculate ship heading $H^*$ from $X^*$ to the next stage position $P(i', k + 1)$, where: $i' = 1, 2, 3..., N(k + 1)$. Iterate step 4 to 5 for each $H^*$. If a ship heading $H^*$ violates the heading constraints or geographic feasibility, calculation for this heading is given up and go to the next heading calculation.

Step 4: Iterate steps 5 for each $u^* = u(m)$, where: $m = 1, 2, 3..., M$. If certain ship speed $u^*$ violates the control constraints or safety constraints, skip out of this loop and go to the next loop.

Step 5: Choose $u^*$ and $H^*$ as the control variables, calculate the fuel consumption $\Delta f_{m,\,i'}$ and the voyage time $\Delta t_{m,\,i'}$ between state $X^*$ on stage $k$ and next state on stage $k+1$ with a geographic position $P\,(i',\,k+1)$ by using the method described in section 3.3, $t_{m,\,i'}$ is denoted as the arrival time at position $P\,(i',\,k+1)$ from the initial state, $t_{m,\,i'} = t_{i,\,j,\,k} + \Delta t_{m,\,i'}$, the position $X' = (P\,(i',\,k+1),\,t_{m,\,i'})$ forms a new possible state on stage $k+1$, The fuel consumption at $X'$ is $f^*$ is determined by $f^* = F_{opt}\,(X\,(i,\,j,\,k)) + \Delta f_{m,i'}$.

Step 6: When the calculation of all possible states $X'$ between time $\bar{t}_{j'}$ and $\bar{t}_{j'+1}$ at position $P\,(i',\,k+1)$ on stage $k+1$ is completed, the possible state which has the minimum fuel consumption $f^*_{min}$ is chosen as the state $X\,(i',\,j',\,k+1)$ Thus, $F_{opt}\,(X\,(i',\,j',\,k+1)) = f^*_{min}$. The departure state $X^*$ on stage $k$, the arrival state $X\,(i',\,j',\,k+1))$ on stage $k+1$ and the corresponding control variables between the two states are saved for tracing the optimum route by a backward procedure at the end of the calculation. During the optimization process states within a time interval are floating. The benefit of using float states is that it eliminates the calculation of the interpolation. Thus, it can save computing time. When the weather in $\Delta \bar{t}$ time does not change much this method will not influence the accuracy of the optimized result.

Step 7: Let $k = k + 1$, then go back to step 2 until $k = K$.

Once the final state on stage $K$ has been obtained, a backward calculation procedure is used to identify the optimized fuel consumption route with the specified arrival time and the corresponding control variables during the entire voyage.

# 4 CASE STUDY

This section presents two case studies with the use of above described 3DDP method. As a simplification, the weather conditions are set artificially. Although the weather conditions used are not real and certain conditions may never happen in the reality, the use of artificial weather conditions will offer the same effect as the real ones in illustrating the methodology and advantages of the 3D dynamic programming. Holtrop method, a regression analysis method, is used to predict the total resistance in calm water. The engine power is calculated by propeller characteristics of the case ship.

Two different sets of weather conditions are used in the case studies with the following common parameters:

- Case ship: a 54,000 DWT container ship.
- Departure from: $\vec{X}_s = (0,\,0)$.
- Arrival at: $\vec{X}_{end} = (90,\,0)$.

- Time interval between states on a stage: $\Delta \bar{t} = 1$ hour.
- Time step for fuel consumption calculation between two stages $\Delta t = 6$ hours
- Ship speed: $u = 5$ to 30 knots.
- Ship speed change step: $\Delta u = 0.1$ knots.
- Total stage number: $K = 16$.
- Total number of stage projection on a stage: $N\,(1) = 1$, $N\,(K) = 1$, $N\,(k) = 17$, where $k = 2, 3, 4 \dots K - 1$.
- Stage space: $\Delta X = 360$ nautical miles.
- State space: $\Delta Y = 75$ nautical miles.

## 4.1 *Case study 1*

The geographic constraints and weather conditions for case 1 study are shown in Fig 3. The geographic constraints are set as a rectangular area which can be islands, rocks or mine fields. The scope of the geographic constraints is longitude from 50 to 70 degree and latitude from -1 to 7 degree. The envelop of the bad weather is set as a rectangular area as well, positioned longitude from 50 to 70 degree, latitude from - 1 to - 9 degree at time $t = 0$. The bad weather stays at this initial area for 60 hours before moving towards south with a speed of 3 knots. The ship is not allowed to enter into the bad weather area for safety consideration. Fig.4 shows the results of fuel consumption vs. time obtained from the route optimization for the case study 1. When the specified arrival time is smaller than 233 hours, both of the 3DDP and 2DDP methods can get the similar strategies which choose the route closed to the dotted line in Fig. 5 and a constant ship speed during the voyage. That means, changing the ship heading can get much benefit than changing the ship speed at this weather condition. When the specified arrival time is more than 272 hours, the time during the voyage is relatively long, so the bad weather already pass away before the ship arrive there, the 3DDP method also get a similar strategy with the 2DDP method which choose the route closed to the solid line in Fig. 5 and a constant ship speed. When the specified arrival time is between 233 hours and 272 hours, the 3DDP method can get a better result than it calculated by the 2DDP method.

Fig. 5 and Fig. 6 show the optimized route and optimized ship speed obtained by using the 2DDP and 3DDP methods under a specified arrival time $t_{end} = 264$ hours. The results have demonstrated that fuel consumption obtained by 2DDP is 1014.54 tons for the given voyage conditions and that is 969.25 tons if the ship follows the route resulted from 3DDP. As a result, route and operation profile optimized by the 3DDP offers about 4.5% of fuel saving compared that with the 2DDP. The reason for the fuel saving is that the 3DDP method permits the ship to change the heading and speed during the route. In the first section of the route, the ship slows down to let the bad weather pass-by first. Once the bad weather has passed, the ship increases her speed to ensure the desired arrival time is achieved.

## 4.2 *Case study 2*

In case 2 study the geographic constraints is the same as that of case 1. Whereas the bad weather area at time $t = 0$ is longitude from 50 to 70 degree, latitude from - 7 to - 15 degree. The bad weather moves towards north with a speed of 3 knots. Fig. 7 shows the geographic constraints and weather conditions. Fig.8 shows results of fuel consumption vs. time obtained from the route optimization for the case study 2. Because of the same reasons with the case 1, when the specified arrival time is smaller than 270 hours or bigger than 303 hours, both the 3DDP and 2DDP methods can get the similar results; when the specified arrival time is between 270 hours and 303 hours, the 3DDP method can get a better result than it calculated by 2DDP method.

Fig. 9 and 10 present the route and ship speed optimized by the 2DDP and 3DDP methods under the same arrival time $t_{end} = 278$ hours. The fuel consumption calculated by the 2DDP is 898 tons and that is 852 tons from the 3DDP calculation. A 5.1% of fuel saving has been achieved by using the 3DDP compared with the 2DDP method. Unlike the case study 1 where ship speed is reduced to wait for the bad weather to pass during the first part of the route, the ship speed is increased to pass the region before the bad weather comes. Once the ship has passed the region where the bad weather is going to pass, the ship speed is slowed down and maintained the desired arrival time.

## 5 CONCLUSION

A newly developed 3DDP for weather routing has been presented. Case studies have shown that compared with the use of traditional 2DDP method, fuel saving can be achieved by using the newly developed 3DDP method in certain constraints and weather conditions. Since the speed of the ship varies according to the weather conditions and movement, the newly developed 3DDP increases the safety of shipping.

The 3DDP method considers optimization of both the ship speed and heading. Its operation and programming are easier and straight forward.

In this paper, real weather forecast is not considered, but this 3DDP method can also give enlightenment for the weather routing problem. In the future, this method will be used based on real weather forecast and ship hydrodynamics.

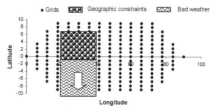

Figure 3. Geographic constraints and weather conditions (case 1).

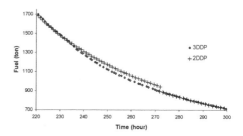

Figure 4. The fuel consumption vs. time curve (case 1).

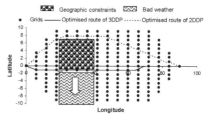

Figure 5. Optimized route (case 1).

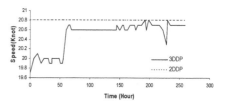

Figure 6. Optimized speed (case 1).

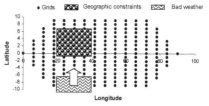

Figure 7. Geographic constraints and weather conditions (case 2).

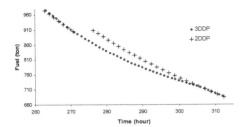

Figure 8. The fuel consumption vs. time curve (case 2).

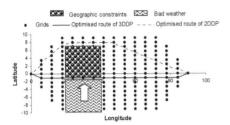

Figure 9. Optimized route (case 2).

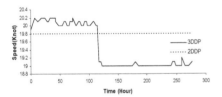

Figure 10. Optimized speed (case 2).

REFERENCE

Aligne F, Papageorgiou M, & Walter E. 1998. Incorporating power variations into weather routing and why it may lead to better results. *IFAC Control Applications in Marine Systems*, Fukuoka, Japan, pp 269-274.

Bijlsma S.J. 1975. *On minimal-time ship routing*. PhD thesis. Royal Netherlands Meteorological Institute, Delft University of Technology.

Calvert S, Deakins E & R. Motte. 1991. A dynamic system for fuel optimisation trans-ocean. *Journal of Navigation* 44(2): 233-265.

Calvert S. 1990. *Optimal weather routeing procedures for vessels on trans-oceanic voyages*. PhD thesis, Polytechnic South West.

Chen H. 1978. *A dynamic program for minimum cost ship routing under uncertainty*. PhD thesis, Dept. of Ocean Engineering, Massachusetts Institute of Technology.

Chinmaya Prasad Padhy, Debabrata Sen, & Prasad Kumar Bhaskaran. 2008. Application of wave model for weather routing of ships in the North Indian Ocean. *Natural Hazards* 44(3): 373-385.

De Wit C. 1990. Proposal for low cost ocean weather routeing. *Journal of Navigation* 43(3): 428-439.

Hagiwara H & Spaans JA. 1987. Practical weather routeing of sail-assisted motor vessels. *Journal oF Navigation* 40: 96-119.

Hagiwara H. 1989. *Weather routing of (sail-assisted) motor vessels*. PhD thesis, *Delft* University of Technology.

Harries S, Hinnenthal J.2004. A systematic study on posing and solving the problem of pareto optimal ship routing. *3rd International Conference on Computer Applications and Information Technology in the Maritime Industries (COMPIT 2004)*, Siguenza, Spain.

Klompstra M.B, Olsde GJ & Van Brunschot Pkgm. 1992. The isopone method in optimal control. *Dynamics and Control* 2(3):281-301.

Kyriakos Avgouleas. 2008. *Optimal ship routing*. Msc thesis, Massachusetts Institute of Technology.

Masaru Tsujimoto, Katsuji Tanizawa (2006) Development of a weather adaptive navigation system considering ship performance in actual seas. *Proceedings of OMAE2006 25th International Conference on Offshore Mechanics and Arctic Engineering*, Hamburg, Germany.

Spaans J.A. 1995. New developments in ship weather routing. *Navigation* 169: 95-106

Aviation and Air Navigation

# 24. Position Reference System for Flight Inspection Aircraft

M. Kubiš & A. Novák
*University of Zilina, Zilina, Slovakia*

ABSTRACT: This paper describes the consistent preview of the position reference system for flight inspection aircraft equipped by flight laboratory capable to flight inspection of radio navigation systems. It focused on description of ground position reference station designed to receive GPS signals and to transmit differential GPS data to the aircraft for DGPS corrective calculation for purpose of improved precision of flight inspection of radio navigations systems such ILS or VOR/DME.

The purpose of a flight inspection is to calibrate and evaluate the performance of aircraft navigation and landing aids to ensure conformance to specifications. This mission requires that the flight inspection platform have a reference position estimate significantly more accurate than that of the facility under inspection, it means tenths of meter accuracy over a region of many kilometers in a dynamic environment. In generally the flight inspection data must be calculated with a higher degree of accuracy than the tested systems. For example, if a DME station is tested the DME receiver itself calculates the data with an accuracy of about 180 m. The position of the aircraft respective to the DME ground station (evaluated by the flight inspection equipment) must have a measured accuracy of less than 180 m. From all navaids the instrument landing system (ILS) requires the strictest accuracy of flight inspection therefore flight inspections systems have to be in compliance with these requirements.

## 1 FLIGHT INSPECTION ILS ACCURACY REQUIREMENTS

The most critical flight inspection accuracy requirements involve checking the alignment and displacement sensitivity of high precision (Category III) Instrument landing Systems (ILS). The alignment values are defined as the average angle from the glide path or localizer antenna to the aircraft, when the ILS signal indicates that the aircraft is on course on path. The ILS alignment errors are defined as the average differences between the instantaneous localizer or glide path angles defined by the ILS receiver and the true angles, measured from the relevant ILS antenna on the ground. Displacement sensitivity is a measure of the scale factor of the associated ILS signal (microamps per degree). Measurement of the glide path displacement sensitivity requires tighter angular accuracy than measurement of glide path alignment.

The International Civil Aviation Organization (ICAO) define system accuracy specifications and flight inspection standards, has specified three categories of ILS runways. These three categories (I, II, III) permit landing under worse conditions of ceiling and visibility, and achieve their purpose by providing more accurate signals in space. The accuracy requirements for flight inspection also become more demanding as the specified facility accuracy requirements are tightened. The most demanding accuracy requirements are imposed for inspecting performance category III ILS runways, specifically for verifying the glide path and localizer alignment and the displacement sensitivity. ICAO requires the inspection device to have a two-sigma (95 %) error that is not more than one third of specified ILS alignment accuracy. The 95 % probability values are shown in Table I.

Table 1. 95 % Flight inspection accuracy requirements(units are in degrees)

| | CAT I | CAT II | CAT III |
|---|---|---|---|
| Localizer Alignment | 0,042 | 0,028 | 0,014 |
| Localizer Displacement Sensitivity | 0,035 | 0,035 | 0,021 |
| Glide Path Alignment | 0,063 | 0,063 | 0,035 |
| Glide Path Displacement Sensitivity | 0,028 | 0,021 | 0,014 |

## 2 HISTORICAL OVERVIEW OF FLIGHT INSPECTION SYSTEMS

The first instrument used for the ILS calibration was a theodolite, an instruments that measures horizontal and vertical angles. Figure 1 shows the old fashioned ILS calibration procedure with a theodolite.

Figure 1. The old fashioned ILS calibration procedure with a theodolite (Bates & Feit, 2002)

This procedure required very skilled people and was time consuming. Then, an automatic light or laser tracker replaced the manual theodolites during 70s of the 20<sup>th</sup> century. Various trackers were used, but they all tracked light or laser from its source installed on the airplane. Flight paths were still estimated on the ground and the overall ILS calibration procedure took a significant amount of time. During 1980s, the Inertial-based Automatic Flight Inspection System (AFIS) was developed. This system used a navigation grade INS (Inertial Navigation System) as a primary sensor with a barometric altimeter, a radar altimeter, a camera system (TVPS), and a pilot event button. A Kalman filter was used to estimate a flight trajectory by using the measurements from those sensors. This system is an automated self-contained system that made the ILS calibration procedure more efficient and convenient. Therefore the current automated flight inspection systems are Inertial-based AFIS and DGPS-based AFIS whose characteristics are quite different in terms of cost, accuracy and efficiency.

## 3 AUTOMATED FLIGHT INSPECTION SYSTEM BASED ON GPS

The current flight inspection systems are capable to perform flight inspection of navaids in accordance with international standards, e.g. ICAO DOC 8071 and FAA OAP 8200 such:
- ILS cat. I, II, III (Localizer, Glide Path)
- MKR
- VOR
- DME
- VHF COMM/VDF
- UHF COMM/UDF
- NDB
- SSR mode A, C, S
- RNAV Visual Landing Aids (VASIS, PAPI)

### 3.1 DGPS AFIS

DGPS (Differential) is an enhancement of the U. S. Department of Defense's Global Positioning System through the use of differential corrections to the basic satellite measurements from the user's receivers. This DGPS approach to flight inspection uses measurements from a receiver in the flight inspection aircraft and corresponding measurements from a second GPS receiver at a location that has been accurately surveyed relative to the flight inspection facility (e.g., the glide path and localizer antennas) to be inspected. The accuracy of the GPS position increases to less than 0, 2° in all directions. However, it requires a time-consuming procedure in setting up a local reference station in each airport, which is the main drawback of the DGPS-based AFIS.

Figure 2. The onboard DGPS AFIS ( www.aerodata.com)

# AFIS Block Diagram

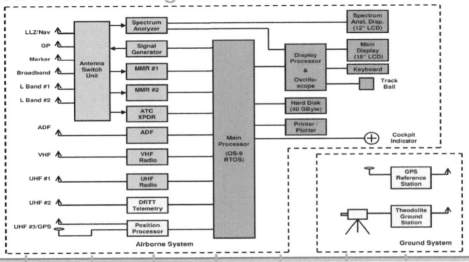

Figure 3. The scheme of current modern AFIS (Haverland, 2009)

In the current tests, an all-in-view receiver at the reference site extracts signals from all visible satellites and measures the pseudorange to range. Since the satellite signal contains information on the precise satellite orbits and the reference receiver knows its position, the true range to each satellite can be computed. By comparing the computed range and the measured pseudorange, a correction term can be determined and used to correct each associated pseudorange measurement in the aircraft. The main advantage of employing DGPS as opposed to stand-alone GPS arises from the improved accuracy that can be achieved through use of this relative navigation technique. In particular, those GPS error components that are common to the two sets of satellite-to-receiver links either disappear or are significantly reduced, especially when the two GPS receivers are in close proximity. These errors include range variations introduced by selective availability, atmospheric propagation delays, satellite clock errors, and ephemeris errors. Differential corrections can reduce navigation errors from 100 meters (95 %) to one meter or less, depending on receiver accuracy and distance from the reference station.

A particular problem that governs the attainable accuracy of DGPS measurements is the positioning (survey) accuracy of the reference GPS receiver relative to the landing aid under inspection. The first order effect of any such survey bias error is a corresponding equal shift in the apparent location of the GPS receiver in the flight inspection aircraft and, hence, in the relative coordinates of the navigation aid. The effect of such a bias error on alignment accuracy has been calculated. A one meter deviation in vertical position results in a 0,0147 degree shift in

the measured elevation angle over the ICAO-specified inspection region. Similarly, a one meter shift in cross runway position yields an azimuth angle shift of 0,0134 degrees for a Category I runway, 0,0153 degrees for a Category II runway, and 0,0198 degrees for a Category III runway (Bates & Feit, 2002). These different impacts are due to the varying inspection regions associated with the three ILS categories. This analysis emphasizes the need for surveying the location of the reference receiver decimeter accuracy relative to the landing aid under inspection. Such survey accuracies are most readily attained if the reference GPS receiver position is within a few miles of the landing aid. Note that the effects of along runway bias errors on the glide path (elevation) angle are reduced by the descent geometry. For a three degree descent angle, a one meter along runway bias error results in an elevation angle change of 0,00077 degrees. The impact on the localizer (azimuth) angle is almost negligible, since the aircraft is nominally on the runway centerline. Displacement sensitivity requires subtraction of two angles. Because of the near linearity of the arctangent for small elevation angles, the first order impact of the bias term disappears, and the effect of bias is very significantly reduced.

### 3.2 Comparison of INS and DGPS flight inspection systems

The Inertial-based AFIS is an onboard system that has a navigation grade INS, GPS, a barometric altimeter, a radar altimeter, and a Television Positioning System (TVPS). In this system, the fusion of a navigation grade INS, GPS, and a barometric altimeter provides high quality velocity during approach. A radar altimeter and a TVPS provide accurate position fixes at the runway threshold and departure end. Those position fixes are used to refine the velocity are used to refine the velocity during approach by calibrating various INS biases. Then, the flight path during the flight inspection approach is estimated by intergrating the velocity backward from the position fix at the runway threshold. On the other hand, the DGPS-based AFIS uses a Real-Time Kinematic (RTK) DGPS system that can provide centimeter level accuracy. An RTK systems differential GPS techniques with two receivers and utilizes GPS carrier phase measurements as ranging sources. This system requires an installation of a local reference receiver near a runway before the flight inspection is carried out. These two different positioning schemes result in substantial differences in the tradeoffs between cost and efficiency of the current flight inspection systems. The Inertial-based AFIS is much more expensive than the DGPS-based AFIS mainly due to the use of navigation grade INS.

On the other hand, the flight inspection procedure with the DGPS-based AFIS takes significantly more time than the Inertial-based AFIS because a flight inspection aircraft first lands on a runway to install a local reference receiver to begin flight inspection. A civil aviation administration (CAA) of a country typically chooses either one of the systems that better fits its own preference. For example, the Federal Aviation Administration (FAA) mainly uses the Inertial-based AFIS due to the large volume of flight inspection required.

### 3.3 Stand alone GPS-based flight inspection system

The standalone GPS-based FIS has a single frequency GPS receiver, a radar altimeter and a TVPS (Tel-

evision Positioning System). The same kinds of radar altimeter and TVPS being used in the current Inertial-based AFIS are taken in the standalone GPS-based FIS. The 95 % accuracy of the radar altimeter is better than 15 cm. The 95 % accuracy of the TVPS is better than 15 cm in cross-track and 30 cm in along-track (Kim & Powell & Walter, 2004). These integrated systems are optimally designed for the ILS calibration problem in terms of accuracy, cost, and efficiency.

Figure 4 illustrates the overall algorithm of the standalone GPS-based FIS. During approach, GPS measurements are collected. Over the threshold of a runway, the radar altimeter measures the vertical distance between the airplane and the runway threshold. At that point, the TVPS measures the cross-track and along-track deviations of the airplane from the center line and the threshold mark of the runway by using its camera images. Since the position of the threshold is accurately surveyed the radar altimeter and the TVPS provide an accurate instant 3D position of the airplane over the threshold. A specialized positioning algorithm Time-Differenced Precise Relative Positioning (T-D PRP) method uses this reference position and the carrier phase measurements to compute precise relative positions. The estimated flight path during approach is obtained by adding the relative positions to the reference position. To ensure sound position solutions, satellite exclusion tests are implemented to discard a satellite that should not be used in T-D PRP. In addition, the integrity of the T-D PRP solutions is checked in the FIS-RAIM (Receiver Autonomous Integrity Monitoring) that protects against possible satellite failures.

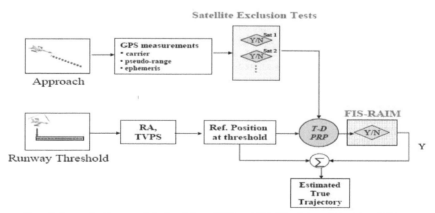

Figure 4: System Architecture of the Standalone GPS-Based FIS (Kim & Powell & Walter, 2004)

### 3.3.1 *Stand alone GPS-based FIS performance*

Considering the errors from the T-D PRP and the accuracy requirements, the most critical regions are around 2200 meters and 2000 meters from the threshold in vertical and horizontal respectively. To see the performance of the standalone GPS-based FIS, it is necessary to consider the total errors caused by both the T-D PRP and the reference position error in cross-track and vertical at the critical regions because those errors may most likely violate the FIS accuracy requirements for CAT II and III ILS calibration.

Treating T-D PRP errors and the reference position errors as zero-mean independent random variables, which is not exactly true but practically good enough the distributions of the total errors can be easily calculated. Taking the accuracies (95 %) of the radar altimeter and the TVPS in the standalone GPS-based FIS to be about 15 cm, the total errors at the critical regions have 9,01 cm standard deviation in vertical and 8,54 cm standard deviation in cross-track. Therefore the 95 % accuracy of the standalone GPS-based FIS about 18,02 cm in vertical and 17,01 cm in cross-track at the critical regions(Kim & Powell & Walter, 2004). Therefore, the standalone GPS-based FIS sufficiently meets the FIS accuracy requirements for CAT II and III ILS calibration whose limits are about 30 cm in vertical and 60 cm in cross-track.

It can be said that the GPS-based FIS provides more optimized performance than the current FIS in terms of accuracy, cost, and efficiency. Its accuracy is between the Inertial-based AFIS and the DGPS-based AFIS, and its cost is significantly lower than the two AFIS. The efficiency of the standalone GPS-based FIS outperforms the two AFIS because it does not need a reference station on the ground nor does it require the FI aircraft to fly level over the whole runway. Compared to the proposed FIS using the WAAS (SBAS), the WAAS-aided FIS and the WAAS-based FIS have better integrity features because they can take advantage of the broadcast integrity messages. However, the probability of satellite failure during FI is expected to be extremely small because a FI is only operated a few days a week and during day time. So a satellite failure will rarely occur in FI. It is also expected that the FIS-RAIM detects most satellite failures that can cause positioning failures. Overall, the standalone GPS-based FIS is a good alternate where WAAS (SBAS) is not available and provides better performance than the current AFIS's.

### 3.4 *WAAS (SBAS)-aided flight inspection system*

It is not the aim of this paper to describe principle of WAAS system. Basically it works on very similar principle like DGPS. WAAS uses a network of ground-based reference stations, in North America and Hawaii, to measure small variations in the GPS satellites' signals in the western hemisphere. Measurements from the reference stations are routed to master stations, which queue the received Deviation Correction (DC) and send the correction messages to geostationary WAAS satellites in a timely manner (every 5 seconds or better). Those satellites broadcast the correction messages back to Earth, where WAAS-enabled GPS receivers use the corrections while computing their positions to improve accuracy. In addition to this system there are also several others space-based augmentation systems (SBAS) like European Geostationary Navigation Overlay Service (EGNOS) in the European Union, Multi-functional Satellite Augmentation System (MSAS) in Japan, and GPS and GEO Augmented Navigation (GAGAN) in India. Korea and Brazil are investigating SBAS, also. At this time, WAAS is the only fully operational SBAS, but EGNOS and MSAS will be complete soon.

The WAAS based AFIS uses a low (tactical or less) grade INS, a certified commercially available WAAS receiver, a radar altimeter and a TeleVision Positioning System (TVPS). The advantages of this system are lower cost and better efficiency than the current AFIS. However, the accuracy of this system is marginal inspection system accuracy requirements far from the runway. The WAAS-aided FIS has some vulnerability to possible accuracy degradation in rare events (e.g., a sharp ionospheric gradient or severe multipath) because it is only utilizing standard positioning outputs from the WAAS receiver. The current WAAS 95 % accuracy is better than 0,935 meters in the horizontal and 1,289 meters in the vertical, which does not meet the ILS calibration accuracy requirements. Although WAAS cannot be directly used for the ILS calibration, WAAS still has useful features because it broadcasts accurate correction messages for GPS errors and integrity messages. The error corrections include satellite clock-ephemeris and ionospheric delay. The integrity messages include satellite anomalies, severe ionospheric disturbances, and the quality of the error corrections. These features play a very important role in helping the WAAS-based FIS have sound positions and firm integrity.

The WAAS –based FIS is a system that has a single frequency WAAS receiver, a radar altimeter, a TVPS, and a computer. This integrated system is optimally designed for the ILS calibration problem in terms of accuracy, cost, efficiency and integrity. Figure 5 illustrates the overall algorithm of the WAAS-based FIS. During approach, WAAS position and raw GPS/WAAS measurements include ephemeris, L1 code and carrier phase measurements and WAAS messages. The ephemeris parameters provide GPS satellite locations at a specific time. L1 code and carrier phase measurements provide range information between an user to satellites. The

WAAS messages provide GPS error corrections and satellite health. Over a runway threshold, the radar altimeter, corrected for roll and pitch angles, measures the vertical distance between the airplane and the runway threshold. At that point, the TVPS measures the cross-track and the along-track deviations of the airplane from the centerline and the threshold mark of the runway by using its camera images for CAT II and III ILS calibration.

However, WAAS can substitute for a TVPS in the WAAS-based FIS for CAT I ILS calibration. Since the position of the threshold is accurately surveyed, the radar altimeter and the TVPS provide an accurate instant 3D position of the airplane over the threshold called a reference position. Again, the reference position can be given from a radar altimeter and WAAS cross-track position in the WAAS-based FIS for CAT I ILS calibration. Similarly as by GPS stand-alone based AFIS a specialized positioning algorithm, Time-Differenced Precise Relative Positioning (T-D PRP) method, uses the reference position and the raw GPS/WAAS measurements to compute precise relative positions. The T-D PRP utilizes the difference of GPS carrier phase measurements over a time interval as ranging sources. It removes the satellite clock-ephemeris errors by using broadcast WAAS correction and the ionospheric effects by using the first order linear regression on the time series of code minus carrier phase measurements during approach. Then, the estimated flight trajectory during approach is obtained by adding the relative positions to the reference position. There are two integrity features for the soundness of the estimated flight trajectory: satellite exclusion tests and validation of the reference position from the radar altimeter and the TVPS. First, satellite exclusion tests are implemented to discard a satellite that should not be used in the T-D PRP. These exclusion tests have the following checks: unhealthy satellite status reported from GPS/WAAS, discontinuity in carrier phase measurements called cycle-slip, severe nonlinearity of ionospheric delay, and satellite outages. If any of these items is reported, the corresponding satellite is excluded in computing position solutions. Second, the integrity of a reference position from a radar altimeter and a TVPS is checked by using both WAAS position during approach and the precise relative position from the T-D PRP. Even though this validation test is limited to the level of WAAS accuracy, it is useful when a radar altimeter or a TVPS introduces an abnormally large error. These features of the WAAS-based FIS provide high performance in terms of accuracy, cost, efficiency, and integrity by taking advantages of WAAS and the near real-time nature of flight inspection.

## 4 CONCLUSION

This paper attempts to offer the reader a consistent overview to the flight inspection in the Slovak Republic and the centre of excellence of air transport at the University of Zilina- ITMS 26220120065. For the flight inspection, special items have to be taken into consideration when selecting the tested aircraft and airborne systems.

For flight laboratory is very important to choose a suitable reference system. The reference system must satisfy the criteria mentioned in this paper as well achieve the required accuracy. Our effort is use for the flight inspection aircraft with MTOW up to 2000 kg and it is a relatively complex task. For our project as the most suitable system appears to be a combination of inertial navigation system and satellite navigation system (DGPS).

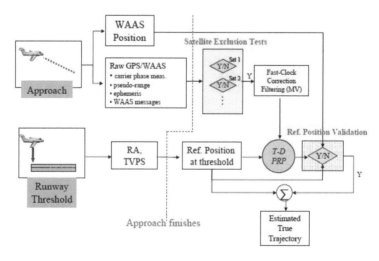

Figure 5: System Architecture of the Standalone GPS-Based FIS (Kim & Powell & Walter, 2004)

## ACKNOWLEDGEMENTS

This project- Centre of Excellence of Air Transport-ITMS 26220120065 is cofinanced from sources of EU, which supports research activities in Slovak Republic.

## REFERENCES

Bates, M. & Feit, C. 2002. Accurate positioning in a flight inspection system using differential global navigation satellite systems. New York

Haverland, M. 2009. New flight inspection equipment for Aerothai. Bangkok

Kim, E. & Powell, D. & Walter, T. GNSS-based Flight Inspection Systems. Stanford university

Novak Sedlackova, A. Selected parts of air law, In: New Trends in Civil Aviation 2009, Brno 28.-29.5.2009 ISBN 978-80-7204-631-7. - S. 84-89.

Topolčány,R.: Safety in General Aviation, Slovak republic. In: Studies of Faculty of operation and economics of transport and communications of University in Žilina. Volume 17. - Žilina : University of Žilina, 2002. - ISBN 80-7100-982-2. - S. 141-145.

# 25. RNAV GNSS Essential Step for the LUN Implementation and the Chance for the Polish General Aviation

K. Banaszek
*Polish Air Navigation Services Agency, Warsaw, Poland*

A. Fellner
*Silesian University of Technology, Katowice, Poland*

P. Trómiński & P. Zadrąg
*Air Force Institute of Technology, Warsaw, Poland*

ABSTRACT: The general idea of this project is to popularize the usage of EGNOS (European Geostationary Navigation Overlay Service) System in Central European countries. The main goal will be achieved by demonstrating key capabilities of the EGNOS System as SBAS (Satellite Based Augmentation System) APV I (Approach Procedures with Vertical guidance) – its main role. After that the National Fly Authority Agency certification procedure will be launched, EGNOS will become fully functional and operational airfield domestic system. Project consists of 4 main phases:

Phase one – Analysis of domestic airfields and Aircraft operator(s) against opportunity of EGNOS installation and utilization of capability for landing purposes;

Phase two – study of demonstrator's technical conception of EGNOS APV implementation connected with blueprint for certification process;

Phase three – on-board and airfield system installation connected with provisional certification;

Phase four – Final demonstration of overall system capabilities, followed by on-going tests, system certification and implementation of Safety Case.

The following entities participate in this project: scientific research centers, government agencies, aircraft operator and local small airfield, which guarantees general outlook analysis, clearly constructed user requirements with correct technical solutions which are in compliance with Polish and European air traffic regulations.

Government agency will keep an eye on realizing a through airfield and aircrafts operators inspection used by research centers for analysis. Additionally government agency will be the supervisor for preparation and certification of the airfield and aircraft.

Scientific research centers will prepare technical solutions for system installation at airfield and aircraft, with certification projects and procedures as well. This includes two types of certification: technical and fly procedures (landing procedures for aircraft and airfield). Safety of fly and Safety Case are especially taken into consideration. Aircraft operator involved in a project will perform the EGNOS system installation on the aircraft and on the airfield, using blueprint for technical solution developed in the second phase of the project, and as result of that in next phases of project will perform test and certification flights. Additionally aircraft operator is allowed to train new and current pilots, which will fully realize plan for preparing all the pilots to be familiarized with EGNOS System, and if so, this system will be widely and commonly used in aviation/air force.

All demonstration, testing and certification flights will be in participation of scientific research center and government agency, which will guarantee the test procedure correctness in accordance to flight safety regulations and rules for new equipment on board aircrafts.

Currently, differential positioning methods, such as EGNOS, ASG-EUPOS, AFREF, WAAS, MSAS, GRAS, GAGAN, SNAS, basing on global satellite navigation systems: GPS, GLONASS and also Galileo in the future, undergo dynamic development. Potential use of these techniques includes, among others, airport traffic control.

## 1 INTRODUCTION

This implementation of GNSS in Poland is „PER ASPERA AD ASTRA". However under the PANSA patronage take action, aiming at the certification of the aviation GNSS application in our country. To distinguish it is possible in this procedure the following phases:

– ARRANGEMENTS TO AVIATION EXPERIMENTS:

- Receiver Septentrio;
- SPAN (Synchronized Position Attitude Navigation)
- ACTIVE PARTICIPATION IN INTERNATIONAL RESEARCH AND IMPLEMENTATION PROGRAMS:
  - HEDGE;
  - EGNOS APV;
  - NPA GNSS;
  - CERTIFICATION;
  - GNSS FUNCTIONING IN AVIATION

The dynamic development of aviation caused huge development of present techniques and technologies in navigation. The demand appeared on entirely new approach in connection with management the air traffic questions in the aim of solution problems connected with enlargement the capacity and transfer function and the skyway as well as existing the far-reaching European formations the ATM. Therefore document was worked out THE AIR TRAFFIC MANAGEMENT STRATEGY FOR THE YEARS 2000, which is the aim the creation of uniform aerospace for Europe. Presented strategy delivers also precise hands and presents effective centers, thanks which is possible to deal all problems and effectively cope with challenges by European ATM in XXI century. In received international solutions mention, that initiation global formation ATM/CNS should take into account present techniques and technologies in wide range and simultaneously build it will make possible the modernization of formations in the future.

Fig. 1. Local Convergence and Implementation Plan Poland 2009 – 2013 and Development Programme network of airports and ground equipment.

The implementation of the EGNOS system to APV-I precision approach operations, is conducted according to ICAO requirements in Annex 10 and of other documents: European Convergence and Implementation Plan 2009 – 2013, Local Convergence and Implementation Plan Poland 2009 – 2013, Program Rozwoju Sieci Lotnisk i Lotniczych Urządzeń Naziemnych - admitted to the accomplishment with Resolution of the Council of Ministers Nr 86/2007 (fig. 1).

Definition of usefulness and certification of EGNOS as SBAS (Satellite Based Augmentation System) in aviation requires thorough analyses of accuracy, integrity, continuity and availability of SIS (Signal in Space). Also, the project will try to exploit the excellent accuracy performance of EGNOS to analyse the implementation of GLS (GNSS Landing System) approaches (Cat I-like approached using SBAS, with a decision height of 200 ft). Location of the EGNOS monitoring station, located near Polish - Ukrainian border, being also at the east border of planned EGNOS coverage for ECAC states is very useful for SIS tests in this area.

According to current EGNOS programme schedule, the project activities will be carried out with EGNOS system v2.2, which is the version released for civil aviation certification. Therefore, the project will allow demonstrating the feasibility of the EGNOS certifiable version for civil applications. Planned demonstration and trials will be provided on 2 - 3 Polish airports (central, eastern and western) chosen based on SIS analysis and EGNOS operational coverage in Poland. For creating and testing software and making other documentations we will use ESA standards like ECSS-E-40, ECSS-Q-80B.

## 2 RECEIVER SEPTENTRIO AND PEGASUS PROGRAM

PEGASUS (Prototype EGNOS and GBAS Analysis System Using SAPPHIRE) is a prototype which allows analysis of GNSS data collected from different SBAS and GBAS systems and using only algorithms contained in the published standards. The tool has been developed in the frame of the GNSS-1 operational validation activity defined in the EUROCONTROL SBAS project and aims to be a first step forward the development of a standard processing and analysing tool to be used for the future EGNOS operational validation. PEGASUS was designed to facilitate the output data handling and interchange. The tool provides several functionalities such as computation of position and GNSS systems attributes like accuracy, reliability, and availability simulating MOPS-compliant receivers, computation of trajectory errors, prediction of accuracy and availability with the required integrity and simulation of GBAS Ground Station processing algorithms. Since June 2003 the GBAS Modular Analysis and Research System (MARS) is integrated in PEGASUS in order to support GBAS data processing needs and activities. The GBAS MARS allows to collect and evaluate relevant data and provide required results and is able to assist Air Traffic Service Providers to aid site approval and obtain operational approval of a GBAS installation for supporting CAT-I precision approach conditions at an airport from their respective safety regulation authorities.

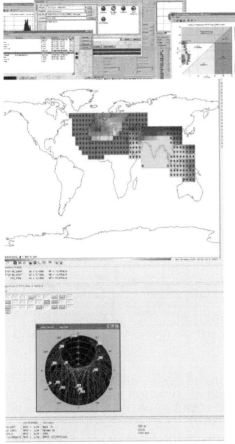

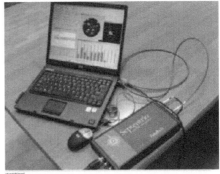

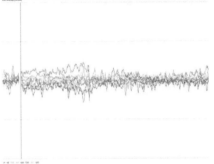

Fig. 3. Workstation configuration Notebook and Septentrio Po-laRx2e (EGNOS L1/L2) receiver and Residuals Plot from RxControl.

Fig. 2. SBAS system coverage, visualization accuracy EGNOS at current time.

## 3 SPAN (SYNCHRONIZED POSITION ATTITUDE NAVIGATION)

The Synchronized Position Attitude Navigation (SPAN) system is NovAtel's Global Navigation Satellite System - Inertial Navigation System (GNSS/INS) solution for applications requiring continuous position, velocity and attitude information. Using Inertial Measurement Unit (IMU) data in addition to GNSS, SPAN provides a high rate position, velocity and attitude solution which seamlessly bridges GNSS outages. The tight integration of the IMU to the receiver core improves GNSS performance by enabling faster signal reacquisition and quicker return to fixed integer status after a loss of GNSS signals. Synchronized Position Attitude Navigation.

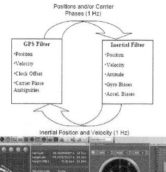

Positions and/or Carrier
Phases (1 Hz)

| GPS Filter | Inertial Filter |
|---|---|
| •Position | •Position |
| •Velocity | •Velocity |
| •Clock Offset | •Attitude |
| •Carrier Phase Ambiguities | •Gyro Biases |
| | •Accel. Biases |

Inertial Position and Velocity (1 Hz)

Fig. 4. Genaral Integration Architecture of SPAN and Novatel CDU panel for receiver configuration and monitoring.

Fig. 5. Visualization postprocesing date(RTK/INS) gathered during the fly over the Mielec airport and GPS/GPRS device for real time monitoring as a second monitoring system on airplane Piper PA-34 Seneca II).

## 4 ACTIVE PARTICIPATION IN INTERNATIONAL RESEARCH AND IMPLEMENTATION PROGRAMS: HEDGE, EGNOS APV, NPA GNSS

The accomplishment of international programs required adopting the following assumptions:
– Final Approach of GNSS landing with "Overlines" method;
– Accomplish of the RNAV GNSS Approach Procedures;
– Certification of the GNSS receivers (on board);
– Test flights - checking assumed solutions;
– Operational of EGNOS System;
– Test flights in frames of programs;
– Collecting indispensable materials and drawing up documents, necessary to do the certification;

– Certification of the gnss approach in Poland

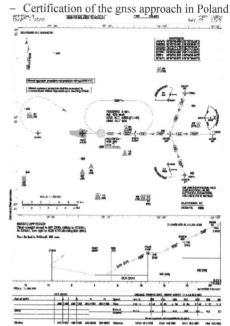

Fig. 6. Example: RNAV GNSS for Katowice – Pyrzowice Airport.

## 5 PROJECT HEDGE - WORK PACKAGE 5: GENERAL AVIATION EGNOS APV DEVELOPMENT AND DEMONSTRATION IN POLAND

Helicopters Deploy GNSS in Europe (HEDGE) is Collaborative Project Response to FP7-GALILEO-2007-GSA-1. The project objectives are to achieve the following by the end of the project to:
– develop the helicopter SOAP (SBAS Offshore Approach Procedure) procedure (and necessary avionics) and then to successfully demonstrate it to the user community;
– develop helicopter PINS (Point in Space) procedures for mountain rescue and HEMS (Helicopter Emergency Medical Services), and to then successfully demonstrate them to the user community.
– demonstrate EGNOS (European Geostationary Navigation Overlay Service) APV (approach with vertical guidance) approaches to general aviation in Spain, Poland;
– (OPTION) To complete EGNOS data gathering that shows the performance of EGNOS.

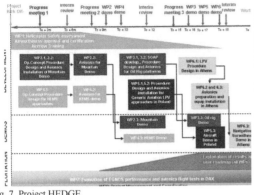

Fig. 7. Project HEDGE.

Coordinator:FP7-GALILEO-2007-GSA-1, for the carrying out of the Helicopters Deploy GNSS in Europe ("HEDGE") Project, hereinafter referred to as "Effective Date" among: Helios Technology Limited, Pildo Consulting, S. L., REGA, TAF Helicopters S.L., P.P.H.U "ROYAL-STAR", Aero Club Barcelona-Sabadell, Polish Air Navigation Services Agency (PANSA), Capital High Tech SARL HELILEO,

The new GNS 430 W (already WAAS enabled) receiver in the Seneca airplane is applied during experiments, becouse:

– already WAAS enabled;
– then use an experimental datacard from Garmin to enable EGNOS;
– perform the necessary tests in the laboratory to validate that the receiver is able to work properly with the broadcasted EGNOS SIS at the time the demonstrations will be performed.

Fig. 8. GNS 430 W receiver and Septentrio receiver in the Seneca airplane and Seneca airplane on the Katowice Airport

Fig. 9. Route of test flights in Mielec 9.04.2010 and 22.04.2010r.

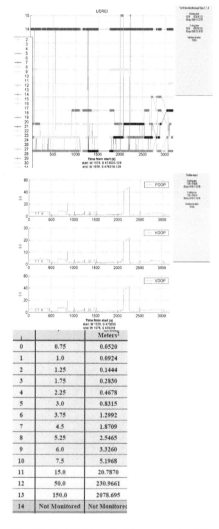

| $i$ | Meters | Meters$^2$ |
|---|---|---|
| 0 | 0.75 | 0.0520 |
| 1 | 1.0 | 0.0924 |
| 2 | 1.25 | 0.1444 |
| 3 | 1.75 | 0.2830 |
| 4 | 2.25 | 0.4678 |
| 5 | 3.0 | 0.8315 |
| 6 | 3.75 | 1.2992 |
| 7 | 4.5 | 1.8709 |
| 8 | 5.25 | 2.5465 |
| 9 | 6.0 | 3.3260 |
| 10 | 7.5 | 5.1968 |
| 11 | 15.0 | 20.7870 |
| 12 | 50.0 | 230.9661 |
| 13 | 150.0 | 2078.695 |
| 14 | Not Monitored | Not Monitored |

Fig. 10. Test flights - checking assumed solutions: UDREI (User Differential Range Error Indicator), PDOP, HDOP, VDOP.

# 6 PROTECTION LEVELS

## 6.1 *General Approach to Protection Levels*

To follow the required path, the aircraft navigation system estimates the aircraft's position and generates commands (either to a cockpit display or to the autopilot). Errors in the estimation of the aircraft's position is referred to as Navigation System Error NSE which is the difference between the aircraft's true position and its displayed position. The difference between the required flight path and the displayed position of the aircraft is called Flight Technical Error FTE and contains aircraft dynamics, turbulence effects, man-machine-interface problems, etc. The vector sum of the NSE and the FTE is the Total System Error TSE. Since the actual Navigation System Error can not be observed without a high-precision reference system (the NSE is the difference between the actual position of an aircraft and its computed position !), an approach has to be found with which an upper bound can be found for this error.

- The Horizontal Protection Level HPL is the radius of a circle in the horizontal plane (the plane tangent to the WGS84 ellipsoid), with the centre being at the true aircraft position, which describes the region which is assured to contain the indicated horizontal position. It is the horizontal region for which the missed alert requirements can be met.
- The Vertical Protection Level VPL is the half length of a segment on the vertical axis (perpendicular to the horizontal plane of the WGS84 ellipsoid), with the centre being at the true aircraft position, which describes the region which is assured to contain the indicated vertical position. It is the vertical region for which the missed alert requirements can be met.

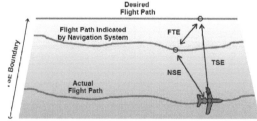

Fig. 11 Navigation System Error, Flight Technical Error and Total System Error

The SBAS protection levels are a function of the satellite constellation and the estimated SBAS performance. Thus, using the SBAS correction data, the protection levels can be determined without using actual pseudorange measurements. The only parameters / items used are:

- the residual error budget as determined for fast-, slow-, ionospheric-, tropospheric and receiver errors
- the satellites received (constellation) and selected (integrity flags) by the receiver in the position solution.

The computed protection levels must be compared to the required Alert Limits AL for the particular phase of flight. If the protection level is smaller than the required alert limit, then the phase of flight can be performed. However, if the protection level is greater than or equal to the required alert limit, then the integrity of the position solution can not be guaranteed in the context of the requirements for that particular flight phase.

$XPL \; < \; XAL$      integrity can be assured

$XPL \; \geq \; XAL$      integrity can not be assured

th   $XPL$      (horizontal or vertical) protection level

     $XAL$      (horizontal or vertical) alert limit

The relevant alert limits, in combination with the required alert limit requirement, are listed in table 1.

In particular, the Integrity Requirements will be used later to derive the protection levels. The corresponding situation in the horizontal plane is depicted in the figure 11.

Table 1. Protection Levels for Flight Phases

| ight runs | Integrity Requrements | Horizontal Alert Limit | Vertical Alert Limit | Note |
|---|---|---|---|---|
| NR | 1·10⁻⁷ per hour | 7400 m 3700 m 1850 m | N/A | different alert limits for domestic and oceanic flight phases |
| MA | 1·10⁻⁷ per hour | 1850 m | N/A | |
| PA | 1·10⁻⁷ per hour | 556 m | N/A | |
| PV-I | 1 - 2 x 10⁻⁷ per approach | 40 m | 50 m | new flight phase defined in the current SARPs |
| PV II | 1 - 2 x 10⁻⁷ per approach | 10 m | 20 m | new flight phase defined in the current SARPs |

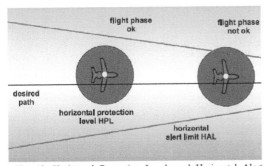

Fig. 12. Horizontal Protection Levels and Horizontal Alert Limit

It should be noted that the main significance using this approach is not the computation of the protection levels and their comparison with the corresponding alert limit. The major interest should be considered to be on the assurance that the computed protection levels represent an upper bound on the NSE with a certain confidence. "Misleading Information" results only, if the NSE is greater than the alert limit and the protection level does not indicate this fact (for a more complete and detailed description of the "overbounding concept" and problems resulting of it, refer to).

### 6.2 Protection Levels

Since the SBAS correction information is applied to each individual pseudorange in different ways (according to the criteria shown in the chapters before), the assumption on one single value for the standard deviation for all pseudorange can not be applied. For the calculation of the protection levels using SBAS corrections, the general approach of the weighted least squares is used . For modes other then precision approach, the weights are undefined. For an unweighted least squares solution, the weighting matrix is a unity diagonal matrix (i.e. the elements on the main diagonal are set to 1). The projection matrix of the position solution is then calculated as:

$$S \; = \; \left(H^T W H\right)^{-1} H^T W \; = \; \begin{bmatrix} S_{x1} & S_{x2} & \cdots & S_{xn} \\ S_{y1} & S_{y2} & \cdots & S_{yn} \\ S_{z1} & S_{z2} & \cdots & S_{zn} \\ S_{t1} & S_{t2} & \cdots & S_{tn} \end{bmatrix}$$

The variances of the model distribution that overbound the true error distribution in each direction of the local tangent co-ordinate system are calculated as:

$$d_x^2 \; = \; \sum_{i=1}^{N} s_{x,i}^2 \, \sigma_i^2$$

$$d_y^2 \; = \; \sum_{i=1}^{N} s_{y,i}^2 \, \sigma_i^2$$

$$d_{xy} \; = \; \sum_{i=1}^{N} s_{x,i} \, s_{y,i} \, \sigma_i^2$$

$$d_v^2 \; = \; \sum_{i=1}^{N} s_{z,i}^2 \, \sigma_i^2$$

th   $d_x^2$      variance of the position solution in the east direction

th   $d_y^2$      variance of the position solution in the north direction

th   $d_{xy}^2$      co-variance of the position solution in the east-north directic

th   $d_v^2$      variance of the position solution in the vertical direction

In the horizontal plane, the error distribution will result in an error ellipse. The principal axes of this error ellipse might not be coincident with the north- and east-directions. The intent is to determine the semi-major axis of that error ellipse and use that error variance in order to determine a protection level. The error covariance in the horizontal plane is:

$$P = \begin{bmatrix} d_x^2 & d_{xy} \\ d_{xy} & d_y^2 \end{bmatrix}$$

Mathematically, an eigenvalue problem of that two-dimensional matrix must be solved. A necessary condition to determine the eigenvalues is:

$$\mathrm{et}(P - \lambda I) = 0 \qquad \text{or:} \qquad \det\begin{bmatrix} d_x^2 - \lambda & d_{xy} \\ d_{xy} & d_y^2 - \lambda \end{bmatrix} = 0$$

us:

$$(d_x^2 - \lambda)\ (d_y^2 - \lambda) - d_{xy}^2\ =\ 0$$

his equation is solved for the eigenvalues $\lambda_{1,2}$, they are obtained t

$$\lambda_{1,2}\ =\ \frac{d_x^2 + d_y^2}{2} \pm \frac{1}{2}\sqrt{(d_x^2 - d_y^2)^2 + 4 d_{xy}^2}$$

Taking the root of the larger eigenvalue, the standard deviation of the horizontal position solution in the semi-major axis of the error ellipse is determined.

$$\sigma_{major}\ =\ \sqrt{\lambda_1}\ =\ \sqrt{\frac{d_x^2 + d_y^2}{2}\ \sqrt{\left(\frac{d_x^2 - d_y^2}{2}\right)^2 + d_{xy}^2}}$$

$\sigma_{maj}$    standard deviation of the horizontal error in the semi-major a:

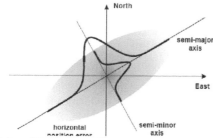

Fig. 13. Horizontal Position Error Ellipse.

The horizontal and the vertical protection level are calculated as:

$$VPL\ =\ K_v\, d_v$$

$$HPL\ =\ K_h\, \sigma_{maj}$$

HPL    horizontal protection level
VPL    vertical protection level

with $K_v$ = 5.33    multiplication factor
$K_h$ = 6.18    multiplication factor (en-route through non-precision approa
$K_h$ = 6.0    multiplication factor (precision approach)

For the precision approach mode, the values of Kh and Kv were selected to bound the user's position in one dimension with a probability of 2 x 10-9 and 10-7, respectively, assuming that the error is characterised by a Gaussian distribution. A Gaussian distribution is used because the aircraft needs to be protected in the vertical and the lateral axes. Only one dimension is used for the HPL, since the along-track tolerance is so much larger than the cross-track tolerance. The worst-case dimension is used. It has been assumed that there is only one independent sample per approach, that half of the total integrity requirement (i.e. 2 x 10-7 per approach) has been allocated to the VPL bounding probability and that the HPL bounding probability has been made negligible. For en-route through non-precision approach modes, the value of Kh was chosen to bound the user's position in two dimensions with a probability of 5 x 10-9 per independent sample, assuming that the error is characterised by a Rayleigh distribution. A Rayleigh distribution is used because the radial error needs to be bounded (both cross-track and along-track errors), using the worst case assumption that the semi-major and semi-minor axes are equal. It has been assumed that there are 10 independent samples per hour, and that half of the total integrity requirement (i.e. 10-7 per hour) has been allocated to the HPL bounding probability.

## 7 SUMMARY

In this article justified the desirability of placing the NPA on Polish airports. Research methods used in the experiments have confirmed that the use of the EGNOS system allows you to safely and continuously carry out the landing maneuver. Presented consortium will be continued its work on the verification of EGNOS for aviation and certification.

## BIBLIOGRAPHY

Helicopters Deploy GNSS in Europe (HEDGE) project documentation,
EGNOS Introduction in European Eastern Region MIELEC project documentation,
ICAO Resolution A32-19,
ICAO Resolution A32-20,
ICAO Resolution A33-15,
ICAO Document 4444,
ICAO Document 7030,
ICAO Document 7300,
ICAO Document 8071,
ICAO Document 8126,
ICAO Document 8168,
ICAO Document 8400,
ICAO Document 8697,
ICAO Document 9161,
ICAO Document 9426,
ICAO Document 9613,
ICAO Document 9660,
ICAO Document 9674,
ICAO Document 9689,
ICAO Document 9750,
ICAO Document 9849-AN/457,
ICAO Document ESARR 1,
ICAO Document ESARR 6.

# 26. Aircraft Landing System Utilizing a GPS Receiver with Position Prediction Functionality

J. Biały, J. Ćwiklak, M. Grzegorzewski & S. Oszczak
*Polish Air Force Academy, Dęblin, Poland*

A. Ciećko
*University of Warmia and Mazury in Olsztyn, Poland*

P. Kościelniak
*Jagiellonian University in Kraków, Poland*

ABSTRACT: The article presents theoretical foundations of the position prediction functionality to be used by a GPS receiver in the event of an instantaneous lack of position data occurring due to various reasons.

## 1 NAVIGATING AN AIRCRAFT IN SPACE BY MEANS OF A POSITION POTENTIAL

The number and character of factors which exert an influence on obtaining the complete navigation information result in the fact that experimental data do not always constitute the complete information regarding the precise assessment of the location of a given aircraft. Even when the satellite system is fully available and its configuration proper, some factors may emerge which prevent its utilization. The proper arrangement of satellite antennas in given points onboard the aircraft is of crucial importance with that regard. Wrong location of those antennas may exert a decisive influence on determining the aircraft's position in space. Factors resulting from abrupt changes of flight parameters, e.g. due to rapid changes of weather conditions, or maneuvers of the aircraft are also to be taken into consideration. The above-mentioned factors necessitate working out additional methods which supplement the process of navigating an aircraft. Mathematical methods are one of the most frequent ways of finding solutions to such problems.

In the presented work, an alternative filter has been developed which relies mostly on the following parameters related to the motion of an aircraft: position, velocity and position error statistics. In reality, a cluster of observed position fixes contains all vital kinematic information of the aircraft. The assumptions of the program use the concept of the "position potential" based on the data received from the navigation services. According to Newton's law of universal gravitation, a free mass particle is attracted by another mass. The acceleration of a particle that has a given mass is directly proportional to the gravitation constant G and inversely proportional to the square of the distance from the attracted mass. Per analogiam, let us regard the statistical confidence regions (position error ellipsoids) of the most probable position fix as a "source of force", which should "attract" a trajectory passing through them. The potential field of a single particle, which in our case is an aircraft, should reflect the observed position fix and the required force to be exerted on the particle, which should monotonically decrease when the particle approaches the assumed position fix.

When the particle enters the position error ellipsoid, it will be attracted with a force whose magnitude will be proportional to the intensity of the potential. The potential monotonically decreases with the decreasing of the distance between the particle and the assumed fix. What is more, in order to allow the particle to continue moving after the position fix appears, the potential of the attracting source will be dissipated exponentially in time. If we select the position density function ("position potential"), which contains the dissipation exponent $\alpha$ and the parameter G (positioning uncertainty, corresponding to Newton's gravitation constant), we assume that the trajectory of the particle will represent the real trajectory of the vehicle.

The first attempt at developing a navigation filter that was based on the above assumptions was made by Inzinga and Vanicek[1]. Their study assumes that the force of the position potential field affecting the particle is connected with the probability which describes the fact that the position fix is included in a two-dimensional error ellipse. Since the motion equation of the particle is extremely difficult to be treated analytically, the following model of operation was selected:

---

[1] T. Inzinga. and P. Vanicek, (1985). "A Two-Dimensional Navigation Algorithm Using a Probabilistic Force Field". Presented at the Third International Symposium on Inertial Technology for Surveying and Geodesy, Banff, Canada, 1985.

– selecting the function of the position potential;
– formulation of temporal dissipation of the potential field;
– describing the motion equation;
– solving the motion equation;
– determining parameters $\alpha$ and G;
– finding the final solution of the navigational problem.

First of all, we have to select the proper potential function. Using the selected function of position density we can establish the time-related position potential field for a sequence of position fixes. Subsequently, we set up a model of motion of an individual particle and find the solution of its motion equation. In order to reflect the changing navigational environment, the potential function contains certain variable parameters, $\alpha$ and G.

When the parameters $\alpha$ and G as well as the initial conditions are not known, the number of possible positions of the particle is infinite. Because of that, further data is necessary in order to determine the proper trajectory of the particle. Therefore, we have to set up a „self-learning" procedure for the filter so that it will determine parameters and conditions based on earlier observations. Parameters and initial motion conditions in the motion model are related to previous observations by means of "motion equations". At this stage, we solve the motion equation with the assumption that the initial conditions are known. Then, we have to determine the parameters $\alpha$ and G and in order to optimize them we will use the least-squares method. The estimation of a given future position of the particle (aircraft), i.e. making a prediction, is possible owing to the model of motion having the determined parameters and on the basis of the present position of the particle. Error estimation is possible at the very moment of obtaining the data concerning a new position fix, by using the values of the differences between the predicted and the actual positions, i.e. checking whether the new position fix is located within the position error ellipsoid.

### 1.1 Navigation Model.

We assume that the density function[2] for a three-dimensional vector of random position is:

$$\Phi_{r_0} = \frac{1}{K}\exp\left[-\frac{1}{2}(r - r_0)^T C^{-1}(r - r_0)\right] \quad (1)$$

where:

$$r = \begin{bmatrix} x \\ y \\ z \end{bmatrix} \quad (2)$$

– particle position vector at the present instant "t"

$$r_0 = \begin{bmatrix} x_0 \\ y_0 \\ z_0 \end{bmatrix} \quad (3)$$

– vector of particle position at the set instant „$t_0$"

$$K = (2\Pi)^{\frac{3}{2}}(\det C)^{\frac{1}{2}} \quad (4)$$

C – symmetrical covariance matrix[3]

$$C = \begin{bmatrix} c_{11} & c_{12} & c_{13} \\ c_{21} & c_{22} & c_{23} \\ c_{31} & c_{32} & c_{33} \end{bmatrix} = \begin{bmatrix} D^2X & \text{cov}(X,Y) & \text{cov}(X,Z) \\ \text{cov}(Y,X) & D^2Y & \text{cov}(Y,Z) \\ \text{cov}(Z,X) & \text{cov}(Z,Y) & D^2Z \end{bmatrix} \quad (5)$$

Let us assume that the particle potential $U_i$ is the basis for the determination of its position fix.

$$U_i(t) = G(r - r_{oi})^T C_i^{-1}(r - r_{oi})e^{-\alpha(t-t_i)} \quad (6)$$

$$(r - r_{oi})^T C_i^{-1}(r - r_{oi}) \quad (7)$$

is the quadratic form of the error ellipsoid at the instant „$t_i$", time $t \geq t_i$. Points located on the error ellipsoid have identical density of probability distribution.

$C_i$ is the positive covariance matrix of the i-th particle, whereas positive parameters $\alpha$ and G (their values are to be determined) represent, respectively:

$\alpha$ - dissipation parameter, with the increasing value of $\alpha$ parameter, the potential of the particle decreases exponentially.

G – uncertainty of the position fix determination. It is a counterpart of the gravitational constant in Newton's attraction.

The potential of the particle is directly proportional to G parameter.

In relation to the potential determined in (6), the time-varying t potential created by "n" position fixes of a particular particle is expressed as:

$$U = \sum_{i=1}^{n} U_i = Ge^{-\alpha t}\sum_{i=1}^{n}(r - r_{0i})^T C_i^{-1}(r - r_{0i})e^{\alpha t_i} \quad (8)$$

where: $t \geq t_n$.

It should be noted that the filter (potential) keeps pace with the kinematics of the particle (aircraft). The time-varying potential field is constantly updated by each new position fix of the particle in order to incorporate the most recent information as soon as possible.

[2] Cz. Platt, Problemy rachunku prawdopodobieństwa i statystyki matematycznej, PWN Warsaw 1977.

[3] A. Plucińska, E. Pluciński, Probabilistyka , Macierz kowariancji, Warsaw 2000, Wyd. Naukowo-Techniczne.

## 2 INTRODUCTION INTO THE ASSESSMENT OF THE ACCURACY OF NAVIGATIONAL SOLUTION

The following parameters are taken into consideration to characterize the error of a position fix that is determined by a receiver:
- SEP (Spherical Error Probable) - 50% of 3D position fixes are located within a sphere with the radius SEP;
- CEP (Circular Error Probable) - 50% of 2D position fixes are located within a circle with the radius CEP;

The assessment of the anticipated accuracy is usually conducted with the knowledge of:
- UERE - estimated standard deviation of the satellite -receiver distance; this parameter is transmitted by the satellite;
- DOP (Dilution of Precision) - coefficients relating the range error with the error of position determination; DOP coefficients are the derivatives of the geometric configuration of the satellites - receiver system.

The errors related to the space and control segments are beyond the influence of the user. All errors are usually treated as random and having normal distribution. Among the DOP coefficients, the following should be distinguished:
- GDOP - Geometrical Dilution of Precision; the coefficient is in inverse proportion to the volume of the solid whose vertices are the positions of the observed satellites and the receiver;
- PDOP - Position Dilution of Precision, 3D dilution of precision,
in the first estimate the value of PDOP is slightly lower than that of GDOP;
- HDOP - Horizontal Dilution of Precision;
- VDOP - Vertical Dilution of Precision;
- TDOP - Time Dilution of Precision.

The preliminary evaluation of the anticipated precision may be obtained by multiplying an appropriate DOP coefficient by UERE. Lower values of the DOP coefficients correspond to better geometrical conditions.

### 2.1 *PARAMETERS OF NAVIGATION RECEIVERS*

The basic set of parameters that characterize a satellite navigation receiver comprises:
- accuracy of position determination; in the autonomous mode accuracy is similar for all receivers because of S/A; in the differential mode it depends substantially on the type of the rover receiver and its base station;
- accuracy of velocity determination, dynamic model parameters;
- accuracy of time determination;

- number of tracked satellites - the minimum number is four satellites, in such a situation, however, losing the signal from one of them (the satellite going beyond the horizon or a terrain feature) causes the interruption of 3D navigation continuity (acquisition of the signal from another satellite and
resuming 3D navigation may take up to 2 minutes), therefore, that solution is not utilized. Practical minimum is 5 satellites - one satellite is always in reserve. The best solution is tracking all visible satellites ("All-In-View" technique), i.e., in practice, up to 12 satellites;
- allowable dynamics (accelerations which the receiver may undergo without interrupting the process of tracking the signal with the carrier- and code-tracking loop) - high dynamics is desirable in aviation applications, whereas in the case of maritime applications that parameter has no importance;
- acquisition time - time to first position fix (TTX);
- re-acquisition time to first fix - time to measurement resumption after a momentary fading of satellite signal;
- receiver sensitivity. Sensitivity of the receiver is of no crucial importance for the proper reception. Almost all receivers have sufficient sensitivity to receive the weakest signals. The parameter that is much more important is the minimum signal-to-noise ratio, at which the receiver is still able to track the satellite signal properly. That criterion is even more important for mobile systems or systems working under the leaf cover. Even though at present there is no the problem with the power level of satellite transmitters there may be some problems related to the reduction thereof in the future. The power output of the present Block II satellite vehicles is four times higher than the power output that is guaranteed by the system specification. It may be expected, however, that for energy saving reasons new Block IIR satellite vehicles will have the power output close to the guaranteed level, which means that their signal-to-noise ratio may be worse by approximately 6dB.
- resistance to jamming signals.

### 2.1.1 *Variation and ambiguity of GPS antenna phase centers*

The problem of the divergence of the real phase center of the antenna and the theoretical point has been known for a long time. Relevant corrections are obtained in calibration procedures and, upon their verification, they are entered into the processed measurement results. The full complexity of the problem became apparent when performing precise, long-chord, GPS measurements and processing their results, with receivers produced by various manufacturers. In extreme instances, when using antennas of various types at the ends of the measured vector, whose length is several thousand kilometers, one may expect a systematic error in the range of 10 cm. The first experiments were followed by systematic laboratory research. All GPS antennas proved to possess variation of the phase center location, depending on the direction from which satellite signal reaches them, i.e. from its azimuth and elevation angle. It was also proved that the phase center shift is different for L1 and L2 frequency of the same antenna. The vertical shift in the case of the best antennas is up to 11 mm for L1 and 8 mm for L2. Horizontal shift is in the vicinity of 1 mm and may, therefore, be neglected. At present, most good quality software used for processing geodesic observations possesses in-built models of constant phase surface for antennas produced by various manufacturers.

The data available with that respect imply that local disturbances (e.g. wind velocity, weather conditions) do not influence significantly the principal directions of the error ellipsoid. Therefore, we may assume that the principal axes of the error ellipsoid are parallel to the coordinate axes (X, Y, Z). Consequently, random variables X, Y, Z are independent and uncorrelated, and the covariance matrix is diagonal

$$\text{cov}(X,Z) = \text{cov}(Y,Z) = \text{cov}(X,Y) = 0 \qquad (9)$$

After using (9) and adopting the following notation:

$$D^2X = \sigma_{11}^2 \qquad D^2Y = \sigma_{22}^2 \qquad D^2Z = \sigma_{33}^2 \qquad (10)$$

the covariance matrix (5) becomes:

$$\mathbf{C} = \begin{bmatrix} \sigma_{11}^2 & 0 & 0 \\ 0 & \sigma_{22}^2 & 0 \\ 0 & 0 & \sigma_{33}^2 \end{bmatrix} \qquad (11)$$

Using (11), we have: [4]

$$\mathbf{C}^{-1} = \begin{bmatrix} \frac{1}{\sigma_{11}^2} & 0 & 0 \\ 0 & \frac{1}{\sigma_{22}^2} & 0 \\ 0 & 0 & \frac{1}{\sigma_{33}^2} \end{bmatrix} \qquad (12)$$

$$K = (2\Pi)^{\frac{3}{2}} (\det \mathbf{C})^{\frac{1}{2}} \qquad (13)$$

$\det \mathbf{C} = \sigma_{11}^2 \sigma_{22}^2 \sigma_{33}^2$ - determinant of a covariance matrix (14)

Positive form determined in (7) is expressed in the coordinates as follows:

$$(\mathbf{r} - \mathbf{r}_0)^T \mathbf{C}^{-1} (\mathbf{r} - \mathbf{r}_0) = \frac{(x-x_0)^2}{\sigma_{11}^2} + \frac{(y-y_0)^2}{\sigma_{22}^2} + \frac{(z-z_0)^2}{\sigma_{33}^2} \qquad (15)$$

Substituting (13) – (15) into (1), the function of position density being Gaussian random reads:

$$f(x,y,z) = \frac{1}{\sqrt{(2\Pi)^3 \sigma_{11}\sigma_{22}\sigma_{33}}} e^{-\frac{1}{2}\left[\frac{(x-x_0)^2}{\sigma_{11}^2} + \frac{(y-y_0)^2}{\sigma_{22}^2} + \frac{(z-z_0)^2}{\sigma_{33}^2}\right]} \qquad (16)$$

Writing U(t) (8) by means of coordinates (see (5) in [3]), for a three-dimensional problem we have:

$$U(t) = \sum_{i=1}^{n} U_i(t) = Ge^{-\alpha t} \sum_{i=1}^{n}\left[\frac{(x-x_{0i})^2}{\sigma_{11i}^2} + \frac{(y-y_{0i})^2}{\sigma_{22i}^2} + \frac{(z-z_{0i})^2}{\sigma_{33i}^2}\right] e^{\alpha t_i};$$
$$t \ge t_n \qquad (17)$$

where:

$$\mathbf{r} = \begin{bmatrix} x \\ y \\ z \end{bmatrix} \qquad \mathbf{r}_{0i} = \begin{bmatrix} x_{0i} \\ y_{0i} \\ z_{0i} \end{bmatrix}$$

Assuming (8)

$$U_i(t) = G e^{-\alpha t} (\mathbf{r} - \mathbf{r}_{0i})^T \mathbf{C}_i^{-1} (\mathbf{r} - \mathbf{r}_{0i}) e^{\alpha t_i}$$

and the equation of the motion (see (6) in [1])

$$\ddot{\mathbf{r}} = -\frac{\partial U(t)}{\partial \mathbf{r}} \qquad (18)$$

By using the potential as described in (8), we will obtain:

$$\frac{\partial U}{\partial \mathbf{r}} = 2Ge^{-\alpha t} \sum_{i=1}^{n} \mathbf{C}_i^{-1} (\mathbf{r} - \mathbf{r}_{0i}) e^{\alpha t_i} \qquad (19)$$

or in an alternative form:

$$\frac{\partial U}{\partial \mathbf{r}} = 2Ge^{-\alpha t} \left[\sum_{i=1}^{n} \mathbf{C}_i^{-1} \mathbf{r} e^{\alpha t_i} - \sum_{i=1}^{n} \mathbf{C}_i^{-1} \mathbf{r}_{0i} e^{\alpha t_i}\right] \qquad (20)$$

After substituting (4,20) into the motion equation (4,18), we have the following equation:

$$\ddot{\mathbf{r}}(t) = -2Ge^{-\alpha t} \left[\sum_{i=1}^{n} \mathbf{C}_i^{-1} \mathbf{r} e^{\alpha t_i} - \sum_{i=1}^{n} \mathbf{C}_i^{-1} \mathbf{r}_{0i} e^{\alpha t_i}\right] \qquad (21)$$

Assuming the following notation:

$$\mathbf{A} = 2\sum_{i=1}^{n} e^{\alpha t_i} \mathbf{C}_i^{-1} \qquad \text{- matrix (3x3)} \qquad (22)$$

$$\mathbf{B} = 2\sum_{i=1}^{n} e^{\alpha t_i} \mathbf{C}_i^{-1} \mathbf{r}_{0i} \qquad \text{- vector} \qquad (23)$$

the motion equation (21) will become

---

[4] Xu. Benlin, A new navigation filter, Calculation of the determinant, inverse matrix and matrix multiplication, 1996.

$$\ddot{r}(t) = e^{-\alpha t} G(\mathbf{Ar} - \mathbf{B}); \qquad\qquad t \geq t_n \qquad (24)$$

It is the motion equation of a particle in the time-varying (t) position potential field, after the occurrence of "n" position fixes.

For simplicity, let us write the matrix $\mathbf{C}_i^{-1}$ as

$$\mathbf{C}_i^{-1} = \begin{bmatrix} \dfrac{1}{\sigma_{11i}^2} & 0 & 0 \\ 0 & \dfrac{1}{\sigma_{22i}^2} & 0 \\ 0 & 0 & \dfrac{1}{\sigma_{33i}^2} \end{bmatrix} = \begin{bmatrix} p_{xi} & 0 & 0 \\ 0 & p_{yi} & 0 \\ 0 & 0 & p_{zi} \end{bmatrix} \qquad (25)$$

Now, computing the product of the matrix **Ar** (see in [2]), the motion equation of the particle (24) in the coordinates can be written as follows:

$$\begin{aligned} \ddot{x}(t) &= -G(A_x x - B_x)e^{-\alpha t} \\ \ddot{y}(t) &= -G(A_y y - B_y)e^{-\alpha t} \qquad t \geq t_n \\ \ddot{z}(t) &= -G(A_z z - B_z)e^{-\alpha t} \end{aligned} \qquad (26)$$

In order to avoid problems with computer calculation, we always move the initial time instant "t" to the time of occurrence of the latest instant „$t_n$". Then, vectors **Ar** i **B** are represented in coordinates as:

$$\mathbf{Ar} = [A_x x, A_y y, A_z z]$$

$$A_x = 2\sum_{i=1}^{n} e^{\alpha(t_i - t_n)} p_{xi}$$

$$A_y = 2\sum_{i=1}^{n} e^{\alpha(t_i - t_n)} p_{yi} \qquad (27)$$

$$A_z = 2\sum_{i=1}^{n} e^{\alpha(t_i - t_n)} p_{zi}$$

and the coordinates of the vector **B** are represented as:

$$B_x = 2\sum_{i=1}^{n} e^{\alpha(t_i - t_n)} p_{xi} x_{oi}$$

$$B_y = 2\sum_{i=1}^{n} e^{\alpha(t_i - t_n)} p_{yi} y_{oi}$$

$$B_z = 2\sum_{i=1}^{n} e^{\alpha(t_i - t_n)} p_{zi} z_{oi} \qquad (28)$$

To solve the system of inhomogeneous linear differential equations (26) of the second order, we will write it in an alternative way:

$$\ddot{x}(t) = -GA_x (x - \frac{B_x}{A_x}) e^{-\alpha t}$$

$$\ddot{y}(t) = -GA_y (y - \frac{B_y}{A_y}) e^{-\alpha t} \qquad t \geq t_n$$

$$\ddot{z}(t) = -GA_z (z - \frac{B_z}{A_z}) e^{-\alpha t} \qquad (29)$$

which is a system of inhomogeneous linear differential equations of the second order.

As an example, let us solve the equation (29)$_1$, to that end making a substitution:

$$x(t) - \frac{B_x}{A_x} = U(t), \qquad (30)$$

then

$$\dot{x}(t) = \dot{U}(t) \qquad (31)$$

and

$$\ddot{x}(t) = \ddot{U}(t) \qquad (32)$$

Equations (29)$_1$ – (32) imply the equation

$$\ddot{U}(t) + GA_x e^{-\alpha t} U(t) = 0 \qquad (33)$$

Substituting the independent variable (time)

$$s = \frac{2}{\alpha} e^{-\frac{\alpha}{2} t} \sqrt{GA_x} \qquad (34)$$

we have

$$\frac{ds}{dt} = \frac{2}{\alpha} e^{-\frac{\alpha}{2} t} \sqrt{GA_x} \left(-\frac{\alpha}{2}\right) = -e^{-\frac{\alpha}{2} t} \sqrt{GA_x}$$

$$\frac{ds}{dt} = -\frac{\alpha}{2} s \qquad (35)$$

$$\frac{du}{dt} = \frac{du}{ds} \cdot \frac{ds}{dt} = -\frac{\alpha}{2} s \cdot \frac{du}{ds} \qquad (36)$$

$$\frac{d^2 u}{dt^2} = \ddot{U}(t) = \frac{d}{dt}\left(\frac{du}{dt}\right) = \frac{d}{dt}\left(-\frac{\alpha}{2} s \frac{du}{ds}\right) = \frac{d}{ds}\left(-\frac{\alpha}{2} s \frac{du}{ds}\right) \cdot \frac{ds}{dt} =$$
$$\left[-\frac{\alpha}{2} \frac{du}{ds} - \frac{\alpha}{2} s \frac{d^2 s}{dt^2}\right] \cdot \left(-\frac{\alpha}{2} s\right) \qquad (37)$$

$$\frac{d^2 u}{ds^2} = \frac{\alpha^2}{4} s^2 \frac{d^2 u}{ds^2} + \frac{\alpha^2}{4} s \frac{du}{ds} \qquad (38)$$

From (29):

$$GA_x e^{-\alpha t} = \frac{\alpha^2}{4} s^2 \qquad (39)$$

Substituting (38) and (39) into (33), we have:

$$\frac{\alpha^2}{4}s^2\frac{d^2u}{ds^2} + \frac{\alpha^2}{4}s\frac{du}{ds} + \frac{\alpha^2}{4}s^2u = 0 \ / : \frac{\alpha^2}{4}s^2$$

$$\frac{d^2u}{ds^2} + \frac{1}{s}\frac{du}{ds} + u = 0 \tag{40}$$

Equation (40) is a Bessel differential equation,[5] whose solution is found to be

$$U(s) = a_1 J_0(s) + a_2 N_0(s) \text{[6]} \tag{41}$$

where $J_0(s)$, $N_0(s)$ are Bessel functions of the first and second kind respectively, with zero index (0).

Returning to the initial variables, i.e. x and t

$$\begin{cases} U = x - \dfrac{B_x}{A_x} \\ s = \dfrac{2}{\alpha}e^{-\frac{\alpha}{2}t}\sqrt{GA_x} \end{cases} \quad \text{see: (30) and (34)}$$

the solution (41) of equation (29)$_1$ is expressed as follows:

$$x(t) = \frac{B_x}{A_x} + a_1 J_0\left(\frac{2}{\alpha}e^{-\frac{\alpha}{2}t}\sqrt{GA_x}\right) + a_2 N_0\left(\frac{2}{\alpha}e^{-\frac{\alpha}{2}t}\sqrt{GA_x}\right) \tag{42}$$

where the constant parameters $a_1$ and $a_2$ are to be determined by the initial condition, which gives the initial position fix x(t) and the instantaneous velocity at the initial instant.

Following the same procedure in solving the equation (29)$_2$ and (29)$_3$ as was employed in solving the equation (29)$_1$, we will obtain the following solution of the system of equations (29):

$$\begin{cases} x(t) = \dfrac{B_x}{A_x} + a_1 J_0\left(\dfrac{2}{\alpha}e^{-\frac{\alpha}{2}t}\sqrt{GA_x}\right) + a_2 N_0\left(\dfrac{2}{\alpha}e^{-\frac{\alpha}{2}t}\sqrt{GA_x}\right) \\ y(t) = \dfrac{B_y}{A_y} + b_1 J_0\left(\dfrac{2}{\alpha}e^{-\frac{\alpha}{2}t}\sqrt{GA_y}\right) + b_2 N_0\left(\dfrac{2}{\alpha}e^{-\frac{\alpha}{2}t}\sqrt{GA_y}\right) \\ z(t) = \dfrac{B_z}{A_z} + c_1 J_0\left(\dfrac{2}{\alpha}e^{-\frac{\alpha}{2}t}\sqrt{GA_z}\right) + c_2 N_0\left(\dfrac{2}{\alpha}e^{-\frac{\alpha}{2}t}\sqrt{GA_z}\right) \end{cases} \tag{43}$$

where:

the first addend $\dfrac{B_x}{A_x}$ is to be regarded as a particular solution of an inhomogeneous differential equation, and both elements containing constants $a_1$ and $a_2$ as general solutions of a homogeneous differential equation.

In order to determine the integration constants $a_1$; $a_2$; $b_1$; $b_2$; $c_1$; $c_2$ the initial conditions have to be used, i.e. they must be added to the system of equations (24) (formulation of Cauchy problem).

To simplify the determination of these constants, $a_1$; $a_2$; $b_1$; $b_2$; $c_1$; $c_2$ the following properties of Bessel function should be used[7]:

$$\begin{cases} \dfrac{dJ_0(t)}{dt} = -J_1(t) \\ \dfrac{dN_0(t)}{dt} = -N_1(t) \end{cases} \tag{44}$$

and:

$$J_1(t)N_0(t) - J_0(t)N_1(t) = \frac{2}{\Pi t} \tag{45}$$

For example:

$$\frac{d\left[J_0\left(\dfrac{2}{\alpha}e^{-\frac{\alpha}{2}t}\sqrt{GA_x}\right)\right]}{dt} = \sqrt{GA_x}\,J_1\left(\frac{2}{\alpha}e^{-\frac{\alpha}{2}t}\sqrt{GA_x}\right)e^{-\frac{\alpha}{2}t}$$

$$\frac{d}{dt}\left[N_0\left(\frac{2}{\alpha}e^{-\frac{\alpha}{2}t}\sqrt{GA_x}\right)\right] = \sqrt{GA_x}\,N_1\left(\frac{2}{\alpha}e^{-\frac{\alpha}{2}t}\sqrt{GA_x}\right)e^{-\frac{\alpha}{2}t} \tag{46}$$

$J_1$, $N_1$ – First order Bessel functions of the first and second kind respectively.

Using (41) in (38) we have:

$$\begin{cases} \dot{x}(t) = \sqrt{GA_x}\,e^{-\frac{\alpha}{2}t}\left[a_1 J_1\left(\dfrac{2}{\alpha}e^{-\frac{\alpha}{2}t}\sqrt{GA_x}\right) + a_2 N_1\left(\dfrac{2}{\alpha}e^{-\frac{\alpha}{2}t}\sqrt{GA_x}\right)\right] \\ \dot{y}(t) = \sqrt{GA_y}\,e^{-\frac{\alpha}{2}t}\left[b_1 J_1\left(\dfrac{2}{\alpha}e^{-\frac{\alpha}{2}t}\sqrt{GA_y}\right) + b_2 N_1\left(\dfrac{2}{\alpha}e^{-\frac{\alpha}{2}t}\sqrt{GA_y}\right)\right] \\ \dot{z}(t) = \sqrt{GA_z}\,e^{-\frac{\alpha}{2}t}\left[c_1 J_1\left(\dfrac{2}{\alpha}e^{-\frac{\alpha}{2}t}\sqrt{GA_z}\right) + c_2 N_1\left(\dfrac{2}{\alpha}e^{-\frac{\alpha}{2}t}\sqrt{GA_z}\right)\right] \end{cases} \tag{47}$$

Equations (43) and (47) with known (to be determined) parameters G and $\alpha$ are the fundamental equations of our study [see (17)].

Using the initial condition for $t_n = 0$

---

[5] E.Kącki, L.Siewierski, Wybrane działy matematyki wyższej z ćwiczeniami, rozdz.1, Warsaw 1975, PWN and in N.M.Matwiejew, Metody całkowania równań różniczkowych zwyczajnych, Chapter VIII, Warsaw 1972, PWN.

[6] E.Kącki, L.Siewierski, Wybrane działy matematyki wyższej z ćwiczeniami, rozdz.1, Warsaw 1975, PWN and in N.M.Matwiejew, Metody całkowania równań różniczkowych zwyczajnych, Chapter VIII, Warsaw 1972, PWN.

[7] W. I. Smirnow, Matematyka Wyższa Chapter V, I, II Part 2, Warsaw 1967, PWN.

$$\mathbf{r}(0) = \begin{bmatrix} x_n \\ y_n \\ z_n \end{bmatrix} \quad i \quad \dot{\mathbf{r}}(0) = \begin{bmatrix} \dot{x}_n \\ \dot{y}_n \\ \dot{z}_n \end{bmatrix},$$
(48)

in equations (43) and (47) and the dependence (45), we compute the constants: $a_1$, $a_2$, $b_1$, $b_2$, $c_1$, $c_2$.

As an example, let us compute the constant $a_1$. To that end, we will write equations: $(43)_1$, $(47)_1$ and (45), after adopting for simplicity, the following notation

$$T = e^{-\frac{a}{2}t}\sqrt{GA_x} \quad and \quad z = \frac{2}{\alpha}T \quad in(4.40)$$

(49)

in the following form:

$$\begin{cases} x(t) = \dfrac{B_x}{A_x} + a_1 J_0\left(\dfrac{2}{\alpha}T\right) + a_2 N_0\left(\dfrac{2}{\alpha}T\right) \\[2mm] \dot{x}(t) = a_1 T J_1\left(\dfrac{2}{\alpha}T\right) + a_2 T N_1\left(\dfrac{2}{\alpha}T\right) \\[2mm] J_1\left(\dfrac{2}{\alpha}T\right)\cdot N_0\left(\dfrac{2}{\alpha}T\right) - J_0\left(\dfrac{2}{\alpha}T\right)\cdot N_1\left(\dfrac{2}{\alpha}T\right) = \dfrac{\alpha}{\Pi T} \end{cases}$$

(50)

By multiplying $(50)_1$ by $-TN_1\left(\dfrac{2}{\alpha}T\right)$ and $(50)_2$ by $N_0\left(\dfrac{2}{\alpha}T\right)$ we obtain:

$$-TN_1\left(\frac{2}{\alpha}T\right)x(t) = -\frac{B_x}{A_x}TN_1\left(\frac{2}{\alpha}T\right) - a_1 TJ_0\left(\frac{2}{\alpha}T\right)N_1\left(\frac{2}{\alpha}T\right) -$$

$$a_2 TN_0\left(\frac{2}{\alpha}T\right)N_1\left(\frac{2}{\alpha}T\right)$$

$$N_0\left(\frac{2}{\alpha}T\right)\dot{x}(t) = a_1 TJ_1\left(\frac{2}{\alpha}T\right)N_0\left(\frac{2}{\alpha}T\right) + a_2 TN_0\left(\frac{2}{\alpha}T\right)N_1\left(\frac{2}{\alpha}T\right) \quad (51)$$

Adding the sides of equations (51), we will obtain:

$$-TN_1\left(\frac{2}{\alpha}T\right)x(t) + N_0\left(\frac{2}{\alpha}T\right)\dot{x}(t) = -\frac{B_x}{A_x}TN_1\left(\frac{2}{\alpha}T\right)$$

$$+ a_1 T\left[\left(J_1\frac{2}{\alpha}T\right)N_0\left(\frac{2}{\alpha}T\right) + -J_0\left(\frac{2}{\alpha}T\right)N_1\left(\frac{2}{\alpha}T\right)\right],$$

and, after using (50), we have:

$$\frac{\alpha}{\Pi}a_1 = \frac{B_x}{A_x}TN_1\left(\frac{2}{\alpha}T\right) - TN_1\left(\frac{2}{\alpha}T\right)x(t) + N_0\left(\frac{2}{\alpha}T\right)\dot{x}(t),$$

hence:

$$a_1 = -\frac{\Pi}{\alpha}\left[\left(x(t) - \frac{B_x}{A_x}\right)TN_1\left(\frac{2}{\alpha}T\right) - N_0\left(\frac{2}{\alpha}T\right)\dot{x}(t)\right].$$

(52)

Because of the initial condition (48) in the time interval $[t_n, t_{n+1}]$ for $t = t_n = 0$ we have:

$$\begin{cases} x(0) = x_n \\ \dot{x}(0) = \dot{x}_n \\ e^{-\frac{\alpha}{2}\cdot 0} = 1 \\ T = \sqrt{GA_x} \end{cases}$$

(53)

Taking into consideration (53) in (52), $a_1$ will read :

$$a_1 = -\frac{\Pi}{\alpha}\left[\left(x_n - \frac{B_x}{A_x}\right)\sqrt{GA_x}N_1\left(\frac{2}{\alpha}\sqrt{GA_x}\right) - \dot{x}_n N_0\left(\frac{2}{\alpha}\sqrt{GA_x}\right)\right]$$
(54)

The remaining integration constants $a_2$, $b_1$, $b_2$, $c_1$, $c_2$ are determined likewise.

Finally, the integration constants are represented by the following formulae:

$$\begin{aligned} a_1 &= -\frac{\Pi}{\alpha}\left[\left(x_n - \frac{B_x}{A_x}\right)\sqrt{GA_x}N_1\left(\frac{2}{\alpha}\sqrt{GA_x}\right) - \dot{x}_n N_0\left(\frac{2}{\alpha}\sqrt{GA_x}\right)\right] \\[1mm] a_2 &= \frac{\Pi}{\alpha}\left[\left(x_n - \frac{B_x}{A_x}\right)\sqrt{GA_x}J_1\left(\frac{2}{\alpha}\sqrt{GA_x}\right) - \dot{x}_n J_0\left(\frac{2}{\alpha}\sqrt{GA_x}\right)\right] \\[1mm] b_1 &= -\frac{\Pi}{\alpha}\left[\left(y_n - \frac{B_y}{A_y}\right)\sqrt{GA_y}N_1\left(\frac{2}{\alpha}\sqrt{GA_y}\right) - \dot{x}_n N_0\left(\frac{2}{\alpha}\sqrt{GA_y}\right)\right] \\[1mm] b_2 &= \frac{\Pi}{\alpha}\left[\left(y_n - \frac{B_y}{A_y}\right)\sqrt{GA_y}J_1\left(\frac{2}{\alpha}\sqrt{GA_y}\right) - \dot{x}_n J_0\left(\frac{2}{\alpha}\sqrt{GA_y}\right)\right] \\[1mm] c_1 &= -\frac{\Pi}{\alpha}\left[\left(z_n - \frac{B_z}{A_z}\right)\sqrt{GA_z}N_1\left(\frac{2}{\alpha}\sqrt{GA_z}\right) - \dot{x}_n N_0\left(\frac{2}{\alpha}\sqrt{GA_z}\right)\right] \\[1mm] c_2 &= \frac{\Pi}{\alpha}\left[\left(z_n - \frac{B_z}{A_z}\right)\sqrt{GA_z}J_1\left(\frac{2}{\alpha}\sqrt{GA_z}\right) - \dot{x}_1 J_0\left(\frac{2}{\alpha}\sqrt{GA_z}\right)\right] \end{aligned}$$
(55)

Because of the initial condition, the computed integration constants $a_1$, $a_2$, $b_1$, $b_2$, $c_1$, $c_2$ represented by (3.55) exist within the time interval $[t_n, t_{n+1}]$. Therefore, the position and velocity at the next instant, $t_{n+1}$ can be determined from equations(43) and (47) for $t = t_{n+1} - t_n$.

Functions present in the above problem are labeled by the following units:

$\alpha[s^{-1}]$ - parameter;
$U[m^2 s^{-2}]$ - potential;
$G[m^2 s^{-2}]$ - parameter present in determining the potential U;
$A[m^{-2}]$ - matrix;
$B[m^{-1}]$ - vector.

$\begin{bmatrix} x \\ y \\ z \end{bmatrix}[m]$ - vector of the particle (aircraft) position;

$\begin{bmatrix} \dot{x} \\ \dot{y} \\ \dot{z} \end{bmatrix}[m\cdot s^{-1}]$ - velocity of the particle (aircraft) position

## 3 OPTIMIZATION OF PARAMETERS α AND G

In order to optimize parameters $\alpha$ and G, we let us analyze the following problem, consisting in solving the motion equation (see: (29)):

$$\ddot{x}(t) = -G(A_x x - B_x)e^{-\alpha t}$$
$$\ddot{y}(t) = -G(A_y y - B_y)e^{-\alpha t} \qquad t \geq t_n$$
$$\ddot{z}(t) = -G(A_z z - B_z)e^{-\alpha t} \qquad\qquad (56)$$

where the vector **Ar** is represented by:

$$A_x = 2\sum_{i=1}^{n} e^{\alpha(t_i - t_n)} p_{xi}$$
$$A_y = 2\sum_{i=1}^{n} e^{\alpha(t_i - t_n)} p_{yi}$$
$$A_z = 2\sum_{i=1}^{n} e^{\alpha(t_i - t_n)} p_{zi} \qquad\qquad (57)$$

and the coordinates of the vector **B** are represented as:

$$B_x = 2\sum_{i=1}^{n} e^{\alpha(t_i - t_n)} p_{xi} x_{oi}$$
$$B_y = 2\sum_{i=1}^{n} e^{\alpha(t_i - t_n)} p_{yi} y_{oi}$$
$$B_z = 2\sum_{i=1}^{n} e^{\alpha(t_i - t_n)} p_{zi} z_{oi} \qquad\qquad (58)$$

with the initial conditions for $t_n = 0$

$$\mathbf{r}(0) = \begin{bmatrix} x_n \\ y_n \\ z_n \end{bmatrix} \quad i \quad \dot{\mathbf{r}}(0) = \begin{bmatrix} \dot{x}_n \\ \dot{y}_n \\ \dot{z}_n \end{bmatrix} \qquad (59)$$

The vector of the particle position is described by the following formulae (see: (43)):

$$\begin{cases} x(t) = \dfrac{B_x}{A_x} + a_1 J_0\left(\dfrac{2}{\alpha} e^{-\frac{\alpha}{2}t} \sqrt{GA_x}\right) + a_2 N_0\left(\dfrac{2}{\alpha} e^{-\frac{\alpha}{2}t} \sqrt{GA_x}\right) \\[2mm] y(t) = \dfrac{B_y}{A_y} + b_1 J_0\left(\dfrac{2}{\alpha} e^{-\frac{\alpha}{2}t} \sqrt{GA_y}\right) + b_2 N_0\left(\dfrac{2}{\alpha} e^{-\frac{\alpha}{2}t} \sqrt{GA_y}\right) \\[2mm] z(t) = \dfrac{B_z}{A_z} + c_1 J_0\left(\dfrac{2}{\alpha} e^{-\frac{\alpha}{2}t} \sqrt{GA_z}\right) + c_2 N_0\left(\dfrac{2}{\alpha} e^{-\frac{\alpha}{2}t} \sqrt{GA_z}\right) \end{cases} \qquad (60)$$

whereas the velocity vector is described by the following equations (see: (47)):

$$\begin{cases} \dot{x}(t) = \sqrt{GA_x}\, e^{-\frac{\alpha}{2}t}\left[ a_1 J_1\left(\dfrac{2}{\alpha} e^{-\frac{\alpha}{2}t}\sqrt{GA_x}\right) + a_2 N_1\left(\dfrac{2}{\alpha} e^{-\frac{\alpha}{2}t}\sqrt{GA_x}\right) \right] \\[2mm] \dot{y}(t) = \sqrt{GA_y}\, e^{-\frac{\alpha}{2}t}\left[ b_1 J_1\left(\dfrac{2}{\alpha} e^{-\frac{\alpha}{2}t}\sqrt{GA_y}\right) + b_2 N_1\left(\dfrac{2}{\alpha} e^{-\frac{\alpha}{2}t}\sqrt{GA_y}\right) \right] \\[2mm] \dot{z}(t) = \sqrt{GA_z}\, e^{-\frac{\alpha}{2}t}\left[ c_1 J_1\left(\dfrac{2}{\alpha} e^{-\frac{\alpha}{2}t}\sqrt{GA_z}\right) + c_2 N_1\left(\dfrac{2}{\alpha} e^{-\frac{\alpha}{2}t}\sqrt{GA_z}\right) \right] \end{cases} \qquad (61)$$

and the integration constants by the relation:

$$\begin{cases} a_1 = -\dfrac{\Pi}{\alpha}\left[ \left(x_n - \dfrac{B_x}{A_x}\right)\sqrt{GA_x}\, N_1\left(\dfrac{2}{\alpha}\sqrt{GA_x}\right) - \dot{x}_n N_0\left(\dfrac{2}{\alpha}\sqrt{GA_x}\right) \right] \\[2mm] a_2 = \dfrac{\Pi}{\alpha}\left[ \left(x_n - \dfrac{B_x}{A_x}\right)\sqrt{GA_x}\, J_1\left(\dfrac{2}{\alpha}\sqrt{GA_x}\right) - \dot{x}_n J_0\left(\dfrac{2}{\alpha}\sqrt{GA_x}\right) \right] \\[2mm] b_1 = -\dfrac{\Pi}{\alpha}\left[ \left(y_n - \dfrac{B_y}{A_y}\right)\sqrt{GA_y}\, N_1\left(\dfrac{2}{\alpha}\sqrt{GA_y}\right) - \dot{x}_n N_0\left(\dfrac{2}{\alpha}\sqrt{GA_y}\right) \right] \\[2mm] b_2 = \dfrac{\Pi}{\alpha}\left[ \left(y_n - \dfrac{B_y}{A_y}\right)\sqrt{GA_y}\, J_1\left(\dfrac{2}{\alpha}\sqrt{GA_y}\right) - \dot{x}_n J_0\left(\dfrac{2}{\alpha}\sqrt{GA_y}\right) \right] \\[2mm] c_1 = -\dfrac{\Pi}{\alpha}\left[ \left(z_n - \dfrac{B_z}{A_z}\right)\sqrt{GA_z}\, N_1\left(\dfrac{2}{\alpha}\sqrt{GA_z}\right) - \dot{x}_n N_0\left(\dfrac{2}{\alpha}\sqrt{GA_z}\right) \right] \\[2mm] c_2 = \dfrac{\Pi}{\alpha}\left[ \left(z_n - \dfrac{B_z}{A_z}\right)\sqrt{GA_z}\, J_1\left(\dfrac{2}{\alpha}\sqrt{GA_z}\right) - \dot{x}_1 J_0\left(\dfrac{2}{\alpha}\sqrt{GA_z}\right) \right] \end{cases} \qquad (62)$$

Our objective is to optimize parameters $\alpha$ and G so that the function of two variables:

$$f(\alpha, G) = \sum_{i=1}^{n} \left[\ddot{r}(t_i) - \ddot{r}_{0i}\right]^2 = \text{minimum}$$

or, written in coordinates:

$$f(\alpha, G) = \sum_{i=1}^{n} \left\{ [x(t_i) - x_{0i}]^2 + [y(t_i) - y_{0i}]^2 + [z(t_i) - z_{0i}]^2 \right\} = \text{minimum} \quad (63)$$

where:

$x_{0i}$, $y_{0i}$, $z_{0i}$ – position of the particle (aircraft) at the instance $t_i$ (data obtained from a satellite);

$x(t_i)$, $y(t_i)$, $z(t_i)$ - position of the particle (aircraft) at the instance $t_i$ computed from (60);

By substituting the optimized parameters $\alpha$ and G into equations (60) and (61) we will obtain from the mathematical model the vector of the particle (aircraft) position and the vector of its velocity at the instant t.

## 4 COMPARISON OF VARIOUS METHODS OF PREDICTING THE AIRCRAFT POSITION

Our task was to create an algorithm, which may be implemented, and then used in practice, in order to predict subsequent positions of an aircraft.

The results of our work are the following:
1 Three different algorithms, written as MS Excel macros, which predict the position of an aircraft in the subsequent instances on the basis of its positions at previous instances.

2 Comparison of the efficiency of the created algorithms when using various values of the parameters.
3 Suggestions concerning the proper selection of an
  algorithm and its implementation.

## 4.1 The methods under comparison

The starting point was for us the method described
in the section *Navigating an Aircraft in Space by
Means of a Position Potential.* Its main advantage is
the fact that it was created on the basis of real physical assumptions. This method is henceforth called
$\alpha Gn$ method - from its two key parameters.

In our studies, utilization of other methods was
also taken into consideration. Having analyzed their
theoretical correctness, we carried out their preliminary tests on the basis of the real data contained in
the MS Excel sheet data.xls. Two of the methods
proved to be worth further testing. They represent
two different approaches: (1) Establishing the relationship between the co-ordinates of the aircraft's
position fix and the time. (2) Establishing the relationship between the aircraft's position at a given instant of time and its position at previous time instances.

In the next part of this report we provide a brief
discussion of the three methods and the results of the
tests conducted in order to assess the usefulness of
those methods.

### 4.1.1 $\alpha Gn$ Method

Technically, its essence lies in expressing the solution of an ordinary second order linear differential
equation as a function of two parameters $\alpha$ and $G$,
and, subsequently, selecting the optimum values of
those parameters, which will minimize the error
function, $F_0$. Knowing those values facilitates the
determining of the predicted values. Bessel special
functions are present in the solution of the equation.

Function $f(\alpha, G)$ (formula (63)) represents the
sum of the squares of distances between the theoretical points and the observed points at $n$ previous
time instances. Initial terms of the series expansion
of the Bessel function were used for the approximation of that function.

Due to high complexity of the formulae describing the function $f(\alpha, G)$ the application of traditional optimization methods was not possible. Therefore,
we decided to use for its optimization a certain version of an evolutionary algorithm developed by our-
selves. The implementation of that algorithm is,
however, time-consuming and increasing the number of previous time instances, n, causes a considerable increase of the duration of its operation. The parameters affecting the course of action of the
evolutionary algorithm are: population size, number

of discarded candidate solutions (individuals), width
(step) of the local method, number of steps and, finally ranges: $\alpha$, $G$.

### 4.1.2 linear n Method

In the opinion of the members of our team and
according to the data that we obtained, predicting the
height h is of crucial importance whereas predicting
the remaining co-ordinates, $X$ and $Y$, is of considerably lesser importance. Therefore both this method
and the next are being discussed in the context of
predicting height, but the discussion may, if necessary, be repeated separately for $X$ and $Y$.

This is one of the simplest methods. It assumes
linear relationship between height and time:

$h(t) = bo + bit.$

Coefficients $b_o, b_i$ are determined with the least
squares method on the basis of the course of the
flight so far (n measurements, $n > 2$) and used for
making a prediction. After preliminary tests, other
methods of determining functional relationship between height and time were discarded as being substantially less effective.

### 4.1.3 baryc n Method

This method is in fact one of the versions of the
*ARIMA (p, d, q)* method, namely $A(1,1,0)$, and its
description is expressed as: $hi = \beta_1 h_{i-1} + \beta_2 h_{i-2}$ with
$\beta_1 + \beta_2 = 1$

Coefficients $\beta_1$, $\beta_2$ are determined with the least
squares method on the basis of the course of the
flight so far *(n measurements, $n > 2$)* and used for
making a prediction. After preliminary tests, other
methods representing the *ARIMA* type were discarded as being substantially less effective. For similar
reasons, two other methods of a similar type were
discarded after preliminary tests.

### 4.1.4 Results for the data set from the MS Excel
   sheet dane. xls (data.xls) (flight of the Cessna
   aircraft)

The data in the tables was used for comparing the
methods mentioned above and for optimizing the $\alpha$
and G parameters.

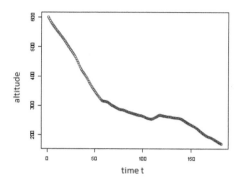

Figure 1: Height in the dane.xls.(data.xls) sheet

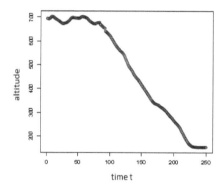

Figure 2: Height in the 4 minuty 92.xls. (4 minutes 92.xls.) sheet

### 4.1.5 *Determined parameters α and G for αG2 method*

|                 | α     | G        |
|-----------------|-------|----------|
| mean            | 13.01 | 0.97     |
| stand. deviation| 5.78  | 1.00     |
| min.            | 1.00  | 0.010003 |
| max.            | 19.97 | 2.993    |

**For αG3**

|                 | α     | G        |
|-----------------|-------|----------|
| mean            | 14.22 | 0.12     |
| stand. deviation| 5.25  | 0.22     |
| min.            | 1.19  | 0.010003 |
| max.            | 19.98 | 1.95     |

## 5 SUMMARY AND CONCLUSIONS

On the basis of the conducted numerical experiments, we hold the view that the methods to be implemented in further, in-flight tests are:
1 α *G2* and α G3
2 *linear 2*

When there are no sudden changes in the course of the flight, the accuracy of the methods mentioned above is considerably higher than the accuracy of the remaining methods. However, only after implementing them in a real instrument, a GPS receiver, the real operational time efficiency of each of them may be assessed. According to our predictions, the *linear2* method is a little faster than the αG2, which in turn is faster than αG3.

In the event of substantial flight disturbances, none of those methods, or any other method known to us, may be used without a running a great risk of making a serious error.

During the flight, making the position prediction ought to be repeated once every second on the basis of the known position fix of the aircraft in the previous 2 or 3 seconds. The prediction may be used as necessary (lack of GPS data) provided that there are no sudden changes of flight parameters. In the event of such changes, the prediction should not be utilized.

The accuracy of the prediction decreases with the time it has been used. Therefore, the pilot ought to be warned after a given time (we recommend 3 seconds) that the position of the aircraft being shown may differ considerably from its real position.

## REFERENCES

[1] Grzegorzewski M.: Navigating an Aircraft by Means of a Position Potential in Three-dimensional Space – Post-doctoral dissertation, 2005.
[2] Grzegorzewski M., Biały J.: Navigating an Aircraft by Means of a Position Potential in Three-dimensional Space – The Journal of Navigation, Great Britain 2007.
[3] Inzinga T. and Vaniček P. „A Two-Dimensional navigation Algorithm Using a Probabilistic Force Field". Presented at Third International Symposium on Inertial Technology for Surveying and Geodesy, Banff, Canada, 1985.
[4] Kącki E., Siewierski L.: Wybrane działy matematyki wyższej z ćwiczeniami. Warsaw 1975. PWN
[5] Ombach J.: Some algorithms of global optimizations. Post-Conference Materials Jagiellonian University. Cracow 2004.
[6] Platt Cz., Problemy rachunku prawdopodobieństwa i statystyki matematycznej, Warsaw 1977. PWN.
[7] Plucińska A., Pluciński E. Probabilistyka. Warsaw 2000. Wydawnictwo Naukowo-Techniczne.
[8] Smirnow W.I., Matematyka Wyższa Chapter I, II, V Part 2. Warsaw 1967. PWN.

# Author index

Asajima, T., 149

Balobanov, O. O., 33
Banaszek, K., 199
Banaś, P., 59
Biały, J., 207
Brandowski, A., 13
Brčić, D., 133
Breitsprecher, M., 59

Ciećko, A., 207
Ćwiklak, J., 207

Dinu, D., 37

Fellner, A., 199
Frackowiak, W., 13

Goerlandt, F., 65, 93, 101
Górnicz, T., 45
Grzegorzewski, M., 207

Hu, Q.Y., 109

Jaworski, B., 85

Kaczorek, T., 41
Kobayashi, E., 149
Kolendo, P., 85
Kopacz, P., 123
Kos, S., 133
Kościelniak, P., 207
Krata, P., 165
Kubiš, M., 191
Kujala, P., 65, 93, 101
Kulczyk, J., 45

Lammi, H., 65
Lenart, A.S., 141
Lisowski, J., 75
Liu, R.R., 109

Mielewczyk, A., 13
Montewka, J., 65, 93, 101

Neumann, T., 3, 9
Nguyen Cong, V., 115
Nguyen, H., 13
Novák, A., 191

Oszczak, S., 207

Pipchenko, O. D., 157
Postan, M. Ya., 33

Scupi, A., 37
Shi, C.J., 109
Ståhlberg, K., 101
Sueyoshi, N., 149
Szlapczynska, J., 165, 173
Szlapczynski, R., 173
Szyca, G., 51
Śmierzchalski, R., 85

Tomera, M., 21
Trómiński, P., 199

Wei, S., 181
Weintrit, A., 3, 9, 123

Xiang, Z., 109

Zadrąg, P., 199
Zhou, P., 181